Understanding Modern Electronics

Richard Wolfson, Ph.D.

THE
GREAT
COURSES

PUBLISHED BY:

THE GREAT COURSES
Corporate Headquarters
4840 Westfields Boulevard, Suite 500
Chantilly, Virginia 20151-2299
Phone: 1-800-832-2412
Fax: 703-378-3819
www.thegreatcourses.com

Richard Wolfson, Ph.D.

Benjamin F. Wissler Professor of Physics
Middlebury College

Richard Wolfson is the Benjamin F. Wissler Professor of Physics at Middlebury College, and he teaches in Middlebury's Environmental Studies Program. He did undergraduate work at the Massachusetts Institute of Technology and Swarthmore College, graduating from Swarthmore with a bachelor's degree in Physics and Philosophy. He holds a master's degree in Environmental Studies from the University of Michigan and a doctorate in Physics from Dartmouth.

Professor Wolfson's books include *Nuclear Choices: A Citizen's Guide to Nuclear Technology* and *Simply Einstein: Relativity Demystified*, both of which exemplify his interest in making science accessible to non-scientists. His textbooks include three editions of *Physics for Scientists and Engineers*, coauthored with Jay M. Pasachoff; three editions of *Essential University Physics*; two editions of *Energy, Environment, and Climate*; and *Essential College Physics*, coauthored with Andrew Rex. Professor Wolfson also has published in *Scientific American* and writes for *The World Book Encyclopedia*.

Professor Wolfson's current research involves the eruptive behavior of the Sun's corona, as well as terrestrial climate change. His other published work encompasses such diverse fields as medical physics, plasma physics, solar energy engineering, electronic circuit design, nuclear issues, observational astronomy, and theoretical astrophysics. His knowledge of electronics stems not only from his professional activities as a physicist but also from a lifelong interest in electronics as a hobbyist.

In addition to *Understanding Modern Electronics*, Professor Wolfson has produced four other lecture series for The Great Courses: *Einstein's Relativity and the Quantum Revolution: Modern Physics for Non-Scientists*; *Physics in*

Your Life; *Earth's Changing Climate*; and *Physics and Our Universe: How It All Works*. He also has lectured for One Day University and *Scientific American*'s Bright Horizons cruises.

Professor Wolfson has spent sabbaticals at the National Center for Atmospheric Research, the University of St. Andrews, and Stanford University. In 2009, he was elected a Fellow of the American Physical Society. ■

Disclaimer

This series of lectures is intended to increase your understanding of the principles of modern electronics. These lectures include experiments in the field of modern electronics performed by an experienced professional.

These experiments may include dangerous materials and are conducted for informational purposes only, to enhance understanding of the material.

Warning: The experiments performed in these lectures can be dangerous. Any attempt to perform these experiments on your own is undertaken at your own risk.

The Teaching Company expressly disclaims liability for any direct, indirect, incidental, special, or consequential damages or lost profits that result directly or indirectly from the use of these lectures. In states that do not allow some or all of the above limitations of liability, liability shall be limited to the greatest extent allowed by law.

Table of Contents

Table of Contents

Table of Contents

Note: Some images in this guidebook include screen captures courtesy of CircuitLab, Inc., or powered by DoCircuits, www.DoCircuits.com.

Understanding Modern Electronics

Scope:

Electronic devices are woven into the fabric of 21ˢᵗ-century life. Some—your smartphone, your laptop computer, your flat screen TV—are obvious. Others—the myriad electronic sensors, computers, and controls that keep your car running at top efficiency; the microcontrollers that increasingly find their way into the most humble of kitchen appliances; the smart electric meters that help manage your household energy consumption—ply their electronic magic largely hidden from view. And devices that were once purely mechanical—thermometers, bathroom scales, clocks and watches—are now largely electronic.

In the context of this electronic world, *Understanding Modern Electronics* has two purposes, either one of which provides ample reason to view and participate actively in this course. First, the understanding this course provides can enhance your appreciation and effective use of electronic devices. Second, for those who might want to go on to build their own electronic circuits, the course lays the groundwork of basic electronic concepts, introduces the building blocks of electronic circuits, and develops many important circuits used in modern electronics.

The course begins by distinguishing electronics from electricity, introduces basic electrical and electronic concepts, and quickly moves into electronic circuitry. You will learn the language of electronic circuit diagrams and meet new and useful electronic components. After a look at AC versus DC, power supply circuits, and electronic filters (for, among other things, bass/treble controls and noise reduction), the course introduces the semiconductors that are at the heart of modern electronics. Focusing on transistors—among the most important inventions of the 20ᵗʰ century—the course then describes how these devices are used as amplifiers. By the end of this section, you will have a strong grasp of the workings of the audio amplifiers that are in everything from phones to televisions to stereo systems.

The course then turns to operational amplifiers—versatile, inexpensive, integrated-circuit devices that function not only as amplifiers for audio and for electronic instruments but can also add, subtract, accumulate, and serve as oscillators to generate useful electronic signals and sounds. Op-amps achieve their versatility through the magic of negative feedback; thus, this section of the course begins with a look at the feedback concept and the many systems—electronic and otherwise—that make use of negative feedback. Practical applications discussed here include an electronic thermometer, a light meter, and servomechanisms that allow us to control, with exquisite precision, the movement of massive machinery.

To this point, the course has emphasized analog electronics—circuits in which electrical quantities are analogs of continuously varying physical quantities, such as sound, temperature, and brightness. But today's electronic circuits are increasingly digital, encoding information as quantities that can take on only two values—the binary digits 0 and 1. The penultimate section of the course begins with digital basics, introducing the logical operations that are at the heart of all digital processing, including computers, then explores contemporary electronic realizations of these digital logic operations. The course goes on to build more sophisticated digital circuits, including computer memory circuits, electronic counters, frequency dividers (used in digital watches), and circuits that let computers exchange information with peripheral devices or help cell phones stream your voice over the airwaves as a sequence of digital 1s and 0s.

The final section of the course shows how electronic circuits convert information from the analog world of everyday life to the world of digital processing—and back again. Important examples include cell phone communications and contemporary music recording, distribution, and playback; these examples bring together material from throughout the entire arc of the course.

The course ends with a lecture on electronics in your future, showing how continued miniaturization and the associated increase in the power of electronic circuits will bring sophisticated electronics into virtually every aspect of 21st-century life.

You can learn about electronics as a purely intellectual exercise, and this course will help you do so. But electronics is also an active pursuit, whether at the hobbyist's workbench or in the engineer's laboratory. Throughout *Understanding Modern Electronics*, you'll get many practical tips about building electronic circuits and working with electronic instruments. Although the course doesn't provide you with actual hands-on experience, it comes close—through the use of electronic circuit simulation software that's web-based, independent of what type of computer or even tablet you use, and free or at low cost. If you choose to do the projects at the end of each lecture, you'll use this software to build, explore, troubleshoot, and even design electronic circuits that are almost like the real thing. ■

Abbreviations and Symbols

AC	alternating current
A	ampere
β	beta (transistor current gain)
DC	direct current
F	farad
Hz	hertz
kHz	kilohertz (k means 1000)
μF	microfarad (μ means one-millionth)
mA	milliamp (m means one-thousandth)
Ω	ohm
V	volt
W	watt

Electricity and Electronics
Lecture 1

The most obvious reason for learning about modern electronics is that electronic devices are ubiquitous in our world. Consider some of the things you use that are electronic: an iPad, a smartphone, a laptop, a TV, the instrument panel of your car, a GPS system, a camera, the loudspeaker in your audio system, and so on. We'll start in this lecture by distinguishing electricity—the more mundane topic—from electronics—the more sophisticated. Then, we'll cover a bit of background on basic electricity. Key topics in this lecture include the following:

- Electricity and electronics
- A brief history of electronics
- Moore's law

- Volts, amps, and watts
- Resistance and Ohm's law.

Electricity and Electronics

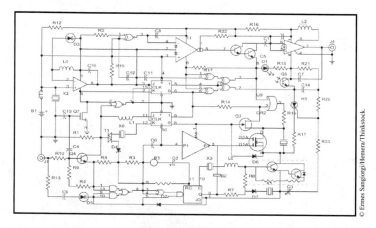

© Ermes Sangiorgi/Hemera/Thinkstock.

Electronics involves the control of one electrical circuit by another, and it involves using devices that allow that kind of control, often with weaker electrical quantities controlling stronger ones. An *electrical circuit* is an interconnection of components intended to do something useful, such as display video or amplify sound. Usually—but not always—a circuit includes a source of energy.

A Brief History of Electronics

For the first half of the 20th century, the device used to control electric current was the vacuum tube. These were evacuated glass tubes containing metal electrodes, between which electrons flowed. The tubes were designed in different ways for different functions. By applying different electrical signals to the *grid* (the *control* electrode in the tube), the amount of current that flowed through the tube could be controlled.

© Alessandro Felizian/iStock/Thinkstock.

© theIIPEN/iStock/Thinkstock.

In the 1950s, a revolution occurred in electronics with the invention of the transistor. The transistor did exactly what the vacuum tube did—that is, it allowed one circuit to control another—but it was much smaller, and it was a solid-state device.

© pigphoto/iStock/Thinkstock.

In the late 1960s and early 1970s came yet another development: the technological ability to interconnect many transistors and other electronic components on a single tiny wafer of the element silicon. That was the integrated circuit. Today, integrated circuits contain anywhere from a handful to billions of individual transistors.

Moore's Law

In 1965, Gordon Moore, a cofounder of Intel Corporation, made the following prediction: As we learn to make integrated circuits increasingly compact, the number of transistors on a single chip will double every 18 months. Actually, the number of transistors has doubled about every 2 years. Moore's prediction is known as *Moore's law*, and it will probably hold for at least another 10 years.

© Umberto Pantalone/iStock/Thinkstock.

Electric Charge

- Likes repel; opposites attract
- Protons carry positive charge
  - So do holes in semiconductors
- Electrons carry negative charge
  - Metals contain vast numbers of free electrons

Electric charge is a fundamental property of matter. It comes in two varieties: positive and negative. The protons that are in atomic nuclei are the main carriers of positive charge, and electrons are carriers of negative charge. There are vast numbers of free electrons in metals.

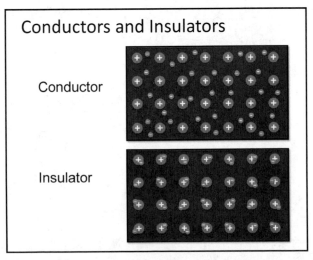

Conductors and Insulators

Conductor

Insulator

We can talk about free charges as a way of distinguishing two kinds of materials: conductors and insulators. An electrical conductor is a material containing charges that are free to move. In metals, the most common conductors, those free charges, are electrons. Typically, one or two electrons at the outermost periphery of the metal atoms become free, not bound to individual atoms. They're free to move throughout the metal, and that's what makes the material a conductor. An insulator is a material lacking free charges. In an insulator, all the individual electrons are bound tightly to the atoms and can't be moved.

Current and Voltage

- Current: a flow of charge
  - Amount of charge per time crossing a given area
  - Unit: ampere (amp, A)
    - 1 A is about 6 x 10^{18} electrons per second
    - 1 A is about the current in a 100-watt (W) incandescent light bulb
    - Common in electronics: milliamp, mA (0.001 A)
    - Direction of current is that of positive charge flow
      - Even though most current is carried by electrons
- Voltage: the "push" that drives current
  - A measure of energy per charge
  - Unit: volt (V)
  - AAA, AA, C, D batteries: 1.5 V; car battery: 12 V; wall outlet: 120 V
  - Typical in electronic circuits: 5–15 V

Electric current is a flow of charge, which is measured in amperes (A). Voltage is the push that drives current through a wire or an electrical or electronic device. It's a measure of the energy per charge, and its unit is volts (V).

Electric Power

Electric current: charge per time

Voltage: energy per charge

Multiply them together:

$$\text{current} \times \text{voltage} = \left(\frac{\cancel{charge}}{time}\right)\left(\frac{energy}{\cancel{charge}}\right) = \frac{energy}{time}$$

= POWER (watts)

Power (watts) = amps x volts

Multiplying current (flow of charge) by voltage (push) gives us a third important quantity, energy per time, or electric power. Power is the rate at which a system delivers, consumes, loses, produces, or transfers energy from one form to another. Power is measured in watts (W).

Resistance and Ohm's Law

Electrical Resistance

- Measures resistance to the flow of *current*
  - Property of a material and its geometrical size and shape
  - Ohm's law:
    $$\text{current} = \frac{voltage}{resistance} \qquad I = \frac{V}{R}$$
  - Via algebra:
    $$V = IR, \quad R = V/I$$
  - Unit of resistance: ohm (Ω)
    - $1\ \Omega = 1$ volt/amp

Typically, conductors have resistance; they don't let current flow easily. The resistance of a particular component is a function of both the material it's made of and its geometrical size and shape. There's a simple relationship, known as *Ohm's law*, between current and voltage: current = voltage/resistance, or $I = V/R$. The unit of resistance is the ohm (Ω). Ohm's law can be written in three different forms, but they're all equivalent: $I = V/R$, $V = IR$, or $R = V/I$.

Suggested Reading

Introductory
Brindley, *Starting Electronics*, 4th ed., chapter 1 through p. 28

Lowe, *Electronics All-in-One for Dummies*, book I, chapters 1–2.

Platt, *Make: Electronics*, chapter 1.

Shamieh and McComb, *Electronics for Dummies*, 2nd ed., chapters 1–2.

Advanced
Horowitz and Hill, *The Art of Electronics*, 3rd ed., chapter 1, sections 1.1–1.2.1.

Scherz and Monk, *Practical Electronics for Inventors*, 3rd ed., chapter 1; chapter 2 through section 2.5.

Projects

Stimulate the circuits for the projects in these lectures online at CircuitLab (www.circuitlab.com) or DoCircuits (www.docircuits.com).

Hard Starting
The connection between a car's battery and the wire supplying the starter motor has corroded to the point where its resistance is 0.05 Ω. When you try to start the car, the starter motor draws 100 A. What's the voltage across the bad connection?

LED Lamps

An LED lamp operates at 120 V and puts out the same amount of light as a standard 100-W incandescent lamp. It draws 150 mA of current. Compare the power consumption of the two lamps.

Questions to Consider

Answers to starred questions may be found in the back of this guidebook.

1. What is Moore's law, and how is it responsible for the proliferation of increasingly powerful and less expensive electronic devices?

*2. An electric stove burner has a resistance (when it's on) of about 40 Ω. It operates at 240 V (as do some large household appliances, such as stoves, water heaters, and electric clothes dryers). Find (a) the current through the burner and (b) its power consumption.

Electricity and Electronics
Lecture 1—Transcript

Welcome to modern electronics, a course in which I'm going to try to get you to understand the electronic devices that are so much a part of your world that are around you everywhere.

Why should you want to understand modern electronics? The most obvious reason is that electronics are ubiquitous these days. If you're a member of our modern technological society—and virtually everybody is, and you certainly are if you're watching this course because you're using electronics to do that—then electronics is everywhere. Look at some of the things that you use that are electronic: your iPad, your smartphone, your laptop, your TV, the instrument panel of your car, your GPS system, your camera, your loudspeaker in your audio system, the microphone that allowed us to record the music, and so on, and so on, and so on; in fact, the distinction kind of blurs because even your refrigerator or your toaster may have electronics in it.

This lecture today is going to distinguish electricity from electronics. Two different things: Electricity being a little more mundane, electronics being a little more sophisticated; I'm going to try to give you that distinction. Then I'm going to give you a little bit of background on basic electricity, not nearly at the deep level I would do in a physics course, but enough to get you equipped to handle what we're doing in here.

Let's look at some definitions. If you take a formal definition of electronics, this one from Wikipedia: Electronics is the branch of physics, engineering and technology dealing with electric circuits that involve active electrical components—and I'm going to emphasize that one—such as vacuum tubes, transistors, diodes, and integrated circuits, and associated passive interconnection technology, which means wires, plugs, and other things that connect these. That's a great definition, but I like a simpler definition. My definition, which isn't quite accurate but I think is simpler and captures the spirit: Electronics involves the control of one electrical circuit by another, and it involves using devices that allow that kind of control. Some would quibble. They'd argue there are electromechanical devices—things called relays, for example—that allow one circuit to control another, and we probably wouldn't call those electronic. But the essence of electronics is we have devices that allow one circuit to control another; often it's a situation where a weaker circuit with weaker electrical quantities controls stronger electrical quantities. That's what happens, for example, in an amplifier.

What's an electrical circuit? An electrical circuit is some kind of interconnection of things called components—and I'll describe a number of electronic components as we go on—and it's intended to do something useful. It could amplify; it could display video, as in a television, or a video monitor, or a screen on your computer or on your smartphone; it could make sound; it could do whatever. It usually, but not always, includes a source of energy. I'm going to give you just a couple of examples of when it doesn't include a source of energy: I drove down to the Teaching Company studio, the Great Courses studio, yesterday with a carload of electronic equipment and I went zipping through a number of toll booths, and I did so thanks to my EZ Pass, which is basically an RFID device in which the source of energy came from outside. It was beamed at me by the tollbooth, energized my circuit, and my circuit sent a signal back to it. So the circuit doesn't have to include a source of energy, but usually it does.

Our definition of electronics is basically looking at how one device or one circuit controls another circuit, and we're going to look at the devices that do that. I'd like to begin by giving you a simple history of electronic devices. This is about modern electronics, and if you look at the set behind me you see many eras of electronic devices, from fairly old to fairly modern. I want to distinguish what's different about modern electronics, and surprisingly—to me, at least, this always surprises me—there's less of a distinction than you might think. I've been teaching electronics for 30 years, and I teach basically the same fundamental ideas. What's changed is the devices we use, and more importantly, what's changed is the number of those devices we can cram into a given circuit. Let's take a look at a quick history of electronics.

I want to begin by talking about the devices that use one circuit to control another, because they've undergone a remarkable evolution since the early years of the 20th century, and yet their function is still the same. Let me begin: Some of you old timers may remember vacuum tubes. Vacuum tubes were evacuated glass tubes containing metal electrodes, and electrons flowed in the empty space between these electrodes. The tubes were designed in different ways for different functions, but the basic function was there was an electrode in there. It was called the grid, and by applying different electrical signals to that grid, you could control the amount of current—the amount of electric current, which I'll introduce soon—that flowed through the vacuum tube. This vacuum tube is probably from about the 1920s. It's a big clunky thing; they're obviously very fragile with that glass envelope; they consume a lot of power because they have a hot filament like a light bulb filament that has to be heated and glow to give off electrons; but there they are and that's what we had for most of the 20th century. Vacuum tubes got smaller as the 20th century evolved; there's one from probably the 1940s or 1950s. They got smaller still; here's a miniature tube from the 1960s probably, a little tiny thing. It still has a whole lot of pins coming out the bottom to connect all kinds of things to it.

Then a revolution occurred in electronics in the 20th century: In about the 1950s to 1960s came the invention of what's probably one of the most important devices ever invented in the 20th century, and that's the transistor. The transistor did exactly what these vacuum tubes did—that is, it allowed one electronic circuit to control another—but it was much smaller. More importantly, it was a solid-state device, meaning it was built out of solid materials, not a glass envelope with vacuum in it and electrodes stuck in there for electrons to flow back and forth. Here is a transistor, a typical transistor; this is a modern transistor. The early transistors looked a lot like this. It has only three wires, which is the minimum number of wires it could have for one circuit to control another. This particular transistor probably costs less than $1 and can do the same functions as this large vacuum tube, so we obviously had an enormous miniaturization of electronics with the invention of the transistor. In the 1960s and 1970s, everything was listed, "This is solid-state, solid-state"; it means it no longer had vacuum tubes, it had transistors. Here's a power transistor. It's built on a metal case that bolts down to a heat sink because it gets very hot. This could be used as the output amplifier of a stereo amplifier; it might be able to tolerate 100 watts of power or something like that.

Then in the late 1960s and early 1970s came yet another development, which was the technological ability to put lots of transistors and other electronic components interconnected on a single tiny wafer of the element silicon. That was the integrated circuit, and integrated circuits contain anywhere from a handful to billions—literally, these days, billions—of individual transistors; that is, billions of these functions. The first computer contained about 20,000 of these tubes, took up a whole room, and generated an enormous amount of waste heat. The integrated circuits in your personal computer, in your smartphone, and so on have literally hundreds of millions to billions of transistors on a little chip. Here's an integrated circuit; this particular one probably has a few dozen transistors in it. Here's a much bigger integrated circuit with probably thousands of transistors. Here's a memory module from a computer that consists of a whole bunch of integrated circuits, each of which probably has on the order of a few million transistors. Here—this is already obsolete—is an Intel Pentium 3 chip used in an earlier generation of personal computers, and, again, probably has several hundred million transistors.

We've gone from this device, which allows one circuit to control another, to integrated circuits that contain hundreds of millions, to, nowadays, billions of transistors that function similarly to this device. What's happened in electronics isn't so much a change in function over all these years, but a miniaturization and a huge increase in power because we're able to cram more and more of these control devices on a single circuit.

I just want to illustrate that with one example of that evolution. Here's a rather archaic device; we don't actually think about these devices so much because they're built into all kinds of things: radios. Here's a radio from probably about the 1950s; it's just an old AM radio. I'm going to turn it around and show you what's inside it. Inside it you can see several tubes in their glass envelopes; there's a special tube that has metal around it to shield it from electronic noise. That's the insides of that radio, and the company that manufactured this radio would boast about how many tubes there were in the radio. Next to the radio—the tube radio from perhaps the 1950s—is a transistor radio, an early transistor radio, circa about 1970. It's obviously a lot smaller. It can be powered easily by batteries or by being plugged into the wall; more on that in future lectures. If I turn it around, you can see it's crammed with components—a whole lot of different components that we'll learn about—but in there are nine transistors, and the company that built this radio boasts about, "This radio has nine transistors," carrying on the tradition of how many tubes did it have? My next radio I can't even take apart. If I did you wouldn't see much in it, but a lot of black rectangles that are the integrated circuits that make it up, many of which are so complex they perform almost all the functions of a radio in one chip. There it is, and it probably has in it—I'm guessing; I don't know what the circuits are in it, but I'm guessing—probably thousands to millions of transistors, even in something like that; probably more like thousands in that on its integrated circuits. There's the evolution of the radio: tubes, individual discrete transistors, and finally integrated circuits.

What's powered this revolution in electronics? Again, a revolution in the power of the electronic devices and the number of components we can cram into them, but not a revolution in the basic functions of these devices. Part of what's powered it is an idea that was promulgated by Gordon Moore, who was a cofounder of the Intel Company that we all know about that makes the processor chips in our personal computers, and in 1965 he made a prediction. He said: As we learn to make integrated circuits more and more compact, the number of transistors on a single chip—that is, a single piece of silicon that makes up one of these complete integrated circuits—would double every 18 months. He was off by a little bit; actually, the number of transistors has been doubling about every two years. The graph here shows you that actual data up until 2015 and then it's an extrapolation beyond that to where we're headed. We now have in today's integrated circuits—today's computer chips and so on—literally billions of transistors, and that number is doubling every couple years. That's the reason almost any electronic device you buy is obsolete almost as soon as you buy it and you envy the people who have the newer one just six months later.

That's Moore's law, and Moore's law seems to be holding and will probably continue to hold for probably at least another 10 years, and then some fundamental limitations imposed by quantum mechanics may come in and make us have to rethink how we make very small, miniature devices.

That's a quick history of electronics. Let me say a little bit now about this course, and then I'll move on and give you a little bit of a background on electricity. First of all, I'm not going into depth here in detailed physics. I'm not going to explain much of the physics of how these devices work; you can see other Great Courses for that, including ones that I've done. I'm giving you instead of kind of intuitive feel for electronic devices. I'd like you to be able to build basic simple circuits; more on that in lecture three, how you'll build them. You won't come out of this course an electrical engineer, but you'll have a sense of how the devices you use work.

Let me also say a little bit about myself as your teacher here. I've been teaching electronics, as I mentioned, for about 30 years. I teach a course called "Electronics for Scientists" at Middlebury College. It's not just for physics majors; it's for pre-meds, chemists, others who are going to use electronic instrumentation in their careers. We've had a course called "Electronics for the Intimidated," for the non-scientist. Although I'm a theoretical astrophysicist by research, I actually love electronics. I've been an electronics hobbyist since I was 12 years old, and I was actually hired at Middlebury College because somebody left in the middle of the year who was teaching their electronics course and I was hired. That's how I got my job: by being electronics hobbyist, actually. ✓

Let's talk a little bit about the basic things we need to know about electricity to understand electronics. There's a fundamental property of matter called electric charge, and it comes in two kinds called positive and negative. They don't mean the presence and absence of something; they're just names that Ben Franklin gave them. We know that like charges repel each other, so the diagram shows two positive charges repelling each other and two negative charges repelling each other; and that opposites attract, so we see also a negative charge and a positive charge attracting each other. The protons that are in the atomic nuclei are our main carriers of positive charge, although there's something called holes in semiconductors that we'll see a little bit later that are really important in the way transistors and other semiconductor devices work; and then the electrons carry negative charge, and there are vast numbers of electrons, free electrons, in metals. ╱

We can talk about free electrons as a way of distinguishing two kinds of materials: An electrical conductor is a material containing free electrons. An example is the electrons in metals; another example is the electron and these positive mysterious holes—which I'll talk about later—in semiconductors.

But the point is there are charges that are free to move, and that's what makes the material a conductor. So the diagram at the top shows a bunch of positive charges; these are the cores of the atoms, and in a metal, typically, one or two electrons at the outermost periphery of the atom, when the atoms come together to form a solid piece of metal, they become free, they aren't bound to individual atoms. They're bound to the entire metal, but they're free to move throughout it, and that's what makes this thing a conductor. In an insulator, on the other hand—shown in the diagram at the bottom—the individual electrons, all of them, are bound tightly to the atoms and they can't be removed. That's a very quick description of the difference between conductors and insulators. There's a much more detailed description in terms of quantum physics, but I'm simply not going to go into that.

So we have electrical conductors, and what can happen in electrical conductors is those charges can move. I'd like to introduce two quantities now—two electrical quantities, important quantities—that we'll deal with throughout this course that speak to the movement of electric charge through conductors. We have two things: We have electric current. Current is a flow of charge. You sit and look at a wire and you count the amount of charge that goes down the wire in a given time, and the amount of charge that goes by in a given time is a measure of the electric current. It's like watching a river and saying, "How many gallons per minute are flowing by? How much charge per, usually a second, is flowing by in this circuit?"

The unit of electric charge is the ampere. I'm not going to go into the detailed definition of the ampere, but an ampere consists of about 6×10^{18}. Wow. That's lots and lots and lots of charges, of elementary charges, the charges on the electron or on the proton—the charges are the same, but they have opposite signs—going past a given point every second. An ampere is a fairly familiar level of current. If you have an old-fashioned 100 watt incandescent light bulb—the kind of thing that's going obsolete pretty rapidly, as we'll see in a moment, and why that is—that's about an ampere; a little bit less. In electronics, we'll be dealing mostly with milliAmps, thousandths of amps, abbreviated lowercase "m" for "thousandths" and capital "A" for amperes—it's capital, by the way, because it's named after the Frenchman, Ampere, who worked a lot with electric current and its relation to magnetism—so we're going to be dealing often with milliAmps or mA in this course, and so I'll often use the term milliAmp or label a <u>current in mA</u>.

Here's an important point about electric current: The direction of the current is that in which positive charges flow. If you have electrons moving through a wire from left to right, the current is actually flowing from right to left, and that's a little bit confusing; you have to scratch your head about that. You can blame Ben Franklin, who assigned those terms not realizing that the dominant charge carriers in metal wires are electrons.

Then we have the voltage: the push that drives the current through a wire or through any other electronic device. If you have taken any kind physics course, you've probably learned that voltage is a measure of the energy per charge. The unit is the volt. If you want to know what a volt is, it's a joule of energy, the SI unit of energy per coulomb, the unit of electric charge; but we don't need to go into that. For example, common batteries: If you have an AAA battery, or an AA battery, or a C battery, or a D battery, those are all 1.5 volt batteries. That tells you how much push they have; that tells you how much energy they give each amount of charge. A car battery is about 12 volts. The wall outlet in North America supplies typically 120 volts. In the electronic circuits we're going to be developing in this course, typical voltages are in the range of 5–15 volts, sometimes a little lower in some really new circuits, and sometimes a little higher maybe in a power amplifier. We have a flow of charge, current, and we have a push, the voltage, which pushes current through devices.

Now let's develop a couple of relationships that will help us to understand how current and voltage behave in circuits and how they're related. I'm going to move over to our big monitor, which we'll be using in this course as a kind of electronic blackboard. If I ever have to do mathematics—and this isn't a heavily mathematical course, but there will be a little bit of math in it occasionally—I'll do that on the big screen. We'll also use the big screen to display the output of certain instruments so we can see them a little more carefully.

Let's go and look at the quantity that we're going to call electric power, a third quantity in addition to current and voltage. Electric current is the amount of electric charge flowing per a given time. Voltage, as I indicated, is the energy each charge has or gets from the energy source in whatever circuit it's in. If I take and multiply these two together, current times voltage is going to be charge per time (that's current), energy per time (that's voltage)—a little algebra here; you notice charge on the top and charge on the bottom—I can cancel them, and I end up with energy per time. If you've had a physics course, energy per time, you know, is power. Power is the rate at which a system delivers energy, consumes energy, loses energy, produces energy, transfers energy from one form to another, whatever. It's a rate; it's the rate of energy consumption, generation, whatever. In electric circuits, this little sort of word equation tells us that current times voltage gives us power. Power has the units of watts, named after James Watt. You may think of watts, kilowatt hours, and things like that as uniquely electrical. They aren't; watts measures energy rates in any kind of system. But in electricity, watts are particularly convenient because multiplying the current in amperes times the voltage in volts gives you the power in watts.

Let's do just a quick example where we worry about electric power; power being watts, amps, times volts. We'll do some power examples, and we'll remember that power in watts is simply current times voltage. Later we'll use symbols for these, but right now I'm just going to call it amps times volts. Here's a question: Your car's 12 volt battery delivers 125 amps while it's cranking the starter motor, trying to get the car started. What's the power it supplies? Power is amps times volts, and here we have 125 amps times 12 volts. That's going to be a little bit more than 10 times 125; it's going to be somewhat over 1,000. It comes out 1,500 watts or 1.5 kilowatts, where k means 1,000. That's, by the way, roughly the same power that a stove burner consumes. If you have a hot stove burner on high, it's consuming power at the rate of about 1,500 watts, and that's the rate at which your car's starter motor is extracting energy from your car's battery when it's cranking. When you're just running the headlights, or running the GPS, or running the instrument panel, or the ignition system, it's less; but while you're starting the battery's got to be able to supply a lot of power, over a kilowatt.

Here's another example, perhaps one you worry about more in these days: Your cell phone has typically a 3.7 volt battery; it consumes about half a watt while you're talking—significant for that tiny little device—and about 25 milliwatts, thousandths of a watt, when it's on standby. What's the current delivered by the battery in each case? We can handle that one with some algebra. We can rearrange that equation, dividing both sides by volts, and we have amps is power divided by volts. When we're talking, we've got half a watt, 0.5 watts, divided by 3.7 volts, and that comes out 0.135 amps or 135 milliAmps. Again, that's a significant amount of current for a relatively small battery to deliver. When we're on standby it's 25 milliwatts over 3.7 volts, and notice the "milli" in milliwatts makes the answer just come out in milliAmps. I don't need to do a conversion; we'll often be doing little tricks like that. In standby mode you only consume current—charge flowing out of the battery—at a much lower rate, and we could calculate, by multiplying these currents by the voltage, what the actual power is in that case. There are a couple of examples of power as volts times amps.

Let's now look at one other relationship between current and voltage, and it has to do with this pushing by voltage of current through a conductor. Conductors, typically—with the exception of things called superconductors, which we won't go into in this course—have resistance; they don't let current flow easily. The resistance of a particular component is a function of both the material it's made of and its geometrical size and shape. There's a simple relationship between current and voltage when we know the resistance of a material. It's called ohm's law, and it says current is voltage divided by resistance. "I" is going to be our symbol for current, "V" for voltage, "R" for resistance; so we can write it $I=V/R$, or we can manipulate it algebraically to write $V=IR$ or $R=V/I$. Those aren't different laws; they're just different ways of writing the same thing. The unit of resistance is called the ohm. If you have one volt and it pushes one amp through a material, a conductor, than that conductor has a resistance of one ohm.

By the way, ohm's law isn't a deep fundamental law like the laws of electromagnetism, Gauss's law and Faraday's law, Maxwell's equations and so on; it's not like the second law of thermodynamics; it's not something fundamental. It's something that happens to be true, at least approximately for a number of materials. Resistance measures the resistance to the flow of electric current, and it's measured in ohms. We're going to just use ohm's law to do some examples; but before that, let me just show you some resistors.

I have over here some devices that are made to have electrical resistance. They typically look like a little tiny cylinder with some stripes on them. They're designed to have particular resistances. They're typically made of carbon that's been compressed. Here's a slightly bigger one. Here's a variable resistor that might be a volume control in an old radio like this one. Here's a big, heavy resistor that can dissipate a lot of power. Here's an enormous variable resistor that I'll use later in the course. Here's is a big resistor: The hot plate burner is simply a material that's got a lot of resistance—or not a lot, but enough resistance—that current flows and heats this device up. Those are resistors.

Over here I have an ohm meter, which I'm going to use to measure resistance. In a subsequent lecture we'll come to understand how this device works. I'm going to grab this resistor, which happens to be a slightly bigger one than the little one I showed you, but that's because it can dissipate more power. I can read the color code off it; I'm not going to teach you about that, but that says this is a 2,700 ohm resistor, and it says 2.64 kilo ohms. There's a little gold band at the end of this resistor that says it's good to within 5%, and that's within 5% of 2.7 kilo ohms or 2,700 ohms. There's a resistor and I've just measured its resistance; so many kinds of resistors.

Let's take a look briefly at how we might use ohm's law to do some experiments here. Let's talk about using ohm's law just briefly. Again, ohm's law written in three different forms but they're all equivalent; and so here's the question: A typical 120 volt household outlet can supply up to 20 amps; after that the circuit breaker goes, to keep so much current from flowing that wires would heat up and may be risk of fire. What's the minimum resistance you can connect across it, and what's the maximum power? Let's use the form resistance equals volts over amps: 120 volts over 20 amps, 6 ohms. You can connect 6 ohms, 7 ohms, 8 ohms, 100 ohms, a million ohms, no problem. You connect 5 ohms, you're going to blow the circuit breaker. Power is I times V. That's 20 amps times 120 volts; that's 2,400 watts or 2.4 kilowatts. More than that car starter, but not much more; more than that electric stove burner, but not much more. An electric stove burner would typically be operated off a bigger circuit. Here's another one; more abstract: What's the voltage across a 1.8 kilo ohm resistor when it carries a current of 5 milliAmps? Voltage: I'm going to use the form V=IR. V=IR; 5 milliAmps times 1.8 kilo ohms is 9 volts. A little 9 volt battery, the kind with the little snaps at the top, you put a 1.8 1,000 ohm, 1.8 kilo ohm resistor across it; 1,800 ohms, and you'd get 5 milliAmps flowing.

There's how we use ohm's law. Notice something: When we have "m" in milliAmps, which we're going to use a lot in electronics, and we have "k" in kilo ohms, which we're also going to use a lot, those two cancel. The thousandths and the 1,000 cancel and we get the voltage right out in volts.

Let me wrap up by just showing you a few interestingly special resistors. Inside your computer's hard drive, if you still have a computer with an old-fashioned mechanical hard drive, is a resistance that depends on how much magnetic field it's exposed to. As the hard disk spins and the magnetized information goes by that, its resistance changes, and that's how you read the information. Here's another interesting resistor: This is a photo resistor. Its resistance changes with light, and we're going to use photo resistors and related devices in several circuits we'll be building later. I'm connecting the photo resistor up, and you'll notice it has 0.123 kilo ohms, 123 ohms. Now I simply cover it up with my hand so it can't see the light and that number goes up; it's already up to about 50-something thousand ohms. This is a light-sensitive resistor, a photo resistor. Again, we'll use that in a lot of useful applications. I imagine you can even think of some right now that you might want to use.

Let's wrap this lecture up. First of all, we've seen the essence of electronics: It's about one circuit controlling another. We have electric current, the flow of electric charge; we have that occurring in conductors. Its unit is the ampere. We have voltage, the push that drives the current through a conductor; it's the energy per charge. Its unit is the volt. We have resistance: Resistance R resists the flow of current, and the unit is the ohm. Electric power is voltage times current. Finally, we've worked with ohm's law and understood how voltage and current are related and related to resistance.

That's all I have to say now except I want to remind you that each lecture in this course has with it a project, which you are welcome to do. It's separate, but you can do the project. The project is sometimes a simple calculation, but more frequently it's going to be you're designing or working with a circuit. Here's the project; if you'd like to go on and do the project part of this lecture, you're welcome to, but you certainly don't have to: The first is about a real situation in which your car battery gets corroded. There's a little too much resistance, and you can't start your car. The second one is about an LED lamp, and that will show you why an LED lamp is so much more energy efficient than that old-fashioned 100 watt incandescent lamp.

Circuits and Symbols
Lecture 2

In this lecture, we'll look at electronic circuits, that is, interconnections of electronic components. As you recall, in the last lecture, we looked at a particular electronic component, the resistor. We'll look at resistors in more detail in this lecture, but we'll also learn how to represent many other kinds of electronic devices in circuits and how to read circuit diagrams. In addition, we'll begin to see how to interconnect electrical systems. Key topics we'll cover are:

- Components and their symbols
- The ideal battery
- Voltage-current characteristics
- Series components: resistors

- A simple circuit: the voltage divider
- Real batteries
- Parallel components: resistors.

Components and Their Symbols

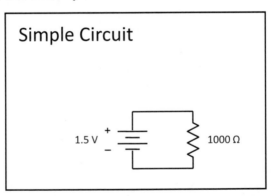

A simple electrical circuit might consist of a 1.5-V battery connected with a 1000-Ω resistor. This circuit has a source of electrical energy, a battery, and a *load*, something to which that energy is supplied. That load might be a loudspeaker, a motor, or as here, simply a resistor.

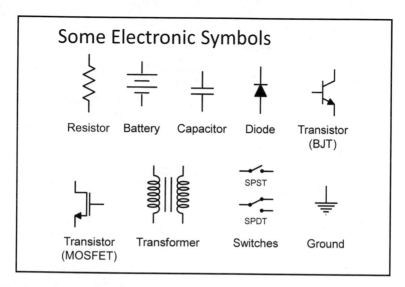

Some Electronic Symbols

Resistor   Battery   Capacitor   Diode   Transistor
(BJT)

Transistor   Transformer   Switches   Ground
(MOSFET)

SPST

SPDT

The symbols that represent circuits constitute a language to learn.

The Ideal Battery

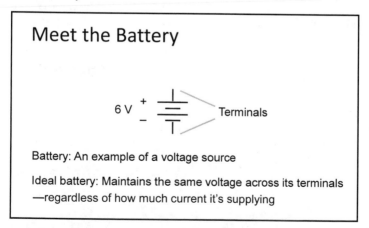

Meet the Battery

6 V Terminals

Battery: An example of a voltage source

Ideal battery: Maintains the same voltage across its terminals
—regardless of how much current it's supplying

A battery is a simple example of a *voltage source*. It converts chemical
energy into electrical energy, and it produces, in principle—if it's ideal—
exactly the same voltage across its terminals, regardless of how much current
it supplies.

Voltage-Current Characteristics

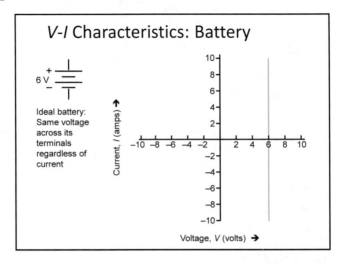

To describe the voltage-current (*V-I*) characteristics of an ideal 6-V battery, we can draw a graph with voltage (V) on the horizontal axis and current (A) on the vertical axis. The resulting *V-I* characteristic curve is a straight vertical line—6 V, regardless of current.

We can draw a similar graph for a 1-Ω resistor. The resulting curve for the resistor is a straight diagonal line with a slope of 1 A for every volt.

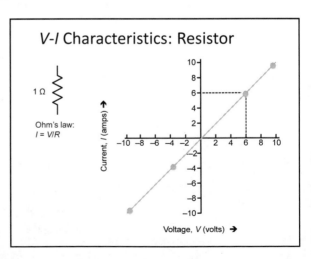

Series Components: Resistors

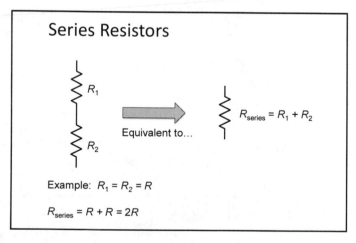

Series Components

- Two components are in series if current flowing through one of them has nowhere to go but into the second component
- Same current (I) through both components

I   Component 1

I   Component 2

Two electronic components are in series if the current flowing through one of them has nowhere to go but into the next one. Because the current that's flowing through one device goes into the other, the same current flows through both components.

Series Resistors

R_1

R_2

Equivalent to...

$R_{series} = R_1 + R_2$

Example: $R_1 = R_2 = R$

$R_{series} = R + R = 2R$

Imagine we have two resistors, R_1 and R_2. Because current through R_1 can flow only through R_2, the two are in series. Resistors in series simply add; thus, the series resistance is the sum of the two resistances.

A Simple Circuit: The Voltage Divider

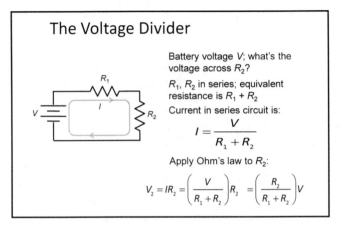

A *voltage divider* is basically two series resistors, R_1 and R_2, across some source of voltage, V. This device divides the voltage in proportion to the resistances.

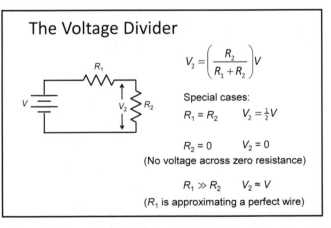

One special case of a voltage divider is when R_1 and R_2 are equal. In that case, the output voltage, or the voltage across R_2, is ½ V. If R_2 is 0, we get no voltage across it; it's a perfect wire. If R_2 is much greater than R_1, then V_2 is almost equal to the battery voltage; R_1 becomes almost a perfect wire if R_2 is much greater than R_1.

Real Batteries

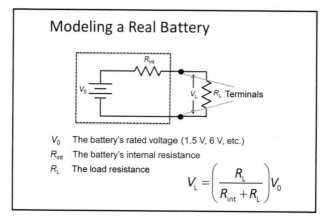

Modeling a Real Battery

V_0 The battery's rated voltage (1.5 V, 6 V, etc.)
R_{int} The battery's internal resistance
R_L The load resistance

$$V_L = \left(\frac{R_L}{R_{int} + R_L} \right) V_0$$

To model a real battery, imagine an ideal battery in series with a resistor, representing the battery's *internal resistance*. Real batteries or any real sources of voltage have internal resistance, which must be taken into account to use them correctly. The loads put across them can't be too low so we don't drop a voltage across that internal resistance, which would make the voltage across the battery terminals much lower than the battery's rated voltage.

Parallel Components: Resistors

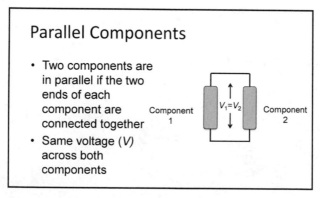

Parallel Components

- Two components are in parallel if the two ends of each component are connected together
- Same voltage (*V*) across both components

Component 1 $V_1 = V_2$ Component 2

Two components are in parallel if the two ends of each component are connected, and therefore, they have exactly the same voltage across either of them.

Parallel Resistors

R_1, R_2

Equivalent to...

$$\frac{1}{R_{parallel}} = \frac{1}{R_1} + \frac{1}{R_2}$$

Example: $R_1 = R_2 = R$

$$\frac{1}{R_{parallel}} = \frac{1}{R} + \frac{1}{R} = \frac{2}{R}$$

$$R_{parallel} = \frac{R}{2}$$

Parallel combination is sometimes called a *current divider* because current coming into the combination splits between R_1 and R_2, with more current going through the lower resistance.

Suggested Reading

Introductory

Brindley, *Starting Electronics*, 4th ed., chapter 1, p. 28 to end.

Lowe, *Electronics All-in-One for Dummies*, book I, chapter 5.

Platt, *Make: Electronics*, chapter 1.

Shamieh and McComb, *Electronics for Dummies*, 2nd ed., chapter 3.

Advanced

Horowitz and Hill, *The Art of Electronics*, 3rd ed., chapter 1, sections 1.2.2–1.2.3; appendix B.

Scherz and Monk, *Practical Electronics for Inventors*, 3rd ed., chapter 2, sections 2.11–2.13.

Resistors

Using only 1000-Ω resistors (as many as you want), design a voltage divider that produces 4 V when connected across a 6-V battery.

Suppose you connect a 10,000-Ω resistor (10 kΩ) across your voltage divider between the points where you're supposed to have 4 V. Will the voltage across this resistor be exactly 4 V? If not, explain why not and determine what the actual voltage will be.

Questions to Consider

*1. Identify pairs of series and parallel components in the circuit below.

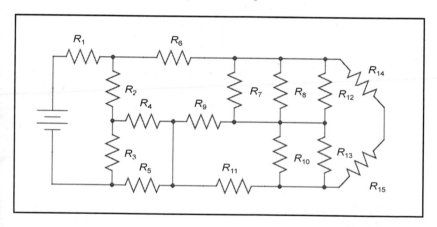

*2. A 6-V battery has an internal resistance of 0.25 Ω. If you connect a 5-Ω resistor across the battery, what will be the voltage across the battery's terminals?

Circuits and Symbols
Lecture 2—Transcript

Welcome back. In today's lecture we're going to be looking at electronic circuits; that is, entire interconnections of electronic components. You may recall in the preceding lecture, we actually looked at a particular electronic component, the resistor, in some detail. We'll look at it a little more detail here, but we'll also learn how to represent many other kinds of electronic devices in circuits. We'll learn how to read diagrams for circuits, and we'll learn a little bit about how to interconnect electrical systems.

Let me give you an example of a complete circuit. Over here I have a board from a computer, probably a decade old by now, but it still illustrates the main idea here that we're talking about devices that are interconnected. In modern circuits, they're almost always interconnected by putting them on a board made of insulating material and wires are etched out of copper that was originally applied to the board in a solid piece, and that's how the interconnections are made. In the old days, they were interconnected by running wires from one component to another and soldering them together. But the point is an electronic circuit is simply an assemblage of interconnected electronic components, a big interconnection or a very small interconnection. I have another circuit here or part of a circuit consisting of a couple of light bulbs, and they're simply connected together by this one wire, and later we'll connect those up into another complete circuit.

We'd like to represent circuits in a kind of symbolic way, so let me give you an example of how we're going to do that. Here's a complicated-looking circuit; it happens to be the circuit diagram for an audio amplifier. If it looks daunting to you right now, come back after a few lectures and you'll actually be able not only to identify the individual electronic components in this circuit diagram, but you'll also be able to kind of follow through what the circuit diagram does and what each component does. Just to give you a few examples: On the left you see an input. You see a number of zigzag structures; those are resistors that we talked about in the preceding lecture. You see some devices with three wires coming out of them or three connections coming out of them, and one of them has a little arrow on it; those are transistors. You see some capacitors, and you see on the right hand side a loudspeaker, because this is an audio amplifier. What we have here is a language to learn: the language of electronic circuit diagrams, schematic diagrams. We could draw pictorial diagrams—I could draw a picture of a transistor, and a picture of a battery, and a picture of a resistor—but different resistors look different, and different transistors look different. But their circuit symbols look the same, and that gives us a common language for representing electric circuits.

Let me give you some other examples. Here's a much simpler example of perhaps the simplest electronic circuit—or electrical circuit in this case—that you might be able to build. This one consists of a battery, typically a C or D flashlight battery, connected in series with a resistor. It's 1,000 ohm resistor—I can tell that by reading off its color code—it's a 1.5 volt battery. On the left is a picture of what that circuit would actually look like if you put it together. The white lines represent wires, which we consider to be perfect conductors; I'll say a little bit more about that later. We ignore any resistance they might have, but there's plenty of resistance in that resistor: 1,000 ohms.

On the right I've shown the circuit diagram, and in that diagram are two electronic circuit symbols: the symbol for a battery, which consists of this series of alternating long and short lines. Technically, a battery consists of several cells. This is actually just one cell, but we're going to use this symbol for batteries in general; I'll talk a little bit more shortly about what a battery does. That battery in the circuit is marked with its voltage; it's a 1.5 volt battery. Remember, a volt is a measure of the push that that source can supply in terms of pushing current through the circuit. On the righthand side, you can see connected directly across the battery is the 1,000 ohm resistor as represented by that zigzag kind of symbol.

There's about the simplest circuit you could imagine making: A simple circuit consisting of a source of electrical energy, a battery, and a load, something to which that energy is going to be supplied; sometimes called a load, maybe it's a loud speaker, maybe it's a motor. In this case, it's simply a resistor that's going to heat up with the power that's dissipated in the resistor.

What we'd like to do is look briefly at some symbols for other circuit elements. I won't introduce them all now, but I'll simply show you them, and then throughout the course of the lecture series you'll see each of these devices introduced. Here are some common electronic symbols; we'll use every one of these in this course, some of them coming much later, some of them coming much sooner: On the left at the top, you see the wiggly symbol that's the resistor that I just introduced. We see the battery. We see a capacitor, which isn't the same as a battery; it consists of two equal-length lines with a gap between them that reflects the electrical nature of the capacitor. Sometimes you'll see one of those two horizontal lines drawn as a curve, but that's still a capacitor. Next element is a diode, a kind of one-way valve for electric current. We'll be using those shortly. On the top right is a transistor, a particular kind of transistor called a bipolar junction transistor, which is commonly used in, for example, audio amplifiers. On the bottom left is another kind of transistor that we'll have more to say about when we get to logic circuits, digital circuits in particular. Next is a transformer, a device for changing voltage levels in electric circuits. We'll be using that when we build a power supply in lecture four. Right after that come a pair of switches. Switches are simply devices for opening or closing an electric circuit. In the simplest case of the single pole, single throw switch at the top. In the case of the switch at the bottom, a single pole, double throw switch. It's a switch that can be moved to one of two different possible positions, and we'll have occasional uses for these switches.

At the bottom right is a really important symbol: It's a ground symbol. Ground is a slightly ambiguous meaning in electronic circuits, but in most electronic circuits there's a common point to which many things are connected. It's often, in a battery powered circuit, the negative terminal of the battery. In a car, for example, the metal body of the car is called ground, and it's usually—not always, but usually—connected to the negative terminal of the battery. In electric power wiring, the electric power system is actually, for reasons of safety, connected directly to the physical earth ground. Outside your house there's a copper stake—typically a copper stake—driven many feet into the ground, and your electric power system is tied to ground there.

I mentioned ground because we'll often use ground as a shortcut. Instead of drawing all the wires that come back to the negative terminal of the battery or the negative terminal of the power supply, we'll show a ground symbol. Many circuits we'll talk about—and particularly some of the simulations we'll do of circuits in the project—will require us to have a ground symbol in there to have the circuit be grounded. That ground symbol is going to be a common, important one. Sometimes you'll see it not drawn as this series of decreasing-length horizontal lines, but instead as simply a little open triangle pointing down at the bottom; so we're going to deal with ground symbols. Those are some electronic symbols.

What we'd really like to do is to be able to characterize the electrical characteristics of all these different kinds of devices. We can't do all of them in exactly the same way, but a remarkable number of them we can characterize in a similar way. The way we're going to characterize them is by drawing graphs that show their characteristics. I'm going to start with the battery. I have that battery symbol again. I'm marking the plus and minus terminals of the battery—the longer bar on the outside edge is the plus terminal of the battery, and the lower one, the shorter one, is the minus terminal of the battery—and we'll mark what the battery's voltage is. That's a 6 volt battery, for example. Over here I have a 6 volt battery, a common battery; it's called a lantern battery. It actually consists of four 1 1/2 volt cells inside it. These two spirally wire things at the top are the two terminals. If I look at it, the middle one says minus; so that's the minus terminal and that's the plus terminal. Every battery has two terminals and a rated voltage.

The battery is one example of what's called a voltage source. There are plenty of other examples: An electric generator could be a voltage source; a so-called power supply is a voltage source; the wall outlet is a voltage source, which is a varying voltage source, as we'll see in another lecture. But a battery is a very simple example. It converts chemical energy into electrical energy and it produces, in principle, if it's ideal, exactly the same voltage across its terminals—in the case of this one I'm showing, that would be 6 volts—regardless of how much current it's supplying. That turns out to be an impossibility; if we could have an ideal battery, we would've solved all the world's energy problems (I'll talk a little bit more about that later). But that would be an ideal battery.

Now I'd like to try to characterize an ideal battery by drawing a graph that shows how the voltage and current are related in an ideal battery. As I do that, remember, an ideal battery maintains the same voltage across its terminals, regardless of how much current you or the external circuit it might be connected to is asking it to supply. For example, that 6 volt battery I just showed you is sitting on the table with nothing connected to it. It's not being asked to supply any current right now; it's supplying zero current. But even if it were supplying 100 amperes of current, if it were ideal it would still have 6 volts across those two terminals.

The key idea here is going to be what we call VI characteristics, voltage current characteristics. I want to show you how we describe VI characteristics first for a battery, then for a resistor. Then, as I introduce new components, we'll introduce their VI characteristics, which will be more interestingly complicated.

Here's our battery again. It's an ideal battery, the same voltage across its terminals regardless of how much current is flowing, and I happen to have chosen it to be a 6 volt battery. To describe the VI characteristics, I want to draw a graph in which voltage is on the horizontal axis, in volts in this case, and current is on the vertical axis, in amperes in this case. Voltage can be positive or negative, depending on which sign is in which place, and current can be positive or negative, depending on which way the current is flowing. That distinction isn't terribly important for a simple circuit like a battery in a resistor, but it becomes very important when we talk about some unusual semiconductor devices whose VI characteristics look very different depending on which way the current is flowing.

Remember our ideal battery: It has the same voltage across the terminals, regardless of the current. In this case it's a 6 volt battery, so it has 6 volts regardless of the current. The VI characteristic curve—it's not a curve, it's a straight line—of that battery is simply a straight vertical line. Go up or down that line and you're changing the value of the current—in fact, you're even changing the direction as you go through zero—but if the battery's ideal, it maintains 6 volts regardless.

What about a resistor? We're going to use the same graph. We're going to put up in the upper left corner a 1 ohm resistor. Remember ohm's law: The current is the voltage divided by the resistance. That's a 1 ohm resistor; so that says if I put, for example, 6 volts across that resistor, 6 amps will flow. The 6 volts, R is 1 ohm, I is V/R, 6 amps. There's one point on the VI characteristic of that resistor. Here's another point: If I put 10 volts across the resistor, 10 amps will flow, because I is V/R. If I put 0 amps, 0 volts across it, no current will flow. That would be another point. If I put -4 volts across it, in this case—put a 4 volt battery across it the other way—I'll get -4 amps of current flowing. If I put -10 volts across it, I'll get -10 amps flowing. You can connect the dots, and you could see that the VI characteristic for a resistor is simply a straight diagonal line. In this case, that diagonal line has a slope of 1 amp per every volt because this is a 1 ohm resistor.

What would happen if I made it into a 2 ohm resistor? Two ohms has twice the resistance, so for a given voltage only 1/2 as much current would flow. A 2 ohm resistor would have the same kind of characteristic curve—that is, a diagonal line—but the slope of that line is less, reflecting the fact that we get less current for a given voltage through the 2 ohm resistor. Think a minute about what a 1/2 ohm resistor would look like. A 1/2 ohm resistor has less resistance than the 1 ohm or the 2 ohm. What would happen in that case? In that case, I'd get more current flowing. The VI characteristic for the 1/2 ohm resistor, 0.5 ohms, is another diagonal line that is in this case steeper because more current flows for a given voltage. That's the VI curves, the VI characteristics, for two devices we've looked at now, namely a battery and a resistor.

Now I want to move on, not to characterizing individual devices, which is interesting in itself, but asking: How do we connect devices together? In fact, there are two important ways of connecting electronic devices, and let me define those and talk about them a little bit. Two components, any components, any kind of electronic components, are in series if the current flowing through one of them has nowhere to go but into the next one. Here's component one, here's component two. If electric current flows through component one, the charge that's flowing has no place to go but into component two. That defines series resistors, and it has to be that way. There can be no other place for the current coming out of one component to go but into the second component. If it has a choice, if there's more than one possibility, they aren't in series. Consequently, because the current that's flowing through one device goes on into the other, the same current flows through both components. I have to qualify that a little bit. I'm talking about circuits where things are happening in kind of a steady state. If electric charge were somehow building up on that wire between the two components, this would no longer be true. But that would be a very impossible situation physically, because soon a big voltage would build up and it would drive a current through the second component, and it would be the same. When you see components in series, they have the same current through both of them.

Here's a little question: Are components one and two in series? How about one and three? Look at what happens to the current through one: It has the possibility of splitting, some of it through two, some of it through three. None of those resistors are in series; there are no series combinations in that particular situation.

That's generally what series means. Let's talk now about connecting resistors in series. Here are two resistors in series; I'm labeling them R1 and R2. Current through R1 can only flow through R2, so they're in series. Without going through the physics of deriving this—and it's not terribly difficult— you could convince yourself that the series resistors is the sum of the two resistances. Resistors in series simply add. Think about a garden hose: The pressure is like the electrical voltage, the push; the flow is like the current. If you attach an extra 100 feet of hose, it's going to be harder to get water through that hose. The resistance goes up, and the same thing happens when you put resistors in series. Those are series resistors. They simply add. Here are some examples: If R1 and R2 are the same—and just call it R—then R series is simply twice R.

That is series resistors. Let me do a little bit of mathematics with series resistors over here on the big screen because I want to describe a particular circuit configuration that's going to be really important in this course. It's called a voltage divider, and it's basically two series resistors across some source of voltage. The battery voltage is V; we have resistor R1 and R2. We want to know: What's the voltage across resistor R2? Those resistors are in series, and the equivalent resistance we've just seen is the sum R1+R2. We know ohm's law, current is voltage over resistance, so the current that flows in that circuit—and here it is going around the whole circuit; everything's in series, same current through everything—is $I=V/R1+R2$. Let's apply ohm's law to R2, because we're asked about the voltage across R2. V is current times resistance. That's this current that we've just derived multiplied by R2. If I do a little tiny bit of algebra on that, I see that what happens is the voltage across the lower resistor, R2, is simply this fraction: R2 over the total resistance times the battery voltage. That's why it's a voltage divider; it divides the voltage in proportion to the resistances.

Let's just take a look at some special cases of the voltage divider. One special case is when R1 and R2 are equal. In that case, got two R down there and one R up there, and the output voltage or the voltage across R2 is half; the voltage divider divides the battery voltage in half. If R2 is 0, on the other hand, we get no voltage across it. It's a perfect wire. Wires don't have voltage across them; no voltage across zero resistance. Finally, if R2 is really big compared to R1, then V2 is almost equal to the battery voltage. I'll have a lot more to say about that in a minute when I talk about ideal batteries. R1 becomes almost a perfect wire if R2 is much bigger than R1. That's the voltage divider; a really important configuration that we'll see again and again and again throughout this course.

I have an example of a voltage divider over here. Here are a couple of circuits consisting of a couple of resistors. Are they resistors? Yes; they're lightbulbs, and incandescent lightbulbs are just resistors. I'm going to connect it across this 6 volt battery. I'm going to come around and get my voltmeter fired up. My voltmeter is set to read DC voltage. I'm going to read the voltage across this second light bulb, and you can see it's just about 3 volts. We're across a 6 volt battery and we have two identical resistors; they happen to be lightbulbs. Consequently, we're good with 3 volts. We've seen that that voltage divider is working; so there's an example of a voltage divider.

I want to take this voltage divider idea and talk a little bit about real batteries, because real batteries are important to understand. Here's how we think about a real battery. This isn't really what's inside a battery, but a way to model a real battery is to think it has an ideal battery—that impossible thing—in series with a resistor, which we call the internal resistance. You can't get inside the real battery; all you have access to are those two terminals on the end of it. What you typically do is connect something—a lightbulb, a whole complicated circuit, a resistor, whatever—across the battery, across its two terminals. The battery has some intrinsic voltage, V sub 0. It has an internal resistance, R sub internal. The battery's V sub 0 is the battery's rated voltage: 1.5 volts, 6 volts, whatever it is. R int is this so-called internal resistance. It's not really a resistor. The battery companies don't take ideal batteries; if they did, they'd solve all the world's energy problems and slap a resistor in series with them. The resistor represents, rather, some physical resistances of the materials the battery's made of. But more importantly, it represents the speed with which the chemical reactions in the battery can replenish, charge, and move positive charge to the positive terminal, and negative to negative. This is sort of a model for a battery.

Now we put a load resistance across it; again, the word *load* meaning what we want to supply power to. If you look at this circuit, you'll see it's exactly like voltage divider circuit we just had. RL is the load resistance, and we want to know what the voltage is across the load. I connect the 6 volt battery, and I connect the resistor or a lightbulb across, and I may not get 6 volts across it. It depends on that voltage divider; it depends on R internal and R load. There's that voltage divider fraction that we saw before: V load is that fraction; R load, that second resistor—I called it R2 before—over the sum of the resistances times V naught. If you want V load to be equal to or approximately equal to the internal rated voltage of the battery, then as we saw when we derived that voltage divider equation, that second resistor, R2 or R load in this case, has got to be much bigger than the internal resistance.

Let me give you a quick demonstration of that. Here I have a simpler circuit, just one lightbulb, one resistance, connected across a 6 volt battery. It's a beefier lightbulb, which means it takes more current and draws more power from the battery. There we go: The lightbulb is lit; the battery seems to be doing its job well. I'm going to disconnect the lightbulb, and I'm going to connect the voltmeter so I can read the battery voltage. There it is, 6.1 volts. That's a reasonably healthy 6 volt battery. It might actually ought to be about 6.3 volts, but that's OK. Now I'm going to connect the lightbulb across it at the same time, and you'll notice the battery voltage has dropped to 5.1 volts. It dropped by a whole volt. That's an indication that this battery isn't doing too well; or to put it a different way, the internal resistance of this battery is high enough that we can't supply current to this particular lightbulb without the battery voltage dropping because we're losing energy in that internal resistance, current is flowing through the internal resistance. Ohm's law says there's therefore a voltage across it, and so that battery isn't working as well; it has too high an internal resistance for that particular application.

This is an old battery. If I had a brand new fresh battery, it'd probably be OK. You may have noticed a similar phenomenon when the furnace comes on in your house or the oven cycles on or something and the lights dim. What you're seeing there is the same effect, but caused by extra current flowing through the internal resistance of the household wiring.

You might ask: What's different about different batteries? Here I've got a bunch of 1.5 volt batteries: a little tiny button battery, a AA battery, a C battery, a flashlight battery, and a D cell flashlight battery. They're all 1.5 volt batteries; they all charge the same amount of energy per unit of charge. What's different is the bigger battery has a lower internal resistance and therefore has more current. Here I've got a little tiny 9 volt battery; it has a rather high internal resistance. Here's a car battery, 12 volts, not much different from 9. What's different is it has a vastly lower internal resistance. Real batteries, or any real sources of voltage, have internal resistance, and you have to take that into account if you're going to use them correctly. You have to put loads across them that aren't too low so you don't drop a voltage across that internal resistance.

Here are devices connected in parallel. Two components are in parallel if the two ends of each component are connected together. There component one and component two are in parallel, and because they're in parallel they have exactly the same voltage across either of them. You connected a battery across there of 6 volts; you'd get 6 volts across both components; same voltage V across both components.

That's parallel components. Are one and two in parallel? They look like they're in parallel; they're right beside each other. But they aren't, because their bottom ends aren't connected together; component three comes in the way. Those aren't parallel components.

Parallel resistors: It's sort of like adding an extra lane to a highway. If you add an extra lane to a highway, it's easier for cars to flow down the highway, it's easier for traffic to flow, and the highway offers less resistance to the flow of traffic. It's the same thing when you put resistors in parallel: The combined resistance goes down. It goes down in a rather funny way—which, again, we could derive, but we're not going to bother to in this course—and it turns out that what adds is the reciprocal of the resistance, one over it. So 1 over R parallel is 1 over R1 plus 1 over R2. Here's an example. If the two resistors are equal and they're equal to R, then 1 over R parallel is 1 over R plus 1 over R. That's 2 over R; and if you invert that, that says R parallel is R over 2. If you put two equal resistors in parallel, you get half the resistance of either one. Remember, if you put them in series, you got a resistance that was twice that of either one.

Two ways of connecting components: series and parallel. Not all circuits exhibit series and parallel connections, but these are easy connections to understand. Series connections, in particular, give us voltage dividers with that fraction we talked about. Parallel combination, by the way, is sometimes called a current divider, because current coming into that combination at the left-hand picture of the diagram is coming in at the top. It can split and go between R1 and R2, and more current will go into the lower resistance; the "Electricity takes the path of least resistance" quote. It doesn't take the path of least resistance; it divides in proportion to the resistance, and more current would flow through the lower resistance.

Let me give you a quick quiz here. I'd like you to take a look at this circuit—it's a fairly complicated circuit; it's got one, two, three, four, five, six, seven, eight resistors and a battery—and I'd like you to think about which pairs of resistors are in series and which pairs are in parallel. Press your pause button and think about this for a minute.

OK, come back out of pause and let's see what we see here. You might think, "Well, it looks like R2 and R4 there are in series or something; or maybe I don't know what's in series, but let's look around." I see actually only two resistors that are in series. There's only one series combination. By the way, when you're looking at circuits, it doesn't matter how they're drawn. Series resistors don't have to be in a straight vertical line; they don't have to be in a straight horizontal line. In this particular circuit, the actual series combination is going around a corner. The resistors that are in series are R6 and R7, because if you come out of R6, the only possible place you can go, if you're electric current flowing, is through R7. Those are in series. R7 and R8 below it aren't in series because there's that path, that horizontal path, at the bottom of R7, and that path could take current that way to the left as well as down through R8. Those aren't in series. The only series combination in that circuit is R6 and R7.

Which pairs are in parallel? Again, there's only one parallel combination: It's the combination of the lower right of R5 and R8. You can see they're connected together at the top by that single wire, and they're connected together at the bottom by a single wire. That's all I need to know to know that those two resistors are in parallel. I don't care what else is connected to that parallel combination. Those two are in parallel, and I'd calculate the equivalent resistance of that parallel combination by using that formula I just described.

There's a quick quiz. It's easy to get this wrong. It's easy to look at a circuit and say, "Oh, these two are in series," but they aren't in series unless there's no place current going through the first component can go except into the second component. It's easy to think things are in parallel when they're not. They're only in parallel if their tops are absolutely connected together and their bottoms are absolutely connected together. In the case of parallel resistors, that means they have the same voltage across both of them. In the case of series resistors, it means they have the same current through both of them.

Let's wrap up what we've seen in this lecture. We've looked at electronic circuits as complicated interconnections of components, or simple interconnections as interconnections of electronic components. We can describe these components by graphs, like we show here with the VI characteristics. We schematize the components with their circuit symbols. Components we found can be connected in either series or parallel, two possible common ways of connecting them. Series resistors simply add. Series resistors form a voltage divider with that combination of voltages; the voltage across the lower resistor being that fraction. It's resistance over the total resistance times the source voltage, in this case a battery. Parallel resistors, we found, add reciprocally.

That's a quick summary of what we've done in this lecture, and we're now set to begin building some simple circuits. Before we go on to another lecture, though, if you're interested there's a project associated with this one that you can do separately. I'm going to ask you to use only 1,000 ohm resistors, but you have as many as you want. Design a voltage divider that producers 4 volts when connected across a 6 volt battery. We'll actually use this particular project circuit—even if you don't do it, I'll describe it—in the next lecture.

Here's a question now: Suppose you connect a 10,000 ohm resistor (10 kilo ohms) across your divider between the points where you're supposed to have 4 volts. Will the voltage still be exactly 4 volts? Or if it won't, explain why it won't be and determine exactly what that voltage would be.

Instruments and Measurement
Lecture 3

In this lecture, we will look at a practical matter, namely, how we measure electrical quantities, especially voltage and current. We need to understand measurement for a number of reasons—to verify that components and circuits are working correctly and to read the output of electronic instrumentation. In addition to describing procedures for measuring electrical quantities, we will also look at the instruments used to measure them. Key topics we'll explore include the following:

- Voltmeters: characteristics and use
- Ammeters: characteristics and use
- Ohmmeters: characteristics and use
- The oscilloscope.

Clarifying Vocabulary

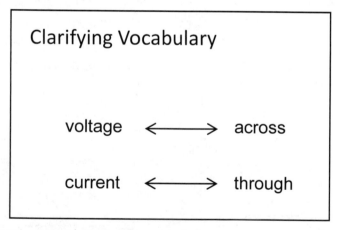

In working with electrical meters, it's important to understand two ideas: (1) Current is a flow *through* electrical components, and (2) voltage is a difference in energy per charge and appears *across* two points in a circuit.

Voltmeters: Characteristics and Use

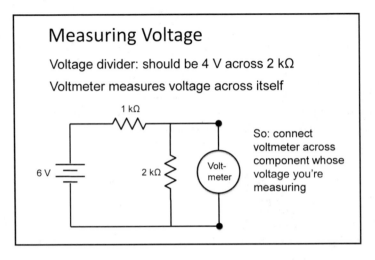

Measuring Voltage

Voltage divider: should be 4 V across 2 kΩ

Voltmeter measures voltage across itself

So: connect voltmeter across component whose voltage you're measuring

A voltmeter measures the voltage between its two terminals—the voltage across itself. To exploit that fact, a voltmeter is connected in a circuit in such a way that the voltage across the voltmeter is the same as the voltage across the circuit component whose voltage is being measured.

The previous lecture ended with the following project: Using only 1000-Ω resistors, design a voltage divider that produces 4 V when connected across a 6-V battery. For this project, we needed to connect the voltmeter to measure the voltage across the 2-kΩ resistor in the circuit, not the battery or the 1-kΩ resistor.

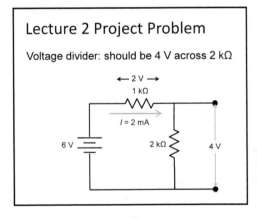

Lecture 2 Project Problem

Voltage divider: should be 4 V across 2 kΩ

With 6 V across a combined resistance of 1 kΩ and 2 kΩ, we have 6 V across 3 kΩ, or 2 mA. We have 2 mA flowing through the 1-kΩ resistor—therefore, 2 V across the 1-kΩ resistor— and the remaining 4 V from the 6-V battery across the 2-kΩ resistor. If we put a 10-kΩ resistor in parallel with the 2-kΩ resistor, we get 3.75 V—less than the 4 V we had at first.

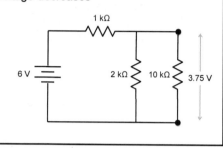

Lecture 2 Project Problem

Voltage divider: should be 4 V across 2 kΩ

Put another resistor in parallel with 2 kΩ: voltage decreases

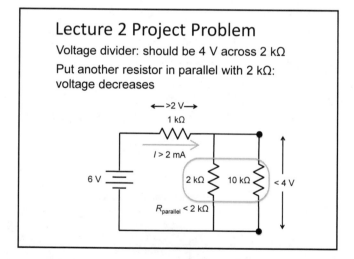

Lecture 2 Project Problem

Voltage divider: should be 4 V across 2 kΩ

Put another resistor in parallel with 2 kΩ: voltage decreases

Current flows through the 1-kΩ resistor, then splits and flows through the parallel combination, but it adds up to a current that's greater than 2 mA. With a current greater than 2 mA flowing through the 1-kΩ resistor, the voltage across the 1-kΩ resistor is greater than 2 V. That leaves a voltage across the parallel combination—remember, parallel resistors have the same voltage—that is less than 4 V. If we work out the math, it's 3.75 V.

Implication for Voltmeters

If voltmeter draws current, it will lower the
voltage it's measuring!

Ideal voltmeter should have infinite resistance!

In practice, much greater than circuit resistances.

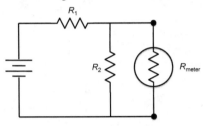

If a voltmeter draws current from the circuit, it will lower the voltage it's
trying to measure. The implication here is that an ideal voltmeter should
have infinite resistance. In practice, it should have a resistance much greater
than the resistances of the circuits being measured.

Ammeters: Characteristics and Use

Measuring Current

6-V battery, 2-kΩ resistor; current should be $I = V/R = 3$ mA

Ammeter: measures current through itself

Series components: current through first component
has nowhere to go but through second component

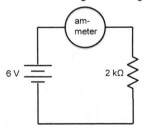

So: Break circuit; put
ammeter in series with
component whose current
you're measuring

An ammeter measures current through itself. Current comes in one lead, goes through the ammeter, and goes out the other lead. We must put the ammeter in a circuit in such a way that the current we're measuring is through the component we're interested in.

If an ammeter has any resistance whatsoever, it will reduce the total circuit current. The implication is that the ideal ammeter should have zero resistance; in practice, it should have a much lower resistance than any resistance in the circuit.

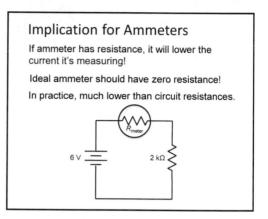

Implication for Ammeters

If ammeter has resistance, it will lower the current it's measuring!

Ideal ammeter should have zero resistance!

In practice, much lower than circuit resistances.

Ohmmeters: Characteristics and Use

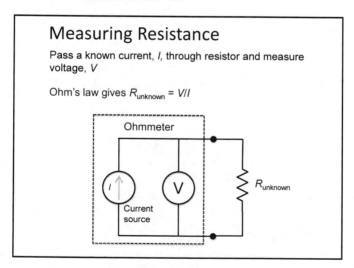

Measuring Resistance

Pass a known current, I, through resistor and measure voltage, V

Ohm's law gives $R_{unknown} = V/I$

Measuring resistance requires an active circuit. We pass a known current through a resistor and measure the voltage, then use Ohm's law to determine the resistance ($I = V/R$ or $R = V/I$).

Don't try to measure the resistance in a circuit by putting an ohmmeter across it because the ohmmeter will try to pass current through the entire circuit—through other resistors and through the battery—and will not properly measure the resistance of the one resistor you're interested in. Instead, disconnect the resistance, then put the ohmmeter across it.

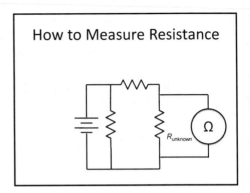

The Oscilloscope

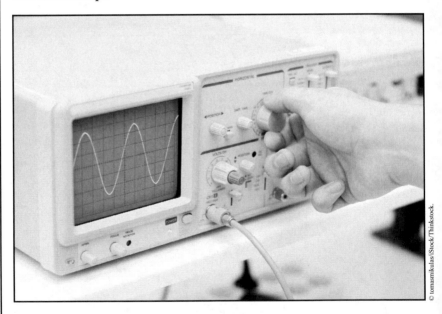

© tomasmikulas/iStock/Thinkstock.

An oscilloscope is a wonderfully versatile instrument that measures voltage as a function of time.

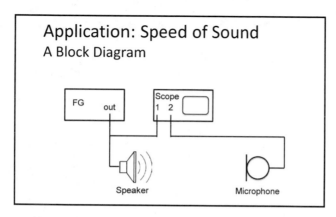

Application: Speed of Sound
A Block Diagram

FG out

Scope
1 2

Speaker

Microphone

To measure the speed of sound with an oscilloscope, we hook up a function generator to a loudspeaker and also to one input channel of the oscilloscope. We then pick up the sound with a microphone and send that signal into the second channel of the oscilloscope. The oscilloscope shows the time difference between when the signal is sent and when it's picked up, and that gives us the speed of sound: a little over 300 meters per second (m/s), or about 700 miles per hour.

Suggested Reading

Introductory
Brindley, *Starting Electronics*, 4th ed., chapter 3.

Lowe, *Electronics All-in-One for Dummies*, book I, chapters 8–9.

Platt, *Make: Electronics*, chapter 1, emphasizing the sections on instruments and measurement. See also p. 231 for a possible inexpensive way to turn your computer into an oscilloscope, although Platt isn't enthusiastic about this approach.

Shamieh and McComb, *Electronics for Dummies*, 2nd ed., chapters 10, 12–13.

Advanced
Horowitz and Hill, *The Art of Electronics*, 3rd ed., appendix L.

Scherz and Monk, *Practical Electronics for Inventors*, 3rd ed., chapter 7, sections 7.3–7.5.

Project

Lecture 2 Project Circuit Simulation
Simulate the circuit from Lecture 2's project. Verify that the voltage across the 2-kΩ resistor is what you expect.

Add the 10-kΩ resistor in parallel and verify that the voltage across the combination decreases as expected.

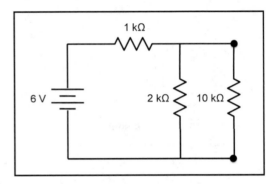

Questions to Consider

*1. An electronics neophyte attempts to measure the voltage across and current through the resistor R_3 in the circuit shown below, simultaneously connecting a voltmeter and an ammeter across the resistor, as shown. Is either measurement successful? Is either meter at risk?

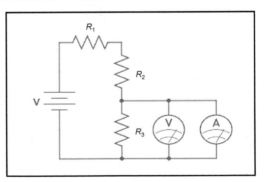

*2. An oscilloscope displays voltage as a function of what other physical quantity?

Instruments and Measurement
Lecture 3—Transcript

Welcome back. In this lecture we're going to look at a very practical matter: Namely, how we measure quantities, particularly voltage and current, in electric circuits, and also how we measure the resistance of individual resistors.

Why do we want to do that? A lot of reasons: For instance, suppose we build an amplifier that's supposed to amplify the input by a factor of 10. We need to look at the input voltage, and we need to look at the output voltage, and we need to verify that the amplifier works. Suppose we got a circuit that isn't working right? The first thing we do is check the various voltages in the circuit and see that they have the values they have. We also need to use voltage measurements and current measurements to read the outputs of electronic instrumentation that may be measuring other quantities—other physical quantities—that get converted like humidity, barometric pressure, temperature, pH, a whole range of quantities we can convert into electrical quantities. Then we need to measure electrical voltages or currents to get the outputs of those quantities.

There's a lot of motivation for measurement. This lecture is going to deal with the process of measuring electrical quantities, particularly voltage, current, and resistance, but it's also going to look at the instruments we're going to use to measure them.

We've had a lot of history behind electoral measurements. Luigi Galvani—for whom the Galvanic cell, the original name for a battery, is named—did experiments with frogs' legs; found that the legs of a dead frog would twitch when they were connected by metal wire, as is shown in the picture here. That was a very early indication of the presence of a potential difference; that is, a voltage. The middle image here shows a 19th-century electroscope, an early device that was able to measure essentially voltage—or actually more like charge. The amount of charge put onto thin gold foil pieces—and that was used for static electricity experiments. Finally, on the right we see an electrician using a modern day voltmeter. We want to come to understand these instruments, how they work, and not only how they work, but more importantly how to use them effectively.

What do we want to do with an electrical measuring instrument? What do we want its characteristics to be? The obvious first characteristic is we'd like it to be accurate; that is, if it reads 6 volts, we want to know that the voltage that that instrument is getting is indeed 6 volts. That's the obvious thing. There's a more subtle thing, and the more subtle thing is we would like the instrument not to disturb the circuit it's in. I want to emphasize that because this is one of the biggest mistakes you can make using electrical instruments. I may have a perfectly accurate meter, and it may read 5.4 volts when I put it in a particular circuit, and it may be 100% accurate. But maybe the voltage before I put the meter there was something different, and so we need to understand how to use instruments and how to specify instruments in such a way that they won't disturb the circuit in which we're putting them.

Let's begin to think about how electrical meters work. Several kinds of electrical meters; and you're probably familiar and have seen both kinds. On the left here I show an analog meter. An analog meter is an old-fashioned meter; there are still plenty of them in use, however. What an analog meter has is a tiny coil of wire—you can kind of see that at the center of the circular part at the bottom end of the needle in this analog meter—and that wire carries a current; so the fundamental analog meter is ultimately a current meter, but we can put resistors in series with it to make it into a voltmeter. That coil of wire is connected to a needle and a spring, and the coil is suspended between the poles of a magnet. If you've taken another physics course from the Great Courses or elsewhere, you know that magnetic forces act on currents, and so that needle twists. The bigger the current, the bigger the twist, and the scale is graduated to read in, for example, in this case, volts. I have over here a very simple analog voltmeter just mounted on a block of wood; that's all it is. It's one of these simple moving coil meters. You can't see the coil, but it's down in under that region. This one is calibrated in volts, up to 10 volts. If I put a voltage across these black and red inputs, the meter will read that voltage. That's a basic analog meter, in this particular case a voltmeter.

Probably the meters you're familiar with are digital meters. Here's a simple digital multimeter; I'll have more to say about that later. Unfortunately, I can't describe as easily how a digital multimeter does its magic; how it takes a voltage or a current that comes in in its inputs and converts it into a numerical readout. You'll have to wait until lecture 23, the next to the last lecture in this series, and then you'll understand how that process of so-called analog to digital conversion works, and we'll actually build an analog to digital converter in that lecture. But we'll be exploiting digital meters and using them long before we understand how they work. We have a primitive understanding of how analog meters work; digital meters are more complicated. A lot of electronics in them, but we'll come to understand them by the end of the time we're in this course.

Let me just point out a few more examples: Again, this is a little inexpensive digital multimeter. You might buy this for $10 or $15. You might have something like this at home. Here's its predecessor; still quite functional, an analog version of basically the same thing. It probably cost about the same, and it's useful for simple home repairs and work and things like that. We'll be using in this course a fancier instrument. This is an Agilent multi digital multimeter; a much more sophisticated instrument. It can do a lot more accurate measurements, a lot more precise measurements; doesn't disturb the circuit you put it in as much. It can actually be controlled or send output to a computer also, but we're not going to do that here. But basically it's the same as the other instruments I showed you, but that's the one we'll be using. I'll talk about this big thing later on in this lecture.

We have a quick look at how meters work, and just some examples of different kinds of meters.

Let me pause and give you a quiz this early in the lecture. Here's a vocabulary quiz: On the left I have two words, *voltage* and *current*, and on the right I have two words, *across* and *through*. I'd like you to think about the meaning of *voltage* as an energy imparted to charge by, say, the terminals of a battery or some other device, or we might ask "What's the voltage between the two ends of a resistor?" Then think about current as a flow, and ask yourself, "Which word from the left column goes with which word from the right column?" Pause and think about that a moment.

OK, coming back from your little quiz. They go right across like that. Voltage goes with across and current goes with through. Current is a flow; it flows through things. Voltage is a difference in energy per charge, and so it goes across between two points in a circuit. Those words are really important. If you don't have that word connection right, you won't understand how to hook up meters; in fact, you may damage the meters, or worse damage yourself. You'll hear misuse of language all the time: "So and so is in the hospital because 20,000 volts went through him." Volts don't go through anything. Volts aren't a flow; they're an energy per charge, a potential difference it's called, which is imposed across two points in a circuit. Current goes through; so get those words right. We'll exploit them again when we talk about how to use these instruments to measure electrical quantities.

Let's talk a little bit about measuring voltage. I mentioned in the preceding lecture that there was going to be a little project and you could do the project or not. Whether or not you did the project doesn't matter, but we're going to talk about the circuit from the project in the previous lecture. Here's the circuit. It's got a 6 volt battery; you were supposed to hook it up in such a way that you had a voltage divider that gave you 4 volts. You'll remember the voltage divider equation, and the voltage divider equation says the voltage across two series resistors gets divided in proportion to their resistance. With 1 kilo ohm, 1,000 ohms, and 2 kilo ohms, the 2 kilo ohms is going to get 2/3 of the voltage across it, and the 1 kilo ohm is going to get 1/3 of the voltage across it. We got the 4 volts across the 2 kilo ohm resistor.

A voltmeter: What does a voltmeter do? A voltmeter, like this simple one I showed you, measures the voltage that you put between its two terminals, its two inputs. That's what it does: It measures the voltage across itself. It doesn't automatically measure a battery voltage or the voltage across a resistor or anything else; it measures the voltage across itself. To exploit that fact, you've got to connect a voltmeter in a circuit in such a way that the voltage across the voltmeter will be the same as the voltage across the circuit component whose voltage you want to measure. Remember from the previous lecture that when we put circuit components in parallel, they have to have the same voltage across them. To use a voltmeter correctly, you connect the voltmeter across the component whose voltage you're measuring; that is, you connect the voltmeter in parallel with the component whose voltage you're measuring. If you connect it in series, it's not going to work because there's no guarantee they have the same voltage across it. You connect a voltmeter in parallel. This particular picture is now showing how you would connect a voltmeter to measure the voltage across the two kilo ohm resistor in that circuit. It doesn't measure the voltage of the battery; it doesn't measure the voltage of the 1 kilo ohm resistor; it's set up to measure the voltage of the 2 kilo ohm resistor. You'd have to put it across the 1 kilo ohm resistor to get that voltage, and you'd have to put it directly across the battery to get the battery voltage.

That's how to hook up a voltmeter. I'm not going to repeat the experiment, but I did that in the previous lecture when I had two light bulbs connected in series and I stuck the voltmeter across one of the lightbulbs. There was a 6 volt battery, two identical light bulbs, and we ended up with 3 volts across one of the lightbulbs. I didn't measure the other one, but if I had it would also have been 3 volts. The voltmeter has to go across or in parallel with the component whose voltage you're trying to measure.

There's something a little bit more subtle, because I made the point that a voltmeter could be perfectly accurate, 100% accurate; it reads exactly the voltage across itself (of course that also is an impossibility). But it could nevertheless disturb the circuit in a way that means the voltage across the voltmeter when you're doing the measurement isn't the same as before you put the voltmeter there. We want to explore what properties a voltmeter has to have in order that it not disturb the circuit in question.

One of the reasons I asked you in the previous lecture's project is to first design this voltage divider that put 4 volts across the 2 kilo ohm resistor was because I then asked: What would happen if you put another resistor across there? Before we put the other resistor, let's notice something. With 6 volts across a combined resistance of 1 ohm and 2 ohms—that's a total of 3 ohms, because resistors in series add—we have 6 volts across 3 ohms. That's, of course, 3 kilo ohms, 3,000 ohms; that's 2 milliamps. Again, kilo ohms, milliamps, the "kilo" and the "milli" cancel out; you can just work things right out in volts. There's going to be 2 milliamps flowing through that 1 kilo resistor, and that 2 milliamps is going to go on through the 2 kilo ohm resistor. Two milliamps is consistent with ohm's law. We have 2 milliamps through 1 kilo ohm. There are therefore 2 volts across the 1 kilogram resistor, and the remaining 4 volts of the 6 from the battery are across the 2 kilo ohm resistor. All this make sense.

What I asked you to do in the preceding lecture was then to put another resistor in parallel with the 2 kilo ohm resistor. You put 10 kilo ohms. If you worked this out, you found that you got 3.75 volts. If you didn't work that out, don't worry; relax. We're going to show you why the voltage is less. I'm not going to go through the mathematics, but I'm going to show you at least why it's going to be less than the 4 volts you had before. Here's why: Here's the circuit with the 2 kilo ohm resistor and the 10 kilo ohm resistor in parallel. Remember that the resistance of parallel resistors is always lower than that of either resistance alone. The reason, again, is we have two pathways, two parallel pathways, for current to flow so it's easier for the current to flow. Without going into the mathematical details, that resistance is less than 2 kilo ohms.

As a result, the current through the whole circuit, the resistance of the whole circuit is less than 3 kilo ohms that we had for the combined resistance, and that means the current is going to be more than 2 milliamps. That current is flowing through the 1 kilo ohm resistor and then it's splitting and flowing through the parallel combination, but it adds up to 1, to this current that's greater than 2 milliamps. But the point is if there's a current greater than 2 milliamps flowing through the 1 kilo ohm resistor, there's a voltage across the 1 kilo ohm resistor that's greater than 2 volts. That leaves a voltage across that parallel combination—remember, parallel resistors have the same voltage—that's less than 4 volts. Consequently, there's no longer 4 volts across that parallel combination, across the 2 kilo ohm resistor, which is part of it. There's somewhat less if you work out the math; the less is 3.75.

What's the implication for a voltmeter? If a voltmeter draws current from the circuit, it will lower the voltage it's trying to measure. Even if it's perfectly accurate, it won't get the voltage right because the voltage will be different with the meter there than if it wasn't there. The implication for voltmeters is an ideal voltmeter should have infinite resistance. In practice, it should have a resistance much greater than the resistances of the circuits you're measuring.

How well do real voltmeters do that? Because analog voltmeters draw current, and they have to draw current, they ultimately don't do as well as most digital meters. This particular meter, I happen to know if it's on the

1 volt full scale range, it has a resistance of 20,000 ohms. That's twice that 10,000 ohm resistor we were throwing into that circuit we were talking about; that would be enough to throw the voltage off in that particular circuit. Digital meters, on the other hand—including inexpensive ones you can buy or fancy instrumentation grade meters typically have millions of ohms—1 million ohm, 10 million ohms, or typically you could even get them up higher, but then they run into various other kinds of troubles. But the point is, in practice, the voltmeter needs a resistance much greater than any resistance in the circuit, and so you can treat the voltmeter as if it were an open circuit, a complete gap, an infinite resistance through which no current can flow. That's the implication for voltmeters. That's how we measure voltage: with voltmeters, which are devices that have effectively infinite resistance.

By the way, that shows what would happen if you put a voltmeter in series with something. It's an open circuit. No current can flow; you'll stop all the current. There'll be no voltage across any of the resistances in the circuit, and you'll see the entire voltage dropped across the meter, but that won't tell you anything. It won't tell you the voltage across, in this case, the resistor R2 that we're trying to measure the voltage across.

Let's move on to measuring current; we'd like to talk now about how you measure current. Again, we want to measure current without doing any disturbance of the circuit. Here's a simple circuit: 6 volts now across a 2 kilo ohm resistance; just one resistor. A 6 volt battery, 2 kilo ohms; the current should be I=V/R, 3 milliamps. The ammeter measures current through itself. Current comes in one lead, goes through the ammeter, and out the other. That's what the ammeter measures. You've got to put the ammeter in a circuit in such a way that that's the current you're measuring through the component you're interested in.

Suppose I want to know the current through that 2 kilo ohm resistor, which in this very simple series circuit is the current through everything. What I have to do is put the two things in series. The current through the meter has no place to go but through 2 kilo ohm resistor or vice versa; it doesn't matter what order. I've got to break the circuit. For an ammeter you've got to break the circuit; you've got to put the ammeter in series with the current, the component whose current you're trying to measure.

There's an implication of that for ammeters, and that's if that ammeter has any resistance whatsoever, it's going to reduce the total circuit current. The implication is the ideal ammeter should have zero resistance, and in practice it should have a much lower resistance than any resistance in the circuit.

This makes ammeters a little trickier, and, in fact, as we'll see in a moment, a little more dangerous to use. Let me do a little demonstration over here. Here I have my digital meter. It's now connected as an ammeter. There's actually a different connection I have to plug into because it's that low resistance connection. It's set to measure amps. Here's my lightbulb that I showed you in the previous lecture, a hefty 6 volt lightbulb. I want to measure the current through that lightbulb. I don't take and stick this ammeter across that battery. To do so wouldn't measure the current through the lightbulb, it would alter the circuit dramatically. It would also probably destroy the ammeter, or in the case of this one, blow the fuse. What I want to do instead is come out of the battery's positive terminal with the wire that's going in to the ammeter. Here's the red wire going into the ammeter, and here's the black wire coming out of the ammeter, and I'm going to connect that to the wire that's going in to the lightbulb. I'll just join these two wires. The lightbulb lights because the ammeter has essentially zero resistance, so the circuit is just the same as before. It thinks the ammeter's just a piece of wire. I look on the ammeter, and it's reading about 1.4 amps; that's a big, hefty current. That's why, in the preceding lecture, when I connected that lightbulb across the battery, the battery terminal voltage dropped considerably because there were 1.4 amps flowing through its internal resistance. That's how we measure current.

I'm going to disconnect that before I do anything that might damage things, and I'm going to remind you once again that you don't want to connect an ammeter, ever, across a voltage source. Here's an example of what you wouldn't do to try to measure the current in that circuit, because suddenly you have a resistance that's almost zero across the battery and an enormous current flows, and the meter either gets destroyed, or if it's a reasonable meter it has a fuse and the fuse blows. When I teach introductory electronics, I have a whole bin full of fuses sitting out there, and many of my students learn the hard way by blowing the fuse and then they have to figure out how to change it. It's not terribly difficult to do. A huge current flows; don't put an ammeter, ever, across a voltage source.

There's one other kind of measurement we want to do, and that's a measurement of resistance. We actually did that in the very first lecture when I showed you, for example, the resistance of a simple resistor, and then I showed you the resistance of a photosensitive resistor whose resistance varied with light. How do you measure resistance? In a common meter, in a multimeter, set to measure voltage, current, or resistance, you have to be a little more active to measure current. You have to have a source of current, somewhere an electronic circuit that generates a constant current, runs it through whatever you connect across the meter—a resistance you're trying to measure—and determines the voltage, and then uses ohm's law to determine the resistance because I=V/R or R is then V/I. That's how you do it. Connect the resistance and away you go. An unknown resistance, you can measure it that way.

Here's what you don't do: You don't measure the resistance when it's in a circuit by putting an ohmmeter across it—that symbol is for an ohmmeter—because that ohmmeter is going to try to pass current through that entire circuit, through those other resistors and through the battery, and it's not going to measure properly the resistance of that one resistor we're interested in. What you have to do instead is to disconnect that resistance and then put the ohmmeter across it. You can sometimes get away with disconnecting just one end of it because then it's not in the circuit. But better, take the resistor and isolate it and measure its resistance separately.

There are the three common things we've talked about—voltage, current, and resistance—and how to measure them. I want to spend the last third of this lecture on another voltage-measuring instrument—it can also occasionally measure current—but it's a wonderfully versatile instrument that measures voltages as functions of time. That's the oscilloscope, and I'm going to begin with a little history of this remarkable electronic instrument.

An oscilloscope is a device that displays, on some kind of screen, voltage versus time fundamentally. I have a wonderful array of oscilloscopes here, starting with one that was manufactured in the 1930s, before there was a very viable oscilloscope industry. It has a tiny little screen about two or three inches across. Most of the rest of it is electronic stuff. It weighs a ton. It's not very useful; it can't do much. In fact, it doesn't even run anymore so I don't have it plugged in. That's an oscilloscope from the 1930s. Here's an oscilloscope, probably built in the 1950s and used into the 1970s and even 80s. There are probably plenty of these so-called 500 series oscilloscopes in operation in research labs today. I actually used this model to teach with when I first started teaching electronics at Middlebury College and it is a great oscilloscope for its date. Its circuitry uses vacuum tubes, like I showed in the very first lecture. More importantly, its display uses a long vacuum tube called the cathode ray tube, which is basically a cousin of the old TV picture tubes before we all had flat screen TVs.

Just to show you what this does: The oscilloscope moves a beam of electrons that hits the screen from inside, causes it to glow, and it moves it across there at a rate that can be varied. There it is slowing down, and you can see that spot moving across the screen. That's actually being supplied with a voltage that's going up and down 440 times a second. That happens to be the note A above middle C. So you see what looks like a line, but it's really going up and down very rapidly. In fact, if I slowed that down, you would see it going much more slowly. This can go very slowly, indeed; it takes many seconds to cross the screen. But even this ancient oscilloscope can be sped up so that that trace is sweeping across the screen in a matter of microseconds (millionths of seconds). That's the versatility of the oscilloscope: It can display signals as a function of time over very wide time intervals, and that particular oscilloscope is displaying a sinusoidal wave form at 440 Hertz.

That's a 1950s-vintage oscilloscope. Here's a 1980s-vintage oscilloscope. It has solid-state electronics—its amplifier is the thing that makes the spot sweep across the screen is basically solid-state electronics, transistors, and related components—but it still has a long cathode ray tube, which is doing the display. It can be sped up, just like the other one, or slowed down; just the same. Same display mechanism; what's different about this one is the electronics is a lot more sophisticated. It takes up less space, is solid-state, so you can see the kind of shrinking in size and weight here. You can also see the screen's getting a little bigger, perhaps.

Here's a modern oscilloscope, probably from the early 2000s. This is the model we use in teaching our electronics course for sort of upper class students at Middlebury; we use a simpler version of the same thing for teaching our most introductory physics courses. It's different. It no longer has a long cathode ray tube for the same reason that your television set doesn't; that's become completely obsolete in favor of flat panel displays. In a sense, this isn't a real oscilloscope. It's a computer that has inputs that turn numbers into digital quantities using the analog to digital conversion process we'll learn about later in the course, and then displays those quantities as time variations on the screen. It actually stores them in a memory and then dumps the memory to the screen. It sort of simulates the old-fashioned cathode ray oscilloscope. I often wonder: If we hadn't ever invented the cathode ray oscilloscope, would we have just jumped to this thing? But then I realized we never would've gotten electronics as advanced without an oscilloscope, because an oscilloscope is absolutely wonderful for looking at electronic signals and what they're doing as a function of time.

Just to give you an example, let me slow down this signal that's going across here. This signal is right now at 440 Hertz. I'm going to drop it down, and you can see that that wave form—we're just seeing a little piece of it—but if I slow down the rate at which the beam is sweeping across these screens, you can see that effect. I'm going to slow down still further. I'm going to drop it to one cycle per second, and now it's going up and down at the rate of one cycle per second. Let's see if we can get it, and there it goes. You can see that spot going across the screen, now at a very slow rate, coming back and returning. It might take a while for it to return. It's going slowly across the screen—there it is—and you can see it tracing that up and down of the voltage as a function of time.

Even these fairly simple—not cheap, but not horribly expensive—instruments, this one can go down to billionths of a second, so it can measure time intervals on the order of billionths of a second. So a contemporary oscilloscope, a 1980s vintage or early 90s vintage oscilloscope, a 1950s and 60s vintage oscilloscope, and a 1930s oscilloscope.

In this course, I'd be perfectly comfortable using this oscilloscope—in fact, it has an analog output that we could supply to a monitor and get a fairly good image on the screen—but instead, in this course we're going to use this instrument. We're very grateful to Agilent Technologies for lending us this very fancy, high-definition oscilloscope. They've just started making high-definition models. You might think that's a gimmick like high-definition television. It does have a high-definition screen, but the essence of the high definition is it breaks the input signals into far finer divisions. In the language of digital electronics, which we'll get to later in the course, most oscilloscopes are 8-bit. They break incoming voltage is 2^8, which is exactly 256 levels. This breaks it into 2^{12} levels, which is about 4,000 different levels. Because the voltage can be displayed much more finely, we need a more high resolution screen. We can take the output of this—we're not going to do it in this lecture, but we will in subsequent lectures—oscilloscope and we'll send it into our big monitor so we'll be able to see exactly what's happening.

I'm going to do a very silly thing with my fancy oscilloscope here: I'm going to simply use it to measure the voltage of that 6 volt battery. Here I have the cable coming from the oscilloscope. You can see that horizontal line that represents 0 volts on that screen. I'm going to connect the oscilloscope across the 6 volt battery; it's a voltmeter. I connect it across, and you see that line jump up. It jumped up—we're getting a little bad connection there; that's why it's wiggling like that, but there it is—to three of these squares on the oscilloscope screen above where it had been. There's a number in the upper left; I don't know if you can see it, it says "2 volts per." That means there are 2 volts for every one of those squares; 2x3 is 6 volts, so this is accurately measuring the 6 volts of that battery. But that is a waste of an enormous instrument; I could do that much more easily with a simple voltmeter.

Let me end by showing you something much more impressive that this oscilloscope, or any of the oscilloscopes I have here, could easily do. I'd like to show you a measurement of the speed of sound. The speed of sound isn't like the speed of light—although we can actually do (and we do this at Middlebury College), a comparable measurement for the speed of light; it's a little more sophisticated—but we're going to measure the speed of sound. Let's look at how we're going to do it.

We're going to hook up a so-called function generator, a source of various electrical signals; and I'll show you what signal I use. I'm going to hook that to one input of the oscilloscope, because the oscilloscope can display several voltages at once, and I'm also going to send it to a loudspeaker. I'm going to pick up with the sound from that loudspeaker with a microphone, and I'm going to send that into the second channel of the oscilloscope. We're going to use that to measure the speed of sound. Here I've implemented the setup that's shown in the block diagram. I have the function generator that makes an electrical signal. It's connected to the channel one input of the oscilloscope; that's the channel that displays in yellow at the top of the screen there. It's also connected to this loudspeaker. A little ways away from the loudspeaker—half a meter to be precise, 50 centimeters—is a microphone, and the microphone is connected to channel two of the oscilloscope, although channel two is not turned on yet so you don't see anything.

This function generator is set to do something interesting: It's set to 1,000 Hertz, 1,000 cycles per second, which if I just ran that would make a kind of high-pitched sound that you could easily hear. But it's set to make only one cycle of that, which will take a thousandth of a second, and then it waits 50 thousandths of a second, 50 milliseconds, and then makes another one. If I turn this on, you first of all hear a buzzing sound. That's the every 50 milliseconds buzzing associated with that one cycle coming on. You can see many of those cycles appearing on the screen of the oscilloscope. I've got the oscilloscope set to sweep rather slowly across the screen. Now I'm going to do two things: First of all, I'm going to turn on the channel two input; so we're going to see what channel two is looking at. There's channel two. Channel two is very noisy; there's a lot of electrical noise in this room. The microphone is picking up other sounds. It's not clean and beautiful. But you can clearly see, in addition to all that noise, little spikes that correspond to the spikes in the input signal that's going to the loudspeaker. The interesting thing about them is you'll notice they're offset horizontally; remember, this direction measures time on the oscilloscope. That offset is going to give us the time it took the sound to travel that half a meter between the speaker and the microphone.

Let me now speed up the trace on the oscilloscope so we can see just one of these spikes. There's fewer, fewer, and there's one. I'll go a little bit further. At this point in time, right there, is the start of that little pulse and this point in time—and it's a little bit hard to see because there are some messy things going on here; this isn't a perfectly accurate measurement, but it'll give us a good estimate—right about there is where that sound is picked up by the microphone, or rather when that sound is picked up by the microphone. If you count between those, there are one, two, three horizontal boxes on the screen of the oscilloscope, and there's a number I'm reading down here that says 500 microseconds, 500 millionths of a second, for each of those boxes. It's taken 3 of those boxes, or 3x500—1,500 microseconds or 1.5 milliseconds—to go half a meter. If it takes 1.5 milliseconds to go half a meter, it's going to take twice that, or 3 milliseconds, to go a whole meter. The speed of sound is 1 meter per 3,000th of a second, and if you work that out, it comes out to a little bit more than 300 meters per second, which is, in fact, the speed of sound under normal atmospheric conditions: About 340 meters a second, about 700 miles an hour, or about 1,000 feet per second. We've used this oscilloscope to do something quite sophisticated and measure a very small time, and measure it right here with this simple tabletop setup reasonably accurately. Just to review, we've gone through a number of instruments and shown how to measure things, and the most sophisticated of those instruments is the oscilloscope.

I could stop there, but I just want to mention the project for this lecture. If you would like in future lectures to actually build circuits by yourself, you can do that, and we're going to do that in this course not with hands on real hardware but with circuit simulators that are available to you on the worldwide web. We'll be trying out two of them, and you can take your pick or decide not to do this at all, but this project for this particular lecture is going to be an introduction to those software packages, those web-based circuit simulators. If you want to get into that, you really want to do the project for this lecture.

AC versus DC
Lecture 4

S o far in this course, we've dealt with batteries as power sources; batteries involve chemical reactions that maintain, basically, a fixed, steady voltage across the battery's terminals. But when you plug something into a wall outlet, you get a time-varying voltage. There are good reasons for using such alternating current (AC) in power systems: Rotating generators naturally make AC power, and it's easy to change voltage levels in AC systems. In this lecture, we'll look at AC with regard to AC power, audio, and other signals that vary with time. Key topics include the following:

- Characterizing alternating current
- Transformers
- Diodes
- Capacitors and regulators.

Characterizing Alternating Current

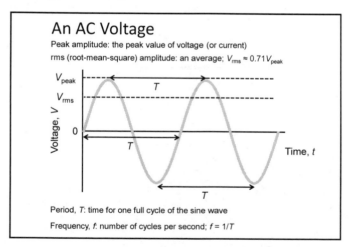

An AC Voltage

Peak amplitude: the peak value of voltage (or current)

rms (root-mean-square) amplitude: an average; $V_{rms} \approx 0.71 V_{peak}$

Period, T: time for one full cycle of the sine wave

Frequency, f: number of cycles per second; $f = 1/T$

The waveform of basic alternating current is called a sine wave. An AC voltage can be plotted as a function of time, with time on the horizontal axis and voltage on the vertical axis; 0 V is the horizontal line through the middle. From an arbitrary start time (designated $t = 0$), the AC voltage rises to a peak, comes back down through 0, and goes to a negative peak.

Transformers

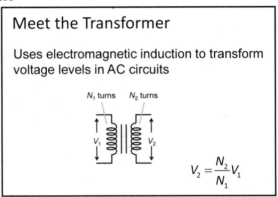

Meet the Transformer

Uses electromagnetic induction to transform voltage levels in AC circuits

N_1 turns N_2 turns

V_1 V_2

$$V_2 = \frac{N_2}{N_1} V_1$$

Transformers use electromagnetic induction to transform voltage levels in AC circuits. Here, V_1 and V_2 represent the voltages that appear across the primary and the secondary. We have N_1 turns on the primary and N_2 turns on the secondary; V_2 is $(N_2/N_1)V_1$. If $N_2 > N_1$, we have a *step-up transformer*; if $N_2 < N_1$, it's a *step-down transformer*.

Diodes

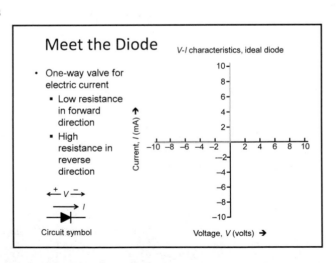

Meet the Diode

V-I characteristics, ideal diode

- One-way valve for electric current
  - Low resistance in forward direction
  - High resistance in reverse direction

Circuit symbol

Current, I (mA)

Voltage, V (volts) →

Practically speaking, a *diode* is a one-way valve for electric current. It has a very low resistance in the forward direction and a very high resistance in the reverse direction. We can characterize the diode by its *V-I* characteristic curve.

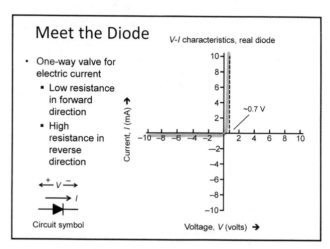

A real diode has a curve that shows a tiny bit of current flowing in the reverse direction and a small voltage drop in the forward direction. For commonly used silicon diodes, that voltage is on the order of 0.6 V to 0.7 V. That small voltage drop will come back to haunt us several times in this course.

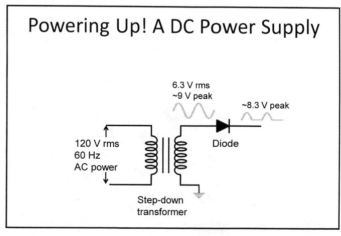

A diode and a step-down transformer are key components in a *power supply*, such as those that supply DC power to electronic devices. Notice that the output of the diode follows that of the transformer while the latter is positive, but as soon as the transformer output goes negative, the diode shuts off; the diode becomes an open circuit and no longer passes current. In a sense, we've made DC, because we've made a voltage that goes in only one direction, but it's far from steady.

Capacitors and Regulators

Meet the Capacitor

- A pair of conductors separated by an insulator
- Stores charge and energy
- Charge proportional to voltage: $Q = CV$
- Capacitance C: ratio of charge to voltage
  - 1 coulomb/volt = 1 farad
- Takes time to charge
  - Introduces delays in circuits
  - Voltage across capacitor can't change instantaneously

+Q +++++++++++

−Q − − − − − − − − − −

V_C

Capacitor symbol

A *capacitor* is a pair of conductors separated by an insulator; it stores charge and energy. The charge and voltage are proportional; thus, if we put a certain voltage across the capacitor, we get a proportional amount of charge on the plates. The capacitance, C, is the ratio of how much charge we get for a given voltage.

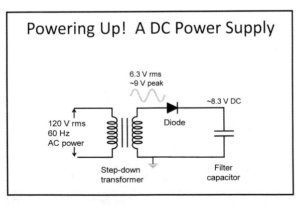

Powering Up! A DC Power Supply

6.3 V rms
~9 V peak

~8.3 V DC

120 V rms
60 Hz
AC power

Diode

Step-down transformer

Filter capacitor

We can use the capacitor as a filter in our power supply to smooth out the variations we see after the diode. After we insert the capacitor into the circuit, we get a nearly steady DC voltage.

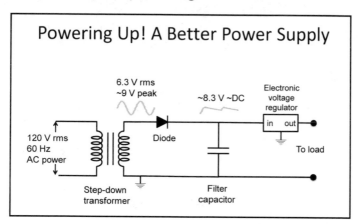

The job of an integrated-circuit voltage *regulator* is to take an unregulated input voltage (that may even be varying somewhat) and produce an output voltage that is fixed by the specifications of the regulator. The regulator we'll use is designed to produce an output voltage of 5 V. Once again, we can measure the output with the oscilloscope, and we see a steady, smooth DC voltage.

Suggested Reading

Introductory

Brindley, *Starting Electronics*, 4th ed., chapters 4; 6; and 7, p. 117 to end.

Lowe, *Electronics All-in-One for Dummies*, book II, chapter 3; book IV, chapters 1–3.

Platt, *Make: Electronics*, chapter 2, pp. 62–63; chapter 3, p. 134.

Shamieh and McComb, *Electronics for Dummies*, 2nd ed., chapter 4 through p. 80.

Advanced

Horowitz and Hill, *The Art of Electronics*, 3rd ed., chapter 1, sections 1.3–1.6.

Scherz and Monk, *Practical Electronics for Inventors*, 3rd ed., chapter 2, sections 2.20–2.23.2; chapter 3, section 3.6.11; chapter 4, pp. 415–420; chapter 11.

Project

A Power Supply

Simulate a simple 5-V DC power supply with 100-Ω load resistance. Use a 120-V AC rms voltage source. Choose an appropriate turns ratio for your transformer:

- Ignore other transformer parameters.
- Don't forget the 0.7-V drop across the diode.

Try capacitors of 1 mF, 10 mF, and 50 mF to see which makes the ripple nearly imperceptible. You'll need to run a time-domain simulation:

- CircuitLab: start/stop times 2, 2.1 s, time step 0.5 ms, "include zero" in advanced graphing.
- DoCircuits: analysis time 0.1 s; use an oscilloscope at the output.

Tweak your transformer to achieve ~5 V DC. For extra learning, add an integrated-circuit voltage regulator.

Questions to Consider

1. A step-up transformer increases voltage. Why, with a step-up transformer, aren't we getting something for nothing?

2. In what way is a diode like a simple on-off switch?

3. Of the components in a power supply, which is most directly responsible for smoothing the varying voltage that the diode would otherwise yield?

4. If you lower the load resistance connected across a simple power supply, such as the one developed in this lecture, will the ripple get worse or better? Explain.

AC versus DC
Lecture 4—Transcript

Welcome back. In this lecture, we'll be considering AC and DC. What's that all about?

So far, we've dealt with power sources that were largely batteries, and batteries have chemical reactions that maintain, basically, a fixed, steady voltage across the battery's terminals. You plug something into the wall and that's not what you get. What you get out of the wall is a time varying voltage, and it'll drive a time varying current through whatever you put there. There are good reasons why we use such alternating current in power systems: One is that rotating generators naturally make AC power. The other reason is it turns out to be easy to transform change voltage levels in AC systems, so we can step the voltage up to hundreds of thousands of volts for long distance transmission line, and then back down to relatively safe levels for use in the home or in industries. We're also going to be looking at AC when we think about audio signals, and a lot of other signals that vary with time. But any electrical signal that represents sound, for example, varies with time.

We want to begin by looking at the waveform of basic alternating current, and that's called a sine wave. One reason a sine wave is useful is because it's simple to deal with mathematically, and also it turns out you can make any complex waveform out of a sum of sine waves. Let's begin by taking a visual look at an AC voltage.

Here's an AC voltage plotted as a function of time. Time is on the horizontal axis, the voltage V is on the vertical axis. Zero volts is the horizontal line through the middle. There's an arbitrary start to time; I'll call time 0, and that happens to be when this voltage is passing through 0 on its way up. The AC voltage rises to a peak, comes back down again through 0, goes negative— in other words, it reverses direction; that's why it's alternating—goes to a negative peak, rises back up, and completes a full cycle.

We want to learn how to characterize this AC voltage. First, we're going to talk about the peak amplitude. That's the peak value of a voltage. That's the voltage at the very maximum; it's also the opposite of the voltage at the very minimum. Sometimes, by the way, you'll see people talk about the peak to peak voltage, which is twice the peak because it's from the bottom to the top, at least for a voltage that swings symmetrically about 0, as this one does. So there's the peak voltage. There's also something called the RMS voltage, the Root Mean Square voltage. That complicated language has to do with squaring this waveform, taking the average of it, and taking the square root. You do that because if you took just the average, you'd get 0 because it's up as much as it is down. It turns out that for a sine wave, the RMS voltage, the Root Mean Square voltage, is 0.71 times the peak voltage. That's actually the peak voltage divided by the square root of 2, but we won't worry about that. There's the RMS; it's about not quite 3/4 of the way up from 0. So two different ways of characterizing AC voltages in terms of how much voltage there is.

We also are interested in the period. What's the time for one full cycle? There's a full cycle from where the wave starts out at 0 rising, goes up, goes through 0, comes down, and it goes back up through 0 rising; that's a full cycle. It's not a full cycle when it reaches 0 again, because at that point it's falling and it has to wait until it comes back, and that's a full cycle. You could measure a full cycle also from the bottommost trough to the next bottommost trough, or from the top peak to the next top peak. All of those are the period "T," the time for one cycle of this sine wave. We also talk about the frequency: the number of cycles per second that the sine wave undertakes. The frequency is simply 1 over the period, the inverse of the period. Those tell you the same thing: Frequency is measured in cycles per second, which is given a special name, the Hertz, after Heinrich Hertz. We'll talk about either frequency or period when we're talking about AC sine waves.

Let me give you a demonstration of an example of what some AC sine waves look like on the oscilloscope and also what they sound like. We'll switch over to the oscilloscope. Here I have the oscilloscope. It's displaying on the big screen; it's also displaying on itself. I'm feeding the oscilloscope's red channel—that's channel four—with the output of this function generator that we used in the previous lecture to generate those pulses for the speed of sound. But now I have it set, basically, to generate a simple signal at a frequency that's audible. There you see the waveform. It's running at a fairly low frequency of 200 Hertz. You can hear a low tone coming out of a loudspeaker. You can see the oscilloscope is having a little trouble dealing with this one with its triggering, but that's OK. There's that waveform. There's one cycle of it displaying on the oscilloscope screen. Now let's switch up to a higher frequency. Two things will happen: You'll see more of the cycles fill the screen. We're going up to 300 Hertz, and now we see about one and a half cycles on the screen and you hear the higher pitch coming out of the loudspeaker. Four hundred cycles, more; four hundred cycles per second; five, six, seven, eight. There I'm adjusting the frequency or period of the waveform. I could also choose to adjust the amplitude, the vertical height, the number of volts being output. Let me do that. There it is coming down, keeping the frequency now constant, but bringing the amplitude down to lower and lower levels.

Amplitude and frequency or amplitude and period are the two things that characterize an AC signal. Before we leave the subject of AC sine waves altogether, let me turn off this function generator and talk about a very familiar AC voltage, at least familiar to people in North America. That's the 120 volt, 60 Hertz AC power line. Frequency 60 Hertz, 60 cycles per second, 160th of a second is the period. That 120 volts is the RMS voltage. The actual peak voltage is about 170 volts; so 120 volts RMS, about 170 volts peak. I say that's familiar to North Americans because if you're in Europe you typically use about 230 volts and your frequency is 50 Hertz, which is one of the reasons you need power converters when you go traveling abroad. There's a familiar example of an AC voltage, and this is the one that's supplied to us by the power companies at all our electrical outlets here in North America.

There's a problem with this AC voltage because electronic circuits like we're studying in this course like, usually, to work on DC. Batteries are great for them, but we don't want to be lugging batteries around all the time, although these days with smartphones and things that have rechargeable batteries that's how we power them. Nevertheless, we need to get DC to charge those batteries. Somehow we need to convert this AC into DC, and that's going to be the subject of the rest of this lecture: Getting practical, making DC from AC. There's another reason we need to do that also, and that's that circuits involving solid state components, transistors typically run on voltage levels of a few to a few tens of volts; 170 volts, that peak of the AC power line, is far too much. Although I will say tube circuits, many of them would've been very happy with that voltage; but transistor circuits, solid-state circuits, no. We've got to do two things: We've got to reduce the voltage and we've got to convert it to DC, to steady DC that doesn't change with time.

We're going to meet three new components that are going to help us to do that, three new electronic components, all of which I showed you circuit symbols for back in the lecture on circuits and diagrams and so on. But now we're going to look at them in a little more detail.

The first of them is the transformer. The transformer is an interesting device. It consists of two coils of wire in close proximity. For the kind of transformers we're talking about that transform power, those coils are typically wound on an iron core. If you've taken any other physics course, either from the Great Courses or in college or high school, you probably learned about the phenomenon of electromagnetic induction, whereby a changing current in one circuit will induce, cause, a changing current in another circuit, even though the two circuits aren't connected. The intermediary is the magnetic field that's created by those currents. I'm not going to go to those details here. Again, in keeping with this being a practical course, I'm simply going to introduce you to the transformer. Here we see its circuit symbol; you can see obviously what it is. It's the two coils of wire, and those vertical bars in between are representing the iron that those coils are typically wound on to concentrate the magnetic field between those coils.

There's what this circuit symbol looks like. We have a V1 and a V2. The left-hand side is called the primary; the right-hand side is called the secondary; although those names could be reversed, it doesn't really matter. We have N1 turns on the primary, and N2 turns on the secondary. What happens is V2, the voltage that appears across the secondary, is the ratio of the turns: N2/N1xV1. If I had a transformer with the same number of turns on both secondary and primary, I would get the same voltage out. You might say, "That's not useful for anything." It's very useful: It isolates the secondary from the primary. A lot of medical applications, for example, use so-called isolation transformers that don't step the voltage up or down, but they keep it roughly constant, but they isolate the circuit that might be connected to the patient, for example.

Here are some examples of transformers. They're ubiquitous. There are huge transformers at power stations that step the voltage up for long distance transmission, and step it back down for safe use for end users in cities and towns. By the way, we don't get something for nothing when we step up voltage. We get stepped up voltage, true, but we get stepped down. Current and power is voltage times current and in principle the power's the same either way. There's a smaller transformer of a type I'll show you in just a moment. There's a typical transformer that you'll see sitting on power poles around your neighborhood and they supply one or more houses with step-down voltages coming from maybe 7,000 or 8,000 volts on the power line down to 240 volts for your home, and that's split into 220 volt circuits. Finally, is a power block, similar to this power adapter, which you see in common electronic equipment. There's a little transformer in there. Sometimes in very small ones there's some other circuitry, but usually there's a small transformer, especially if it's a fairly hefty power block. Then that transforms the AC into a lower voltage, and often it also turns it into DC using devices that we'll see shortly.

The transformer I'm going to use today is this one. It's a small transformer; I've just got it mounted on a plastic block and a couple of binding posts so I can connect to it easily. But the transformer is the thing I'm holding in my hands. It's got this iron core, and then underneath those paper layers are the windings, the two coils. This particular transformer is supposed to take in standard 120 volt power line; there's the power line I'm going to connect it to. By the way, don't go connecting things to power lines, it's dangerous. I know what I'm doing here. Normally, people who are working with electronics as hobbyists never have to touch the power line or go near it; their circuitry already takes care of what I'm about to do here. Here's the power line coming in, and here is supposed to be 6.3 volts RMS coming out. That's a common transformer used in the old days to power the filaments in vacuum tubes, but it's also useful for a lot of other purposes, and I'm going to use it for some of these purposes.

I'm going to begin by doing some changes to the oscilloscope that will allow me to look at the voltage that's being produced by that transformer. Now I have the oscilloscope set up to measure the output of the transformer through this wire, which connects to channel one; the channel display is in yellow. I'm going to connect the two wires to the two outputs of the transformer, and I'm going to plug in the transformer. Let's take a look at what we have. Here we have a clear pattern that looks pretty much sinusoidal. It's not perfect. There's a bunch of electrical noise in the studio, and there are some other issues going on here, but basically it's a nice sinusoidal waveform, approximately. It's running up and down about—here's the 0—10 volts; that's about two divisions. You can see at the upper left it says 5 volts per; that means each of these squares is 5 volts, so we're coming up about 10 volts. I thought I said that transformer was 6.3 volts. It is, but that's 6.3 volts RMS—that's that average—and so the peak is closer to 10, and so it's coming up to not quite 10 volts; and take 0.7 of 10 and you get about 7, and that's a little more than 6.3, so it's working. If you measure the time between these peaks you would find on this oscilloscope, at 5 milliseconds per division, it would come out a 60th of a second.

There's the output of the transformer now swinging about plus and minus 10 volts. It's still AC, but it's now a lower level, more appropriate to the kind of electronic circuits we're going to be building. We've taken the first step in building a power supply to take what comes out of the wall and turn it into a suitable voltage to use to power electronic circuits.

But we've got two more steps to go through before we can do that, and I need to introduce two new components in order to make those steps. Let's take a look at first the diode. I showed you a diode when I gave you a bunch of circuit symbols a while back in the second lecture; now I'm going to say a little bit more about diodes, but we'll really get into them and how they work a few lectures hence. But what a diode is, practically speaking, is basically a one-way valve for electric current. It lets current flow in one direction; in other words, it has a very low resistance in the so-called forward direction, and it has a very high resistance in the reverse direction. There's the circuit symbol for a diode. It's this solid triangle. Sometimes it's drawn hollow with a bar. Current flows through that diode from left to right; that's the direction of the current flow. If you put a positive voltage on the left, you'll get current flowing through it and it'll act like a very low resistance, almost like a wire. If you put voltage the other way, it won't work. We're going to characterize the diode, like we did other circuit components, with a VI characteristic curve. Here's the VI characteristic for an ideal diode. An ideal diode, if you put positive voltage across it or try to, huge amounts of current will flow because it's, after all, just like a short circuit, like a piece of wire: It takes almost no voltage to drive current through it. It's like a switch that's on. But if you put the voltage the other way, you can build up large voltages in the reverse direction and no current will flow because it's like an open switch, and it switches abruptly when the voltage across it changes. Nothing is perfect; nothing is ideal. A real diode has a curve that lets a tiny, tiny bit of current—typically microamperes, nanoamperes, or less—flow in the back direction; so it's very small. It allows a small voltage to build up across it. That voltage for a silicon diode, which is what most diodes are made of, is on the order of 6/10 to 7/10 of a volt. That's going to come back to haunt us many times in this course.

There's the diode; it's like a one-way valve, if you will, for electric current. There's the 0.7 voltage that the diode maintains across itself when it's passing current. Again, if you're dealing with a circuit where voltage is big, that's irrelevant; but if we're dealing with a few volts, you've got to take that into account. That's the diode.

What I'm going to do next is take my power supply now—it's the step-down transformer that I already showed you that I have connected, and I've got out of it 6.3 volts RMS, about 9–10 volts peak—and I'm going to attach a diode. I'm going to look again on the other side of the diode at the output and see what that looks like.

Before I actually demonstrate what the diode does in the circuit, let me say a little bit about the circuit I have here. Here I have a small white board with a lot of little holes in it, and those holes make electrical contact. I've taken one of those boards and taken the back off it so you can see kind of how the contacts go. There are rows of holes that are all connected together by these bars of conducting material, there are rows along the edges that are all connected together, and we're going to wire almost all our circuits from now on on these boards. This is the first example of that; we'll get more sophisticated examples as we move on. I'll be plugging components into these boards, and I'll be doing it in such a way that I connect the parts I want connected, and don't connect the parts I don't want connected by plugging them into the appropriate holes. I have one of these boards mounted here, and it's got connections to these binding posts so I can bring in signals from outside. There's some more stuff on this board; there are actually three things on this board: There's the power supply I'm now describing, there's a regulator circuit that makes it better (and I'll talk about that in a minute), and there's actually an audio amplifier and a loudspeaker, and we'll get to that by the end of the lecture.

Let me begin by connecting the power supply, the input of the power supply, to the transformer. The voltage from the transformer is now coming into there. Now, let me take the oscilloscope, which has this probe that makes it easy to connect to little points on these circuits, and I'm going to connect it right here to the output of the diode. The diode is a tiny black thing that you can barely see, and I'm not even going to bother to hold one up. It looks like a little resistor but without any stripes on it, and it's got a little bar at one end and that indicates the end toward which current flows.

Let's see what we've got. You can see the yellow curve, which is the output of the transformer going up and down and alternating current. You can see the output of the diode, which is almost the same; that little difference is partly due to that 0.6 or 0.7 volt drop across the diode. It's following the AC coming out of the transformer while it's positive, but as soon as the transformer output goes negative, the diode shuts off. It becomes an open circuit and it's no longer passing current, and so we get 0 volts at the output, and then the next cycle comes along and we get a voltage that follows it, and across we go.

You might say, "We've made DC." We have in the sense that we've made a voltage that only goes in one direction, but it's far from steady and it would sound awful—and I'll show you that at the end of the lecture—if we tried to power, say, a piece of audio equipment with that signal. So we need to move on, and we need to look at still another component that we have to add to our power supply.

Meet the capacitor. Capacitor is a pair of conductors separated by an insulator. We'll be using capacitors a lot throughout the course, and I'll be introducing more and more about them. But for now, it stores charge and energy. The charge and voltage are proportional, so if you put a certain voltage across the capacitor you get a certain amount of charge on the plate. The capacitance, C, is the ratio of how much charge you get for a given voltage. One Coulomb per volt is called one farad, which, by the way, is an enormous capacitance. What capacitors do is it takes time to flow charge onto their plates, and so they slow things down; they introduce a time dependence in electric circuits and the voltage across a capacitor can't change instantaneously. We have plus Q on one plate, minus Q on the other plate, and some corresponding voltage between the two plates. That's what a capacitor does. We're going to use the capacitor as a so-called filter in this power supply; we're going to use it to smooth out these variations that we're seeing after the diode.

Let me add a capacitor to this circuit. I'll take a fairly big capacitor—this is a 2,200 microfarad capacitor—and I'm going to put it right into the circuit. So the capacitor, this big cylinder, has been inserted into the circuit; let's take a look at what's happened. What's happened is we have an almost perfectly steady DC voltage. There's a little tiny bit of a blip where the peak of the AC comes in, but it's pretty good, steady DC. By the way, that little blip is called ripple, and you have to decide, if you're an electronic circuit designer, how much ripple you can tolerate. If you buy a power supply it'll be rated for what the ripple voltage will be, what this variation in voltage would be, and typically, it should be a very tiny fraction of the actual output voltage. You'll notice we're producing a little bit less than 10 volts as our peak voltage coming out of this power supply.

We need to do something else to our power supply, and I've actually already got it in there. We've connected the filter capacitor—again, going from the diode, back to ground, back to the return of the power supply—and we've got about 8-something volts DC. Then we connect a load resistance, and I've actually already got a load resistance connected in that circuit. That's the thing to which we're wanting to deliver power. We need to size that capacitor; we need to make it big enough so that during the time the AC has gone negative, and the diode isn't conducting, and there's no more charge or energy coming into the circuit, the capacitor can hold that charge and energy for a long enough time that its voltage doesn't decrease appreciably. This capacitor is doing quite a good job; if I had a smaller capacitor, from here that voltage might've decreased quite a bit and then come back up again, and so on.

There's the power supply with the filter capacitor in place. I'll show you or I'll let you hear soon just why you really want to have that filter capacitor.

Now we've introduced three components: the transformer, the diode, and the filter capacitor; a capacitor in general, a filter capacitor in this case.

Let's improve on this power supply one more time. Here I've got to introduce something whose workings you won't understand for a few more lectures. When we get into the lectures on operational amplifiers, you'll understand how we'll be able to build a device like I'm showing you. Down here at the end I have a little tiny thing. It looks just like a transistor, but it isn't. It's a little black piece of plastic with three wires coming out. What it is is a voltage regulator, an integrated circuit voltage regulator. It probably has something on the order of a dozen transistors in it. Its job is to take an input voltage that may be varying, may not be the voltage you want, and produce an output voltage that's fixed by the specifications of that voltage regulator.

This particular regulator I've chosen is designed to produce an output voltage of 5.0 volts, so let's see how well it does. I have, actually, the output of my power supply also going into the input of that voltage regulator and then I have the output of the voltage regulator. Now I'm going to connect my oscilloscope across the output of that voltage regulator. That output is right here, and there are a few extra components connected to that output. Let's see if we've got the right place, and we want to connect to there. Now we see a very steady, smooth DC voltage, exactly the kind of thing we would like. Just a very slight little wiggle as we go under the peaks here, but hardly noticeable. That's a good power supply. We probably, in fact, with that electronic regulator attached, could've gotten away with a smaller filter capacitor. Since capacitors are expensive, that would've been a good thing.

Our circuit now looks like something with this electronic voltage regulator attached, and then we've got our load resistance following that. It's beautiful, it's smooth, it's DC. Furthermore, if you notice, we're at 5 volts per division on the oscilloscope; that's one division. We're up at exactly 5 volts. If I threw it into my voltmeter, I would see it read 5.01 or 5.02 or something volts, or 4.98. Pretty close to 5 volts. We've built a nice DC power supply.

Now let's do something useful with this power supply instead of just looking at it on the oscilloscope. What you might want to do with a power supply like this is to power some kind of electronic circuit. The circuit I've chosen to power is a little audio amplifier. The last thing I have on this board— and, again, this is something we'll learn about, we'll actually build from individual transistors a few lectures hence, but for now I'm going to simply point it out—a small little rectangular chip. It's an integrated circuit, audio amplifier. It's not a great big power amplifier, but it can put out enough power to drive a loudspeaker. That amplifier is set in that circuit. It's driving this loudspeaker that you see here. It's not yet connected to a power source. This yellow wire brings in its power, but I'm going to connect that yellow wire to the output of my voltage regulator. Then this amplifier needs an input. I've taken as my input my iPod. I've just connected my iPod to some wires that I can plug right into this board, and they're plugged into the input. One of them is grounded, and one's plugged into the input terminal of that amplifier. The iPod is now playing but we don't hear anything. That's because we haven't connected the power or the amplifier. We'll plug that right into the output, and there's our music.

We've got a complete audio system built with this integrated circuit audio amplifier, and we've also built the power supply that's giving it the smooth, steady DC that it wants to have. You'll notice that occasionally there are loud pieces in the music, and they're probably drawing so much current out of the regulator that it can't quite keep up the regulation. If I turn down the volume a little bit it sounds a little better and it's running a little bit smoother, although you see a few glitches. By the way, that lovely music is called "Starlight." It's written and played by my sister Helen Wolfson who's a hammered dulcimer player. That's the music.

Now I raise the question: What good is this filter capacitor in the power supply? What's it good for? What's it all about? Do we really need this big hunking thing there? Let me show you what happens; I'm going to do grievous damage to my sister's music by simply removing the filter capacitor from the power supply. Here's the nice music playing, and that's what we hear: "Blah." What we're hearing are the 60 cycles, the 60 Hertz oscillations; almost the same as the oscillations we heard when I did the speed of sound thing, but this is coming from the fact that the system is unable to regulate. The filter capacitor can't hold enough charge to take care of maintaining the voltage to the regulator during the time that the AC goes negative. The regulator is doing its best to regulate, but it can't regulate if it's not getting enough voltage at its input. What we hear instead of the beautiful music is this "blah" of this every 60 cycles, the voltage coming up, and we get a little bit; you can kind of hear the music playing underneath it, but not very much.

We really do need filter capacitors in power supplies. If we don't have them, we've got a big problem. I'll just stick the filter capacitor back. Back it goes, and we get the beautiful music again.

Let's wrap up. What we've done here is learn how to convert alternating current to lower voltage for solid-state circuits and convert it to nice DC. We needed three more components to do that: the transformer, the diode, and the filter capacitor. We're going to see a lot more of diodes and capacitors. We won't use transformers much more because they're built in to the kinds of things we'll be working with.

If you did the project in the previous lecture, you learned about how to simulate circuits with some of these web-based simulators. If you'd like to extend your prowess to that, I'm asking you in the project for this lecture to design a simple 5 volt power supply. It needs 100 ohm load resistance. You're going to use a source, you're going to have to pick your transformer, you're going to have to try different capacitors and see what happens, and you're going to have to tweak your transformer to achieve about 5 volts. I'm not letting you use one of these electronic regulators. If you want to, as a sort of extra credit or an extra learning, you could add one of the integrated circuit voltage regulators similar to the one I did here that's available in these circuit patterns. I recommend the project for this lecture.

Up the Treble, Down the Bass!
Lecture 5

In the preceding lecture, we looked at AC signals, voltages and currents that varied sinusoidally with time. The emphasis was largely on a single frequency. But life would be pretty dull if we had only one frequency. In fact, we're usually confronted with a mixture of frequencies, as in the audible spectrum—the sounds we hear. We need circuits that can process signals consisting of multiple frequencies—filter circuits. In this lecture, we'll learn how filters can process the different frequencies that appear in electronic signals and, consequently, in other signals, such as sound, that we might then convert to electronic signals. Topics to be covered include these:

- Filters and their uses
- Capacitors up close
- A simple low-pass filter
- A simple high-pass filter

- Filter characteristics
- Band-pass filters
- Application:
  a loudspeaker network.

Filters and Their Uses

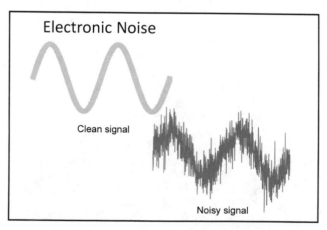

In the real world, there are sources of electromagnetic noise in our electrical signals. We can remove that noise with an electronic filter.

Capacitor Up Close
A conceptual picture of a filter shows an input, consisting of a mix of frequencies, and an output, consisting of an altered mix of frequencies.

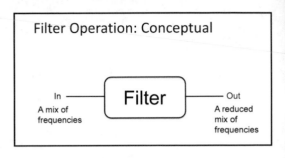

One component that allows the filtering of frequencies is the *capacitor*. A capacitor is a pair of conductors separated by an insulator.

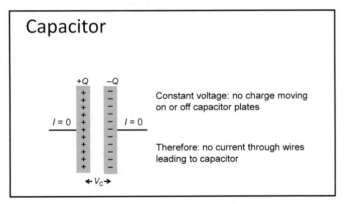

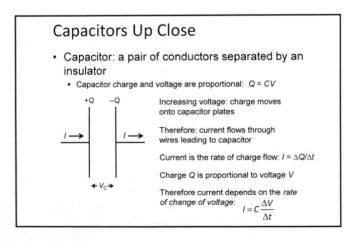

Charge goes onto the capacitor plates when a voltage is put across the capacitor. The voltage and the charge are proportional, and that proportionality defines the *capacitance* of the capacitor.

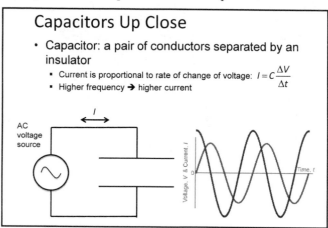

Capacitors Up Close

- Capacitor: a pair of conductors separated by an insulator
  - Current is proportional to rate of change of voltage: $I = C\dfrac{\Delta V}{\Delta t}$
  - Higher frequency → higher current

AC voltage source

I

Voltage, V & Current, I

Time, t

With a sinusoidally varying voltage, charges pile up on the capacitor plates. As the voltage swings negative, charge flows off the plates, and then reverses, with negative charge ending up on the upper plate and positive charge on the lower. This process keeps repeating, so there's always current flowing through the wires leading to the capacitor.

"Ohm's Law" for Capacitors

- Capacitor acts "sort of" like frequency-dependent resistor
  - "Resistance" proportional to 1/frequency
  - High frequency → low "resistance" and vice versa
    - $f \to \infty$ (very high frequency): capacitor acts like short circuit
    - $f \to 0$ (very low frequency): capacitor acts like open circuit
- Why "sort of"?
  - Because capacitor changes phase
  - Current leads voltage in capacitor
  - Better name than "resistance": *capacitive reactance*, X_C
  - X_C proportional to $1/fC$

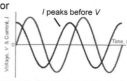

I peaks before V

Voltage, V & Current, I

Time, t

Ohm's law doesn't apply to capacitors because the voltage and current are not proportional in a capacitor. But when there is an AC signal coming to a capacitor, there is a sense in which we can talk about the proportionality between voltage and current—in particular, between the maximum value of the voltage and the maximum value of the current. We can say, then, that a capacitor acts *sort of* like a frequency-dependent resistor.

A Simple Low-Pass Filter

A circuit consisting of an AC voltage source and resistor in series with a capacitor also acts *sort of* like a voltage divider. The capacitive reactance goes down as the frequency goes up; therefore, at high frequencies, the capacitor acts more like a short circuit, or a low resistance. At high frequencies, the circuit is like a voltage divider, with less voltage across the capacitor—and therefore the output—than would be the case with low frequencies. Therefore, this is a low-pass filter.

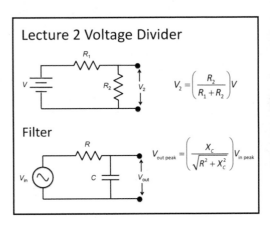

Lecture 2 Voltage Divider

$$V_2 = \left(\frac{R_2}{R_1 + R_2} \right) V$$

Filter

$$V_{\text{out peak}} = \left(\frac{X_C}{\sqrt{R^2 + X_C^2}} \right) V_{\text{in peak}}$$

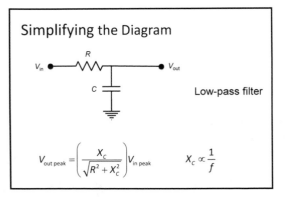

Simplifying the Diagram

Low-pass filter

$$V_{\text{out peak}} = \left(\frac{X_C}{\sqrt{R^2 + X_C^2}} \right) V_{\text{in peak}} \qquad X_C \propto \frac{1}{f}$$

Circuits are frequently drawn with the capacitor going to ground, and it's understood that the input voltage is applied relative to ground and the output voltage is measured relative to ground.

A Simple High-Pass Filter
A high-pass filter is the opposite of a low-pass filter. Again, thinking of this as a sort of voltage divider, the capacitor's "resistance" is small for high frequencies but large for low frequencies. As a result, the output voltage is reduced for low frequencies.

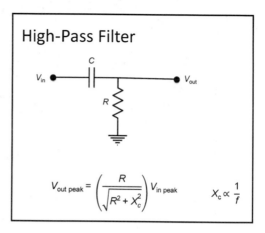

Filter Characteristics

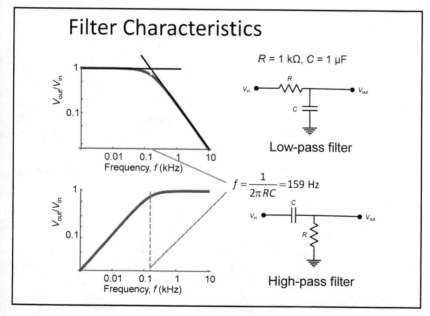

We can analyze high- and low-pass filters a little more carefully and essentially display what they do by looking at curves that describe their frequency response, that is, how much signal they let through as a function of frequency.

Band-Pass Filters

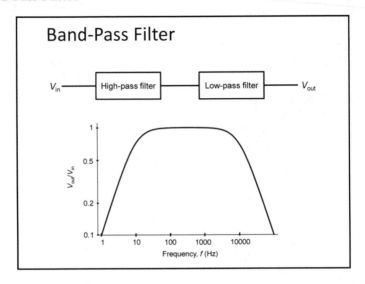

If we combine a high-pass filter and a low-pass filter, it seems as if no frequencies would get through. But if we pick the break points differently (*knee frequencies*), we can build a band-pass filter for a particular band of frequencies. This might be useful in a communication system to allow limited bandwidth to carry information with a limited range of frequencies.

Application: A Loudspeaker Network

In loudspeaker systems, a high-pass filter passes only the high frequencies to the tweeter, and a low-pass filter passes only the low frequencies to the woofer.

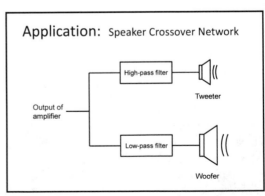

Introductory

Brindley, *Starting Electronics*, 4th ed., chapter 5, pp. 85–91.

Platt, *Make: Electronics*, chapter 2, experiment 9; chapter 5, experiment 29.

Shamieh and McComb, *Electronics for Dummies*, 2nd ed., chapter 4, p. 80 to end.

Advanced

Horowitz and Hill, *The Art of Electronics*, 3rd ed., chapter 1, section 1.7.13.2.

Scherz and Monk, *Practical Electronics for Inventors*, 3rd ed., chapter 2, section 2.33; chapter 9.

Project

Twin-T Filter

Simulate the circuit shown and explore its filter characteristics. Make a log-log plot of the ratio of output voltage to input voltage over the frequency range 10 Hz $\leq f \leq$ 100 kHz.

- Straightforward: Use a voltage function generator (under *Signal Sources*) as the input. Use a fixed voltage but different frequencies. Run a time-domain simulation for each f and plot V_{out}.
- Extra learning: Explore CircuitLab's frequency domain simulation.

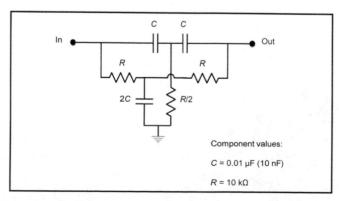

*1. What is an approximate way to characterize a capacitor (a) at low frequencies and (b) at high frequencies?

2. Even though we talk of a capacitor as approximating a frequency-dependent "resistance," that characterization isn't quite accurate. How is the AC voltage-current relation for a capacitor different than that for a resistor?

3. What's the purpose of the crossover network in a loudspeaker system?

Up the Treble, Down the Bass!
Lecture 5—Transcript

Welcome back. In the previous lecture, we looked at AC signals, voltages and currents that varied sinusoidally with time, and the emphasis was largely on a single frequency. In fact, in the end we built a power supply that worked on the 60 Hertz AC power line frequency. Life would be pretty dull if we only had one frequency in the world. In fact, we're usually confronted with a mixture of frequencies. Think of the audible spectrum, the sounds we hear. If we could only have one frequency, all you would hear me do is go [TONE SOUND], and you might reply [TONE SOUND] because we'd only have one frequency. But we have mixes of frequencies, and frequently we want to do things to those mixes. We want to alter the mix of frequencies. We want to emphasize, say, the treble. If you're getting hard of hearing and you need to hear your TV a little better, you might want to boost the treble, the high frequencies. You might have electronic noise; you might want to eliminate it, and so on. We need circuits that can process signals consisting of multiple frequencies. That's what this lecture is about. It's about filter circuits. You've heard that word before because I talked about the filter capacitor in our power supply, and we're going to enlarge on that concept in today's lecture. We're going to learn how filters can process the different frequencies that appear in electronic signals, and consequently in other signals, like sound, which we might then convert to electronic signals.

Let me give you some examples. In the audio world, there's a clear example of the tone controls on perhaps old-fashioned radios and stereos. Nowadays they have more complicated systems called equalizers that can vary the frequencies and different frequency bands. My TV has a menu that shows me able to adjust the treble and the bass up and down various amounts. If you play the electric guitar you've got frequency controllers. You've got some kind of filters in there that you can use to boost or diminish different frequencies. In the audio world, treble and bass controls as they were called in the old days, equalizers and more complicated frequency controls as they're called today. Those are sort of obvious; you're probably familiar with those. But there are many other examples where we have frequency issues.

For example, at the top left in this diagram we see a sort of clean electronic signal. It happens to be a single sine wave. In the lower right it looks the same sort of, but it's got all this noise superimposed on it. The fact about the real world is there are sources of electromagnetic noise that get into our electrical signals and make them noisy, and we'd like to avoid that. We have techniques for avoiding it, but when the noise gets in there we need to remove it, and we can do that with an electronic filter. You use your cell phone; 100 or 1,000 other people are using cell phones and communicating through the same nearby cell phone tower. How on earth does the cell phone system manage to sort all those individual calls out? One way is with very, very fine filters that respond only to a very narrow band of frequencies and are able, therefore, to sort out many, many, many phone calls that are very closely spaced in frequencies but, nevertheless, the filters can handle that.

I want to talk a little bit conceptually about filters, and then about the main component that we're going to use to make a filter work. Here's a conceptual picture of a filter. We have an input; the input consists of a mix of frequencies. We have an output, and the output consists of an altered mix of frequencies. Maybe it's a stereo system and we've boosted the bass, and so the output consists of a greater amount of power at the low frequencies. Or maybe we boosted the treble and it's a greater amount at the high frequencies. Or maybe there was some noise at a particular frequency and we've applied a filter that removed that frequency alone and left the other frequencies unimpeded. We can design filters to do all kinds of things, and we'll be talking about those different kinds of filters throughout this lecture.

But before we do, we need to look at the one component that's going to let us do this filtering, and it's the same component we used when we filtered out the ripple in the power supply that we built in the preceding lecture, and that's the capacitor. I want to get a little bit up close with capacitors and spend a little more time explaining how capacitors work and what they do. Let's go to our big monitor and take a look in more detail at capacitors.

Let's look at capacitors up close. What's a capacitor? I introduced capacitors briefly before. A capacitor is a pair of conductors separated by an insulator. Because of that, charge goes on the plate to these capacitors when you put a voltage across them, and it turns out that the voltage and the charge are proportional, and that proportionality defines the capacitance of the capacitor. There's something funny about a capacitor: It's an open circuit; it's a block; it's an insulating gap. No current can flow through the capacitor, but it can flow in the wires leading to the capacitor. We want to explore what happens under different conditions to that current as a function of what's going on with the voltage you put across the capacitor.

If there's a constant voltage, then the capacitor has got some charge on it. Maybe it's zero, maybe it's some charge. We've got positive charge on one plate and negative charge on the other as we show here with the plus signs and the minus signs, and there's some voltage V sub C across the capacitor. But under these conditions there's no current flowing because there's no charge moving on or off those plates, so with a constant voltage there's no current through the wires leading to the capacitor; I, the current, is 0 in that case. That's not very interesting. Although, it's modestly interesting because under those conditions the capacitor is storing charge and also storing electrical energy, which it could release at a later time through a flow of current out of the capacitor.

But what happens if the capacitor voltage is changing? If the voltage is increasing, then because the charge and the voltage are proportional, that means the amount of charge has to be increasing. Where does that charge come from? It flows in through the wires that are leading to the capacitor plates. Here's a little animation of how that happens. In a real circuit, it would be only electrons moving onto the negative and electrons moving off the positive, but I'm showing it more symmetrically. There was some charge flowing through the wires leading to the capacitor and that brought the capacitor now up to some charge level. Current must flow through the wires leading to the capacitor while the voltage is changing.

Current is the rate of charge flow. It's the amount of charge, which I'm going to call delta Q; a little chunk of charge that flows in, and a little chunk of time, delta T. I is delta Q over delta T, the rate of flow of charge through those wires. The charge, though, is proportional to the voltage. What that means is that the current depends on the rate of change of voltage, so a capacitor has a different kind of relationship between voltage and current than, say, does a resistor. In a resistor, the current is directly proportional to the voltage. In a capacitor, the current—and, again, when I say in a capacitor, it's not through that gap, but in the wires leading to the capacitor—doesn't depend on the voltage, but on how rapidly the voltage is changing. Let's take a little more look at this relationship between capacitors, voltage, and current. In particular, let's look at what happens when we have a sinusoidally varying voltage. There's our current that's going to flow when the voltage is changing.

Here's a little simulation of an AC voltage source; that is, a voltage that's changing in direction in a sinusoidal way: Charges piling up on the plates; then, as the voltage swings negative, charge goes off the plates or swings down again; and then the voltage goes negative, and we end up putting negative charge on the upper plate and positive charge on the lower. As that goes on, what I've done on the right here is to plot the voltage in yellow, which was supplied by that voltage source. We turned a dial on a voltage source and said, "Give us as much volts." We also said, "Give us this frequency." Here's the current in green, and the current depends on the rate of change of voltage. You'll notice where the voltage was rising steeply at the beginning, the current was high. Where the voltage wasn't changing at all, which was happening briefly down here at the bottom and up here at the top, the current's going through 0. The current depends on the rate of change of voltage. You'll notice something else that that leads to: The current and the voltage are what we call out of phase. In this case, the current reaches its peak before the voltage; the current peaks before the voltage peaks, and so on. In a capacitor, the current leads the voltage, and it leads it by, in fact, 90 degrees or a quarter of a cycle of this sinusoidal oscillation.

That was a relatively low frequency; let's move to a high frequency. Things are happening much faster now because we've got a higher frequency. Same voltage; and let's plot it. Two things you notice are different: First of all, the voltage, although it's the same in amplitude, now more cycles fit in here because we've got a higher frequency. The current is bigger because the rate of change of voltage is bigger because the voltage has to change more rapidly to change at this higher frequency.

That's what capacitors look like up close. The key to take away from this is that a capacitor is a device in which the current responds not the voltage, but to rate of change of voltage. We need to think about, also, this phase difference between the voltage and the current in a capacitor.

OK, so that was a conceptual look at how capacitors behave. Now we want to talk about filter circuits in particular and how we'd use a capacitor to make a filter. To do that I have to introduce a little bit more about the voltage and current in a capacitor. I'm going to put in quotes here, "ohm's law for capacitors." Any electrical engineers watching this course, we know ohm's law doesn't apply to capacitors. In fact, I just told you that; I told you that the voltage and the current are not proportional. But when you have an AC signal coming to a capacitor, there's a sense in which you can talk about the proportionality between voltage and current; between the maximum value of the voltage in particular and the maximum value of the current. I want to introduce that.

I'm going to say that a capacitor acts sort of like a frequency-dependent resistor, sort of. I want to keep emphasizing that "sort of" because I'm on thin ice here. This isn't quite correct, but it gives us a feel for what happens. This "resistance" that the capacitor has is proportional to 1 over the frequency. Remember when I showed you the animations: At the low frequency we got a lower current; at the high frequency we got a higher current. It was as if it was easier for current to flow at the higher frequency. The capacitor has a "resistance" proportional to the inverse of the frequency. If there's a high frequency there's a low resistance, and vice versa. If the frequency gets very high toward infinity, the capacitor basically acts like a wire, like a short circuit. A capacitor at high frequency acts like a short circuit. You can already begin to see why our filter capacitor worked in our power supply, because it acted like a short circuit to ground for the high frequencies of the ripple but let the steady, direct current to through without being shunted to ground. At very low frequencies, on the other hand, the capacitor acts like what it is actually: a gap between two conductors. It acts like an open circuit or an open switch; no current can flow. In between, there's some kind of variation.

Why do I say sort of? Because this is not really ohm's law, because a capacitor doesn't just change the amplitude of the current relative to the voltage according to this "resistance," but it also changes the phase, as you saw. The current leads the voltage in a capacitor. That's really important. It makes the mathematics different, and it makes the whole electronics, and it makes the energy considerations all different. I'm not going to go into that; but remember, a capacitor isn't a resistor, and when I say it behaves sort of like one, we've really got to take that with a grain of salt. It does behave sort of like a resistor, but only sort of. A better name for that "resistance" is what's called capacity of reactants. It's given the symbol X, and I put a sub C on it because it's here about a capacitor. That capacitive reactant is proportional to 1 over the frequency times the capacitance.

Now let's start trying to do some things with capacitors. You can see the phase lag there. We're going to try to build filters out of capacitors. Remember in Lecture Two, I introduced the voltage divider. It consisted of two resistors in series, and we asked about the voltage across the lower resistor, R2, and it divided the battery voltage in proportion to that resistor's resistance in relation to the total. We're going to do the same thing with this circuit that consists of a resistor in series with a capacitor instead of a battery. We've got an AC voltage source, a generator, a function generator like I've been using, or whatever. The question is: Does this also act like a voltage divider? The answer is: Sort of, again. The mathematics is more complicated. I'm not going go into it; I'm not going to derive it. If you don't like seeing square roots, you can phase yourself out a little bit. But the point is: There's, in the bottom, the sum of the resistance and the reactants, but they come in kind of with the Pythagorean Theorem, and that's because of that 90 degree phase difference. The answer is: It sort of is like a voltage divider, but the relationship is a little more complicated. That's really all you need to know.

I'm going to stop here and I'm going to give you a quick quiz and ask you to think about this circuit and tell me whether it's a high pass filter or a low pass filter. Remember our conceptual picture of a filter: We had an input, the filter did something to the signals coming in, and the output was different; an altered mix of frequencies. If I put a mix of frequencies into this circuit, will the output consist more of the lower frequencies or more of the higher frequencies? Take a minute, pause, and think about that.

Let's come back and take a look at it. If you want to do the formula you can, but you don't have to. But the fact is that the capacitive reactants—this "like resistance"—goes down as the frequency goes up because it goes like 1 over the frequency. Therefore, at high frequencies, that capacitor acts more like a short circuit, more like a wire, more like a low resistance, and so at high frequencies it's a voltage divider, if you will, with less voltage across the capacitor than would be the case with low frequencies. Therefore, this is a low pass filter. Low frequencies get right through to the output. High frequencies are passed through the capacitor and go to ground and return back through the ground and they're not visible at the output. This is a low pass filter.

Before we do any more with the low pass filter, let's pause and simplify this circuit a little bit because we're frequently going to draw circuits that look simply like that, with the capacitor going to ground, and it's understood that the input voltage is applied relative to ground and the output voltage is measured relative to ground. It's just like the circuit I had before where there was a wire going across, connecting the bottom of the capacitor to the bottom of V in and V out. It's just a shorthand way of writing it, but we're going to do that.

Let's stop looking at conceptual diagrams now and turn to the real world, and let's build a low pass filter and see how it works. What I have here is a very simple circuit, already wired. It consists of a capacitor. It's a 1 micro farad capacitor, a millionth of a farad—that means a millionth of a coulomb for every volt on that capacitor—and I've got a 1,000 ohm resistor, and that's all I've got because that circuit, that filter is that simple. It consists of a resistor in series with a capacitor. The input is this red wire coming into this binding post here; that's connected to the resistor. The output is coming from the junction of the resistor and the capacitor, and that's coming out on this yellow wire and going into this table. I've got the ground wires connected to the black, and that's connected to this whole strip down here that's then grounded. The bottom of the capacitor, you can see, is connected to ground.

I'm going to feed this now with a signal from my function generator, and I'm going to look at both the input—coming in here from the function generator—and the output on the two channels of the oscilloscope. Let me get the oscilloscope set up to handle this. OK, I've turned on both channels of the oscilloscope. You can see on the screen here the traces. Let's turn them on over here so we can see them a little better on our big screen. What you see are two sine waves. They both have 500 millivolts, half a volt per division. They're approximately the same size; they're certainly the same frequency. They're also in phase. There are our input and output, and we're running right now at a fairly low frequency. We're running right now at a frequency of—let me take a look and make sure—100 Hertz.

Now I'm going to begin to turn up to the frequency of the function generator. Up it goes; a little higher; it's a little higher. You can see two things happening: You can see, first of all, the bottom trace, which is the output, getting smaller. You can see both traces getting more compressed because we're going to a higher frequency and so we're seeing more cycles. Go up again and you can see the frequency goes higher but the amplitude of the output drops lower because this is a low pass filter. It's letting the low frequencies through easily. When we started out at 100 Hertz, the two were almost exactly the same amplitude; just a little bit of attenuation in the output relative to the input. But as we go to higher frequencies, the output amplitude drops, and drops, and drops, and drops, and drops, and gets lower, and lower, and lower, and there it is. The oscilloscope is having trouble triggering now because I asked it to trigger off channel two, which is getting a little bit low amplitude. But you can see the idea: We have a filter that's reducing the high frequencies. There I've gone to a very high frequency, and the output is very, very small. This is a low pass filter.

What good is a filter like that? I want to give you one example of an application of that filter other than the one you used in the previous lecture. In the previous lecture, we built a power supply and we basically put a low pass filter in it—that's what the capacitor did—and we filtered out that 60 cycle variation. Remember, when I pulled the capacitor out of the circuit, I was playing beautiful music and all we heard was [STACCATO NOISE] because of that variation. Now I want to consider another situation: I want to consider a situation in which I have a noisy signal, like I showed you at the beginning of this lecture, and we're going to see how well our filter does at removing the noise from that signal. I'm going to change the set ups on the function generator—not the oscilloscope so much, but the function generator—by adding some noise to the signal, and we'll see what the filter does to that noise.

Here we are with a slightly different set up. What I've done, this function generator—and by the way, if you're wondering how this works we'll get to that around lecture 15; you'll actually build one so you'll understand how it works—this is a two channel function generator, which is very convenient, and I've set the second channel to generate random noise. I've set the machine up so that the output combines the random noise with that nice smooth 100 Hertz sine wave that we started with. You see on the big screen the sine wave we had before, same amplitude, same frequency; it's just got all this messy noise piled on top of it. If that were an audio signal that we were trying to play it would sound bad. If that were a signal representing some electrical quantity, that represented some physical quantity we were trying to measure in a scientific lab, that noise would bother us and result in less accurate measurements, for example. We'd like to get rid of that noise.

The reason we can get rid of that noise is because that noise is at much higher frequency than the signal. It's running a bandwidth up to 100 kilohertz, 100,000 cycles per second, although it does include components that are much slower than that. But clearly, as you can see, the variations associated with that noise are much faster than the variations associated with our basic sine wave. This is still connected to the filter, just as it was before. It's going through our low pass filter. Remember how our low pass filter attenuated high frequency noise, or high frequency signals of any kind; in this case, it's the high frequency noise. All I have to do is step over there and turn on the channel two display of the oscilloscope and you'll see what wonders our filter is able to work for us.

This is the same signal that came in, except it's been put through our very simple filter consisting of a 1 kilo ohm resistor and a 1 micro farad capacitor. Nothing else is going on there; it's that simple. We've cleaned the signal back up beautifully; we've removed all that noise. So one very important use of filters is to remove noise in circuits, and we've done a really nice job of it there. Just to remind you, I'm going to let that go a minute or two to show you that that is really doing a great job of removing noise; you can even see the randomness in the noise as the channel one signal jumps around.

There's a beautiful real-world demonstration of an electronic circuit doing just what we want it to do; in this case, an RC resistor capacitor filter removing the noise from a signal and making it beautiful and clean.

Let's think a little bit more about filters. Let me start with a little bit of a quiz for you: What's this circuit? Take a look at this circuit, think about it, pause your pause button, and figure it out. What does this circuit do? It's different from the circuit I showed you before.

What does this circuit do? This circuit; if you want to go through the math you can, but you don't really need to because you know the capacitor acts like a big resistance at low frequencies, and a little resistance at high frequencies, and so if you think of this thing as sort of a voltage divider, the resistor one, the upper resistor, is small for high frequencies. They go through it easily, but it's big for low frequencies. They don't go through it easily, and consequently, what's this? It's a high pass filter; it's the opposite of a low pass filter. We can analyze these high and low pass filters a little more carefully and sort of display what they do by showing curves that describe their frequency response; how much of a signal they let through as a function of frequency. It turns out to be best to do this—and I'm going to do this for the filter I actually have, with a 1 kilo ohm resistor and a 1 micro farad capacitor. I have the low pass filter I actually built, and the high pass filter I just quizzed you on—and it turns out to be good to plot these on logarithmic scales.

Here's the frequency plotted in kilohertz: 0.01 kilohertz, 0.1 kilohertz is our 100 Hertz that we started at, 1 kilohertz, up to 10 kilohertz, and I'm plotting the logarithm on the vertical axis of the output voltage over the input voltage; or actually, I'm plotting the actual output voltage over the input voltage, but I'm doing it on a logarithmic scale also. These individual divisions are a factor of 10, rather than a fixed amount. You see what happens: At low frequencies for the low pass filter, for a long time there's very little variation. Then on this logarithmic graph there is a rather sharp, it's called a knee, where the thing bends, and then it tapers off at a constant rate. By the time we get up to 10 kilohertz in frequency, we've got almost no voltage at the output. That's a filter characteristic curve. We can draw a very similar curve for the high pass filter that used the same components of the same values in this case, and there we have something opposite happening: We have no voltage out at DC at 0 frequency or very low frequencies, and then it gradually rises. At that same break point, which happens to be 159 Hertz for this particular combination of resistor and capacitor, it begins to level off. At frequencies much above that 159 Hertz up to about thousands and thousands of Hertz, it passes all the input into the output, basically unattenuated. Those are characteristic curves for filters.

Now I'd like to take a few more minutes and look at more sophisticated filters we could build. We could clearly put a low pass filter and a high pass filter together. What that's going to do is if we put a high pass filter first, then we're going to let only high frequencies through and we're going to cut off the low frequencies. But then we come to a low pass filter and we're going to cut off the high frequencies. You might say, "What good is that? It lets nothing through." If you pick those "knee" frequencies, those break points, differently, you can build a band pass filter that passes a particular band of frequencies. I want to have a communication system that has limited bandwidth—and bandwidth, that's a term you've heard; it has to do with how much information you can carry on a signal path, like a fiber optic, or the internet, or a cable—maybe you want to limit that bandwidth so you can carry more information, but with a limited range of frequencies, and so you might do that with a band pass filter, letting a particular band of frequencies through.

Here's an application for a band pass filter. If you have any kind of loudspeaker system—and I have an example of one here that does this very nicely—any good loudspeaker system consists of at least two different units and at two different speakers; two different devices that convert AC electrical signals into sound. They do it by running current through a coil of wire. The coil of wire is in a magnetic field and it moves back and forth and moves this cardboard cone, which makes the sound waves in the air. Notice there's a fairly big one in this speaker at the bottom. That's the woofer. That produces bass frequencies easily, but it's not very good at producing high frequencies because it can't move fast enough. Inside here is a tiny little tweeter, which is good for high frequencies. It would be really stupid to have an expensive high power audio amplifier dumping all its power equally to those two speakers. What do we do? We use filters—a high pass filter to pass only the high frequencies to the tweeter, and a low pass filter to pass only the low frequencies to the woofer—and that's how we sort them out. Many speaker systems would have several other drivers, typically a midrange, and there'd be a band pass filter that would pass to the midrange. That's another example of a filter application here used in this loudspeaker system.

Let me summarize and then tell you a little bit about the project for this lecture. We've learned a lot about capacitors; devices that act "sort of" like frequency dependent resistors. We learned how to put them in very simple voltage-divider-like circuits and build filters that did wonderful things for us like extracting a nice clean signal from a very messy amount of noise. If you'd like to go on to the project for this lecture, I have a project called a twin-T filter. A twin-T filter is a very sophisticated filter that has a very narrow frequency response. I'm not going to give away the punchline anymore; if you' like to do this project, I'm going to have you simulate this twin-T filter with your circuit simulation software.

However, I want to tell you how excited the people were here at The Great Courses when I mentioned I was going to do the twin-T filter because they said, "Oh"—it's called a notch filter; I'll give away a little bit, it takes out a very narrow notch of frequencies—"we use notch frequencies all the time." I said, "Well, what do you use them for? They said, "For example, suppose a cell phone goes off in the middle of a recording session or something. We don't want to re-record the whole thing. What do we do? We apply a notch filter at just the frequency of the tone made by the phone ringing. It takes out those frequencies and it doesn't take out anything else, and it leaves the recording nice and clean." They don't use this particular circuit for their notch filter, they probably simulate it digitally, but it's the same idea. Filters do wonders for us because they take a wide range of frequencies and they alter it in a way that we've engineered to do exactly what we want with them.

Semiconductors—The Miracle Material
Lecture 6

In this lecture, we will look at semiconductors, the miracle materials that are at the heart of most modern electronic devices, including transistors, diodes, and a host of others. Semiconductors replaced the vacuum tubes from the first half of the 20th century with tiny solid-state devices that consume very little power; billions of them can fit on a single silicon integrated-circuit chip. In this lecture, we'll discuss diodes, important components in their own right and also important to understanding the transistors that we'll cover in Lecture 7. Main topics in this lecture are as follows:

- Silicon atoms, silicon crystals, and intrinsic semiconductors
- Doping semiconductors: N- and P-type semiconductors
- The PN junction
- Diodes.

Silicon Atoms, Silicon Crystals, and Intrinsic Semiconductors

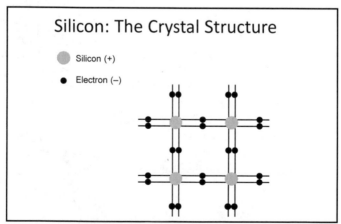

Silicon: The Crystal Structure

- Silicon (+)
- Electron (–)

Silicon is the element at the heart of semiconductor electronics. It has a crystal structure, in which every atom is bonded to its nearest neighbors by the sharing of two electrons.

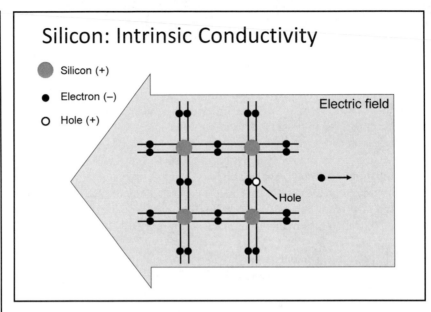

Silicon: Intrinsic Conductivity

Silicon (+)

Electron (–)

Hole (+)

Electric field

Hole

Silicon and other semiconductors conduct electricity by an unusual mechanism: At normal temperatures, random thermal motions might free one of the electrons in a silicon crystal from its bond. Freeing the electron leaves behind a "hole" in the crystal structure where the electron was.

Imagine that we apply an electric field, perhaps by introducing a sheet of positive charge beyond the right edge of the diagram and a sheet of negative charge beyond the left. The positive charge attracts electrons, and a free electron moves in response to this field (opposite the field, since it's negative) and, thus, carries electric current, just as it would in a metal. But something else happens: The field can cause an electron bound in the crystal structure to "jump" into the nearby hole left when the original electron was freed. As a result, the hole effectively moves to the left and acts just like a *positive* charge carrier.

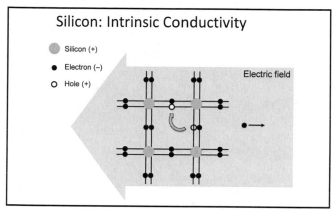

So the intrinsic conductivity of a semiconductor material, such as silicon, is caused not only by the presence of electrons, as in a metal, but also by the *absence* of electrons—that is, by holes. The holes act like positive charge carriers, and we will think of them as particles carrying positive charge. The holes, too, can move through the crystal structure, carrying electric current.

Doping Semiconductors: N- and P-Type Semiconductors

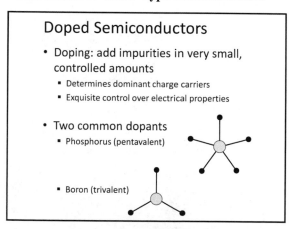

The process of doping determines the dominant charge carrier in silicon and other semiconductors. By doping appropriately, we can make semiconductors that have almost all of their current carried by electrons or by holes. Doping also provides exquisite control over the details of electrical properties, in

109

particular, how good a conductor the material is. Two common dopants are phosphorus and boron.

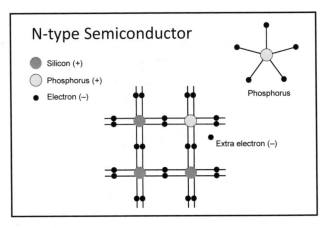

Phosphorus has five outermost electrons, but only four of them can participate in the silicon structure. Once it's introduced through doping, the phosphorus will fit itself into the crystal structure, but that one electron will be free and able to carry electric current. As a result, the dominant charge carriers in this *N-type semiconductor* are electrons.

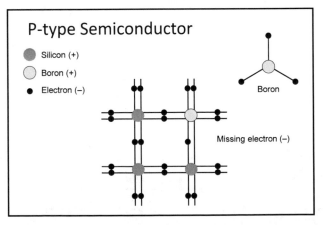

An analogous process occurs with boron. Boron has only three valence electrons, but the silicon structure needs four electrons. Thus, there is a missing electron (a hole) in the bond structure. In a boron-doped material, the majority charge carriers are positive holes. This is called a *P-type semiconductor*.

The PN Junction

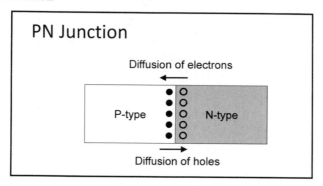

A *PN junction* is the interface between a piece of P-type material and a piece of N-type material. In the N-type material, there are many electrons, and in the P-type material, there are very few free electrons. At the junction, *diffusion* occurs—the process whereby some of the electrons move into the P-type material to ease out the abrupt gradient at the junction. Those electrons then recombine with holes. A similar process goes the other way, with holes diffusing into the N-type material. They, too, recombine with electrons.

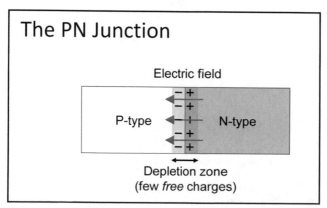

As a result, there are very few free charge carriers in the junction region (here, called the *depletion zone*), making the junction a poor conductor.

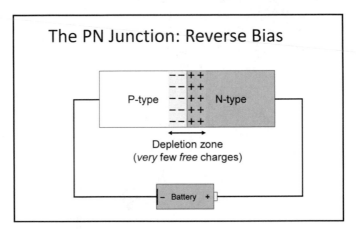

The PN Junction: Reverse Bias

If we connect the positive terminal of a battery to the N-type material and the negative terminal to the P-type material, the electric field at the PN junction is reinforced, causing the depletion zone to grow. Now there is a larger region where there are very few free charges, which means that the device can hardly conduct electricity at all. The PN junction is basically an open circuit. This is called *reverse bias*.

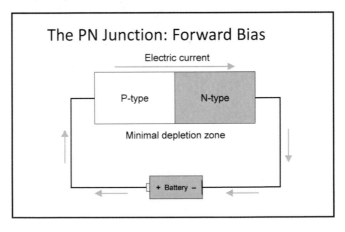

The PN Junction: Forward Bias

If we connect the battery the other way, it can eliminate the electric field and allow current to flow across the junction from the P-type into the N-type material. That's called *forward bias*; under this condition, the junction becomes a fairly good conductor, and electric current can flow around the circuit.

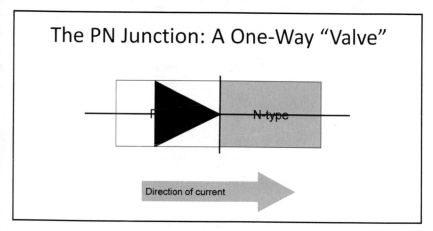

We're left with something that conducts electricity with a battery connected in one direction but not in the other direction. The PN junction becomes a one-way valve for electricity. Current can flow from P to N, but it can't flow from N to P.

Diodes

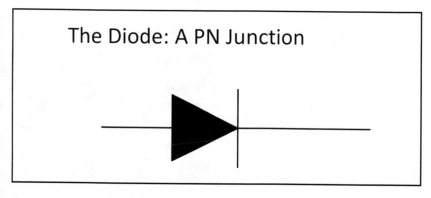

As you recall, we used a diode that passed current in one direction and not the other in our power supply to change AC into DC (the process of *rectification*). That was the step that cut off the negative-going portion of the AC. The PN junction does exactly the same thing; in fact, a diode *is* a PN junction.

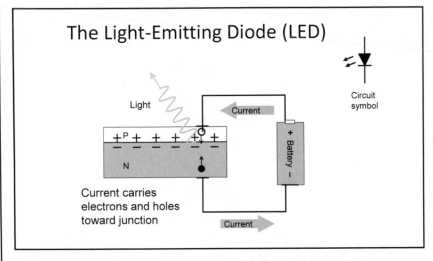

The Light-Emitting Diode (LED)

The obvious use of the diode is in converting AC to DC, but there are many other uses of diodes in electronic circuits. One specialized type of diode is the light-emitting diode (LED), which converts electrical energy to light.

Another specialized type of diode is the photovoltaic cell, which takes in sunlight and produces electricity. It, too, is a PN junction diode.

Suggested Reading

Introductory

Brindley, *Starting Electronics*, 4th ed., chapter 6.

Lowe, *Electronics All-in-One for Dummies*, book II, chapter 5.

Shamieh and McComb, *Electronics for Dummies*, 2nd ed., chapter 6 through p. 125.

Advanced

Scherz and Monk, *Practical Electronics for Inventors*, chapter 2, section 6; chapter 4, sections 4.1–4.2.

Project

Diode Applications

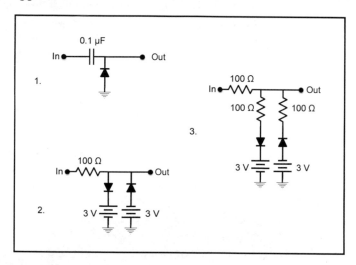

Simulate each circuit below. At the input, use a voltage function generator producing a 10-V p-p, 1-kHz triangle wave. Look at both the input and output with time domain simulation. Describe in words what each circuit does.

Questions to Consider

1. The *N* in *N-type semiconductor* stands for "negative." Does that mean an N-type semiconductor is negatively charged? Explain and answer the analogous question for a P-type semiconductor.

2. Explain the terms *forward biased* and *reverse biased* as applied to diodes.

3. What is the purpose of doping silicon?

Semiconductors—The Miracle Material
Lecture 6—Transcript

Welcome back. In this lecture, we're going to look at semiconductors, the miracle materials that are at the heart of most modern electronic devices like transistors, diodes, and a host of others; the things that replaced the vacuum tubes from the first half of the 20th century with these tiny, tiny devices that are solid-state, consume very little power, and we can cram billions of them on a single chip. This lecture will probably be the most deep physics lecture in the sense of going into exactly how these devices work. In this lecture, we'll talk about diodes in particular and see how they form the basis of the transistors that will come in the next lecture.

I'm going to begin by looking at the element silicon, which is a remarkable element. It's the second most abundant element in Earth's crust after oxygen, and it's the element at the heart of semiconductor electronics. Let me give you a very simplified picture of silicon's crystal structure. It's simplified because silicon is a three-dimensional crystal. It has a more complex structure. But the important point about it is every silicon atom is bonded to its nearest neighbors by sharing of two electrons. In this picture, you see the reddish dots, which are the silicon atoms, and you see the outermost electrons going off and sharing themselves with nearby silicon atoms. That's the structure of silicon. Silicon is one of a number of elements or compounds that act as semiconductors. Their electrical conductivity isn't as good as metals, but they're not bad like insulators. They're somewhere in between, and that's why they're called semiconductors. They conduct by a unique mechanism, and we need to understand that mechanism to see how semiconductor electronic devices work.

Here's a silicon crystal. If were at absolute zero temperature, it would just be sitting there looking like this; but at reasonable temperatures—room temperature for example—it's possible that the random thermal motions, the random motions that we call loosely "heat," might knock one of these electrons out of its bond and make it free, and then that electron would become free to carry a current the way electrons do in metals. An electron in this picture now has been dislodged from the crystal structure; it's free to move. If we apply an electric field—and if you don't know what an electric field is, don't worry; just imagine putting a bunch of positive charge on the right of this picture somewhere off in the distance, and negative charge on the left, and the positive charge on the right is going to attract electrons, and it's going to repel anything that's got a positive charge—so there's our electron in this electric field, and it's going to go happily moving away, and that would be carrying electric current the same way the electrons do in a metal.

But something else can happen here: Notice that the bond that's been broken, if you will, by the loss of that electron is an empty spot. An electron could go there; an electron would like to go there. The tendency of electrons in the presence of that electric field is to try to move to the right. It's possible that the electron sitting right next to that blank spot could move into it. We call that blank spot a hole, and that hole basically acts like something with positive charge because what happens is if the electron moves into the hole to the right, then the hole has effectively moved to the left. Again, in the presence of that electric field, another nearby electron might jump into the hole, and in that process, the hole has moved. What happens in a semiconductor material like silicon, the intrinsic conductivity of that material is caused not only by the presence of electrons, as in a metal, but also by the presence of these—if I can say it this way—the presence of the absence of electrons by these holes. The holes act like positive charge carriers, and we're going to think of them as little particles carrying positive charge. They aren't; they're more complicated than that; they're the absence of an electron. But they, too, can move through the crystal structure. In an intrinsic semiconductor like silicon, there are equal numbers of electrons and holes. Again, the way they get there at all is because random thermal motions can promote an electron out of the bond it's in and let it be a free electron, and that process creates a hole as well, so we have electron hole pairs. In an intrinsic semiconductor, an intrinsic pure semiconductor—that's what intrinsic means: pure silicon, nothing else in there—the current can be carried equally by holes and electrons. But the beauty of semiconductors is we have two kinds of charge carriers instead of one—we have positive holes and negative electrons—and that make semiconductors wonderful materials for building electronic devices.

Intrinsic semiconductors—that is, pure silicon and pure other semiconductors—are actually not very useful. They conduct electricity, but only weakly. One thing they are good for because the amount of electron hole pairs that are being formed depends on the temperature because these electrons are being promoted out of the crystal bonding structure by random thermal motions, the conductivity actually goes up. With increasing temperature, the resistance of the material goes down, and so pure silicon can be used very nicely as a temperature sensor, a resistance-based temperature sensor, and that's a very common use for it.

But most of our semiconductors are not pure. We go to great lengths to make pure silicon. Even though silicon is very abundant, it's very expensive and energy intensive to make it really pure. Then we add impurities in extremely small and carefully controlled amounts, typically on the order of one impurity atom for every 10 million silicon atoms or something in that order; very, very small amounts of these materials. They're called dopants, the process is called doping, and the reason we dope the semiconductors is it does two things: First of all, it determines what the dominant charge carrier is. By doping appropriately, we can make semiconductors that have almost all of their current carried by electrons, or we can make semiconductors that have almost all their current carried by holes. We have exquisite control, not only over which sign of charge carrier operates in that semiconductor, but also over the detail of electrical properties; in particular, how good a conductor this thing is. We can make resistors, for example, by putting the right amount of doping in, and so on. Two common dopants that are used to dope silicon are phosphorus, which is said to be pentavalent—and what that means is every phosphorus atom has five outermost electrons that would like to participate in bonding with other materials or other atoms—and we're going to use boron, which is trivalent. It has three outermost electrons that like to participate in bonding.

Let's take a look at what happens if we add dopants to these semiconductor materials. Here again is our silicon, showing a piece that has four silicon atoms. We're going to replace one of the silicon atoms with phosphorus, the yellowish material. The phosphorus is electrically neutral, but it's got five outermost electrons, and only four of them can participate in the silicon structure. If there isn't too much phosphorus—and remember, we're putting in parts 1 in 10 million, so these are very small amounts of these contaminants, these dopants—the phosphorus will fit itself into the crystal structure, but that one electron will be free to wander around. There's this extra electron. It's free; it's able to carry electric current; it didn't come with a corresponding hole. The material didn't get made more negative because there's extra protons in the nucleus of the phosphorus to accommodate that. The material's still neutral, but it now has free electrons in it. Each phosphorus atom that we put in liberates an electron. As a result, the majority charge carriers—the main dominant charge carriers in this material—are the negative electrons. There are far more electrons coming from the dopants than from that random thermal energy process of promoting electrons and making electron hole pairs. There are some holes, but they don't play a big role. There are a few electrons that got there from thermal agitation, but most of the electrons came from our putting the phosphorus in, and so we call this an n-type semiconductor. Don't get confused. "N" doesn't mean it's negatively charged; in fact, in some conditions, you'll see it's going to be positively charged. The "N" means the dominant charge carriers. The things that are free to move and carry current are, in fact, negative, and they're electrons in this case.

If we do a similar thing, but we put a boron atom in, the boron only has three valence electrons. Only three of them can participate in the bonds, and the silicon structure wants four; so if we put the boron in, we are missing an electron in the bond structure, and that missing electron is a hole. In a boron doped material, a boron doped silicon, there are lots of holes introduced every time we put a boron atom into the structure. The majority charge carriers are positive; they're holes. We call this a p-type semiconductor. Again, it doesn't mean the material carries in that positive charge; it means that the dominant charge carriers, the majority of things that can move to carry current, are holes. Again, remember, a hole is really the absence of electrons, so what's really happening is electrons are bouncing through the crystal structure, filling in the holes and moving that way. But they aren't free electrons, they're still bound electrons; so it's, in a sense, really the hole that's moving, and the hole that is carrying the current.

Now we have two kinds of semiconductors, n-type and p-type, and they differ in what are the majority charge carriers. The whole business of semiconductor electronics consists in getting the doping right and then putting p- and n-type semiconductor in proximity. We'll spend the next chunk of this lecture looking at what happens when we put a piece of p-type material and a piece of n-type material in close proximity, and some really interesting stuff goes on when we do that.

We're going to have what we call a p-n junction. A p-n junction consists of a piece of p-type material and a piece of n-type material and they're sitting right next to each other, and between them is this junction where they meet. This is the essence of semiconductor electronics. Understanding the p-n junction will allow you to understand all the other devices we're going to deal with, the several kinds of transistors, the diodes, and so on. Here's a p-n junction. What happens? In the n-type material, there are a lot of electrons, and there are very few free electrons in the p-type material, as we saw. Whenever you have a big concentration of something over here, and a small concentration of it over here, there is diffusion, the process where some of the electrons move into the p-type material to try to ease out that sharp, abrupt gradient. You know, I could do the same thing by opening a bottle of perfume, and pretty soon perfume atoms would've diffused throughout the room and people would smell them on the other side of the room, because when there's a high concentration of material of any kind, more of it is moving in the direction away from that high concentration, because there's more of it there, than toward it, and so the material spreads out. That's called diffusion. So electrons diffuse into the p-type material, and there are a lot of holes in the p-type material, so holes diffuse into the n-type material. Where the holes coming from p-type material get into the n-type material, they meet electrons. Where the electrons from the n-type material come into the p-type material, they meet holes and they can recombine, and then we lose electron hole pairs.

Something else interesting is happening here: We're getting actually a slight layer of negative charge in the p-type material, and a slight layer of positive charge in the n-type material. You might say, "Why doesn't this diffusion just keep happening?" The electrical repulsion set up by those charges keeps it from happening, and we eventually build up an equilibrium where some electrons have migrated to the p-type material and some holes have migrated to the n-type material. What happens then? Then the process of recombination occurs. We have an electron and a hole. The electron falls into the hole, if you will. It's energetically favorable, and they're all gone. The electron hasn't disappeared from the world; what's happened is it's ceased to be a free electron. It's now part of that crystal binding structure binding the atoms together. It's gone as far as carrying electricity is concerned, and so is the hole. The electron hole pairs recombine and they're all gone.

What happens is in the region right around the junction of the p- and n-type material, there's a zone where there are very few free charges because the electrons have diffused to the left in this picture into the p-type material, the holes have diffused into the n-type material, and they've recombined. There's a zone here where there are very few free charges, and if there aren't very many free charges, the material is a lousy conductor. This p-n junction is, on its own, a bad conductor, and the reason is because of this process of diffusion and recombination that have occurred and depleted the region in the vicinity of the junction of charges. Plenty of holes off to the rest of the p-type material; plenty of electrons off in the rest of the negative; but they can't get through because there's this depletion zone. That's the first thing about p-n junction that's important: It's not a good conductor under these conditions.

There's this electric field established, as I said, because there's now a slight positive charge to the right of the junction and a slight surplus of negative charge to the left of the junction. Remember, I said p doesn't mean it's positively charged; it means the majority charge carriers are holes, positive. N doesn't mean it's negatively charge; it means the majority charge carriers are negative electrons. In this case, we see a situation where, in fact, the n-type material requires a little positive charge and the p-type material requires a little negative charge.

Let's start doing some interesting things to this p-n junction: Let's connect an external circuit. In fact, let's connect a battery, with its positive terminal connected to the n-type material and its negative terminal to the p-type material. Now what happens? That electric field that we had in there that was pointing from the n toward the p gets reinforced by the presence of the battery, and what that does is cause the depletion zone to actually grow. Now there's a bigger region where there are very few free charges, and so the device can't conduct electricity hardly at all. In this configuration, that battery can't push current through that p-n junction, and this thing has become a non-conductor; the p-n conjunction basically is an open circuit. That's called reverse bias. On the other hand, if we connect the battery the other way, the battery goes against that internal electric field that had resulted from the diffusion of holes and electrons into the opposite materials. It can, if it's strong enough that battery—and it needs to be only over about 7/10 of a volt or so—it'll then eliminate that electric field and it'll allow current to flow from the p-type material into the n-type material across the junction. That's called forward bias, and under this condition the junction has become a decent conductor and electric current can flow around that circuit.

We now have something that seems to conduct electricity with a battery connected one way, but not with it connected the other way. It's that one-way valve for electric current that I introduced when we talked about how to build a power supply, for example. The p-n junction becomes a one-way valve for electricity. Current can flow from p to n, but it can't flow from n to p.

Let's consider this p-n junction in the context of a device we've already seen, namely the diode. We used a thing called a diode that seemed to pass current in one direction and not the other in our power supply to achieve this rectification; this changing of AC into DC. It was the step that cut off the negative-going portion of the AC. This p-n junction does exactly that; this p-n junction is, in fact, what a diode is. We can replace our picture of the p-n junction with our symbol for the diode. On the left is the p side, on the right is the n side. This symbol gives you an arrow-like structure pointing here to the right, and that shows you that's the direction in which current will flow. The obvious use of the diode is in converting AC to DC, but there are many, many other uses of diodes in electronic circuits. We'll come across a number of them as we go throughout this course. However, there are also some specialized diodes, and I'd like to take the rest of this lecture to look at some of those specialized diodes, see how they work, and see some real world applications of them.

The first specialized diode I want to look at is the light-emitting diode. In any diode with current flowing through it—so it's forward biased with the positive of the voltage source connected to the p-side of the diode—what happens is holes flow through the p-type material toward the junction in the direction of the electric current, and electrons being negative flow opposite the direction of the current. They flow in through the n-type material, and at the junction they recombine. Normally not much happens out of that recombination—a little bit of heat is generated and so on, so the diode loses a little bit of power perhaps—but in specially engineered diodes, the amount of energy that's released when an electron falls into a hole recombines, joins that crystal structure again, becomes part of a bond between silicon atoms, is that the energy is emitted as a little bundle of light called a photon. Diodes that are specially engineered to do this typically have a very thin p-type layer, so thin that it's essentially transparent to the light. The light is created at the p-n junction, and the light emerges. Typically it emerges at a particular color or a particular frequency, although increasingly we've learned to make wide spectrum LEDs. That's a light-emitting diode. Its circuit symbol is shown at the upper right, it's simply a regular diode symbol with a couple of arrows coming out of it, and those arrows are supposed to represent the light coming out of the light-emitting diode.

Let me show you a few examples of light-emitting diodes. Let's go over here. I have here a setup that I'm going to be using a lot from now on. This is three of those boards I showed you before that I'd built the power supply on, and we built a filter on one of them in the previous lecture. I'm going to build lots of circuits on these boards and, in fact, as you see the board now, it's actually set up with a number of circuits, which we're using in future lectures. But we're focusing on this region right here. This region has a couple of LEDs. One reason we're using this big board, it has a lot of auxiliary components, it has ways to connect things to it, it has switches, it has a built-in function generator, it has variable voltage supplies, and a lot of other things. We're going to use the power supply built in to here to power some LEDs. What I have here are simple green and red LEDs, and as soon as I turn on the power, they light up. For many years the use of LEDs was restricted to very simple low power applications like indicator lights that show you when something is on or off, for example, or what condition it's in. There's a red and a green LED.

You may notice that these two LEDs are actually wired through these points to a resistor, and the resistor is going to ground and the other side of the LEDs is wired to a positive 5 volt power supply. You never connect an LED, or any diode for that matter, directly across a voltage source because remember, a diode, when it's in its low resistance state, when it's in its forward biased state, is essentially a closed circuit, a short circuit, and very large currents will flow. You need a resistor to limit the current that can flow in the diode, and in particular in the LED. These LEDs have a current-limiting resistor connected to them.

If we want to step up a little bit in sophistication, below these two red and green LEDs is a seven-segment display. These are increasingly being superseded by LCD screens, liquid crystal display screens like we have in our smartphones and so on, but they're still used in a number of applications where we need to display numerical or sometimes alpha numerical information. This has a number of straight-lined segments, it has a couple of decimal points, and the straight-lined segments can be lit. What I've done with this particular one is simply taken a wire that I can connect to different points on that display, one at a time. For example, if I plug it in there, the bottommost segment lights and you might use that to make part of, for example, an eight or maybe a two. Let's come up through here; there are a number of connectors. That segment is lit. As I move along, there's the middle segment that would make the cross piece in the eight, or the five, or the six, or the two. We can go on. On the other side, there are additional; there's the decimal point lighting up. There's another segment. By connecting these different segments, we can make displays of different numbers. I'm only connecting one at a time, so I'm not making any numbers. But as we'll see in subsequent lectures, particularly when we get to digital electronics, we'll use these devices to display the count in a digital counter, or in the next to last lecture we'll make an analog to digital converter, basically a digital voltmeter. We'll put in a voltage, and we'll get the voltage displayed with one of these seven-segment displays by carefully lighting the right segments to make the numerical display.

That's all fairly widely established, well-established uses of LEDs, but we can do other things with LEDs. Probably, within a few years of my making this course and maybe by the time you're watching it, all our lightbulbs will be LEDs. We've already gotten to the point where we can replace standard light bulbs—here's a standard 60 watt lightbulb—and we can replace those with LEDs. I have in this lamp an LED. It looks about the same size and shape as the incandescent bulb—I'm not sure we would've designed it that way if we hadn't already had all the infrastructure for incandescent bulbs, but that's what we have—and so that LED lamp is screwed in there. This is a fairly modern LED lamp because it has to be able to do several things that my small ones can't do: It has to be able to tolerate large amounts of power to make large amounts of light, and it also has to be able to produce pleasant-looking light; a spectrum that gives us a sort of warm white light. Both those things were big technological challenges in the development of LEDs, but I think you can see that when I turn that on, we pretty much achieved that. That looks pretty much like the glow of an ordinary incandescent 60 watt lightbulb.

If you did the project for lecture one where I talked about an incandescent lightbulb and what would happen if you replaced it with an LED, you found that the power used by the LED to produce the same amount of light was far, far less than the power consumed by the incandescent light bulb. These are real energy savers, and don't be dissuaded by their relatively high cost, which is still dropping very rapidly as these come into full scale production and we get more and more different varieties of them, because over the lifetime of this thing—and this thing will last almost forever—you'll get back in your electric bills far more than you paid for the LED lamp. Rather than pay a few cents for an incandescent lamp, pay a few dollars for an LED lamp and you won't go wrong.

There's a modern use of LEDs. Within a few years, all of our lightbulbs will be LEDs. You probably have a flat screen TV, and if you do, and if it's a relatively new one, it probably has LEDs as its backlight that sends the light through the pixels that either turn on or turn off to show you the picture on your screen. Some of them still have fluorescent lamps, but most of them now are made with LEDs. LEDs are really everywhere. They're in lots of displays, they're in lots of simple indicator lights, but increasingly they're there to provide illumination.

That's a special kind of diode, the light-emitting diode. But it's a diode; you have to connect it the right way for current to flow

Let's look at another special kind of diode. Here's another diode that's basically the light-emitting diode in reverse: This is the photovoltaic cell. A photovoltaic cell is a device that takes sunlight in and produces electricity. It, too, is a p-n junction diode. In this case, you see the n-type material on the top, the p-type material thicker on the bottom. The n-type material is thin so light can get through it. There's an electric field established in there by that diffusion process that moved holes into the n-type material and electrons into the p-type material, and you can see that electric field. That would want to push positive charge downward and negative charge upward. It's balancing that diffusion effect that's already occurred. What if some light comes in? If light comes in and strikes the junction, a photon—an individual little bundle of light energy; a quantum of light, if you will—can have enough energy to create an electron hole pair. If you design this photovoltaic cell correctly, you'll design it so you'll get that process of electron hole pair creation occurring for photons with the energy invisible light, particularly in sunlight. If an electron hole pair is created, right at the junction there's that electric field. That electric field drives electrons up and it drives holes down. Remember, there aren't usually a lot of holes and electrons at that junction because they've all recombined. But you can, with the light, create an electron hole pair. Sometimes they'll recombine and you've lost that energy, but sometimes they'll be swept away by the electric field. Then, if you connect an external circuit, the electrons can move through the external circuit, and that represents a current flow in the direction opposite the direction the electrons go, and so you can light an external lightbulb, or run a motor, or do whatever you'd like to do with your photovoltaic cell. Today, you see photovoltaic arrays all over the place because photovoltaic cells have become very inexpensive. The prices are still dropping dramatically. They're beginning to supply not yet a substantial but a seriously appreciable fraction of the electricity that we use.

Photovoltaic cells are with us very much, and I have two examples of those over here. This is a little demonstration photovoltaic cell that we use in our physics labs. This is actually a solar battery charger that's used to charge batteries, cell phones, and things like that and it works off ambient light in the room, or sunlight, or whatever. Photovoltaic cells and LEDs are sort of the opposites of each other. One has electricity flowing through it, electric current; takes electrical energy, turns it into light; does so much more efficiently than incandescent light bulbs. The other is the opposite, takes light in and makes electric current.

Before I end, I want to introduce one other interesting kind of diode—and I'm not going to say a whole lot about it, but if you do the project for this lecture you can work with it—and that's called the zener diode. Normally, you put voltage across a diode the wrong way, the reverse biased way, and not much will happen. You'll have no current flowing, you'll build up a voltage across it, and if you make that voltage too big, you'll eventually overwhelm the junction and you'll basically destroy the diode, and current will flow then, and it'll no longer be a diode for you. But there's a special kind of diode that's engineered to be able to tolerate that breakdown in the reverse direction. It's called a zener diode because that process is called zener breakdown, and a zener diode can be engineered to have that breakdown occur at a particular voltage. I've drawn the VI characteristic curve here for a zener diode. It looks at the right like a regular diode; it goes up very steeply at about 7/10 of a volt. But to the left, this particular one suddenly starts conducting current again if you put more than 3 volts across it in the wrong direction, in the backward direction. At that point it maintains 3 volts across it, almost regardless of how much current flows. Zener diodes maintain a constant voltage after you've broken them down in the reverse direction. They aren't destroyed. Take the voltage away, they're back to normal. Put more than 3 volts negative over them or try to, and they'll maintain 3 volts across them in the reverse direction. They're wonderful for voltage regulators. They're often used in power supplies to establish a reference voltage that then determines the output voltage of my power supply. When I built a power supply and used an electronic regulator circuit, there was a zener diode built into the integrated circuit voltage regulator. It wasn't the only thing that was holding the voltage, but it was establishing a reference level voltage.

That's a lot about diodes, many different kinds of diodes, and I'm going to now let you think about diodes more by building some circuits with diodes if you'd like to do the project for this lecture. You don't have to do the project, but again, here are some interesting circuits using diodes, and I'm not going to tell you what they do. I'm going to ask you to build them and experiment with them—again, build them with your circuit simulators—and see if you can figure out what they do. If you do that and want to see the solution to that, I'll actually construct these circuits in the real world as well as in simulations, and we'll have a look at what these interesting diode circuits do.

Transistors and How They Work
Lecture 7

Transistors are among the most important inventions of the 20th century. They came into full-scale use in electronic circuits in the 1960s, quickly replacing vacuum tubes. Transistors are at the heart of all modern electronics because they allow one circuit to control another. In this lecture, an introduction to how ideal transistors work will enable us to understand simple transistor-based circuits. By the end of the lecture, we'll build a circuit that uses a transistor as a switch. Key topics in this lecture include the following:

- Electronic control: the concept
- Field-effect transistors: MOSFETs and JFETs

- Bipolar junction transistors (BJTs)
- *V-I* characteristics for the BJT
- Application: BJT switch.

Electronic Control: The Concept

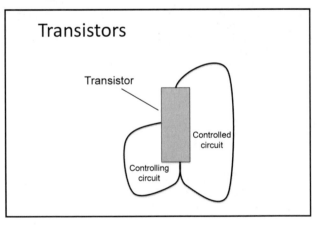

The minimum number of connections to a device that lets one circuit control another is three, and that's the number of connections to a transistor. Because there are only three connections, the two circuits must have one point in common.

There are two main kinds of transistors: field-effect transistors (FETs) and bipolar junction transistors (BJTs). FETs are voltage-controlled devices, while BJTs are current controlled.

Field-Effect Transistors: Metal Oxide–Semiconductor FETs (MOSFETs)

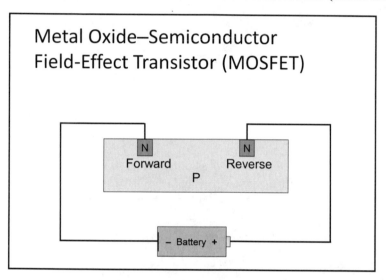

There are two kinds of FETs: metal oxide–semiconductor FETs (MOSFETs) and junction FETs (JFETs). A basic MOSFET consists of a block of P-type semiconductor material embedded with two inserts of N-type material. As we saw in the last lecture, the PN junction is at the heart of almost all semiconductor electronic devices. Here, we see two PN junctions. There is also a battery connected with its positive terminal to the N-type region of the right-hand PN junction. In this configuration, that junction is reverse biased, and no current can flow. Reversing the battery would give a similar picture, now with the left-hand junction reverse biased, so again, no current can flow.

MOSFET

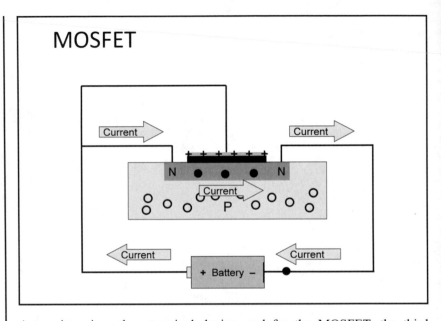

A transistor is a three-terminal device, and for the MOSFET, the third terminal is the *gate*. The gate is a metal electrode separated from the rest of the transistor by a thin insulating layer. This gate structure gives the MOSFET one of its virtues, namely, very high resistance between the metal gate and the rest of the circuit. As a result, the MOSFET gate draws very little current from the circuit it's connected to.

If we then apply a positive voltage to the insulated gate, it acquires a positive charge; that charge can't flow anywhere because of the insulating material, but it attracts negative electrons into the region below the gate. Effectively, the whole *channel* (the region between the two intrusions of N-type material) becomes N-type. Now there are no junctions—just a continuous piece of N-type semiconductor—and the transistor can conduct.

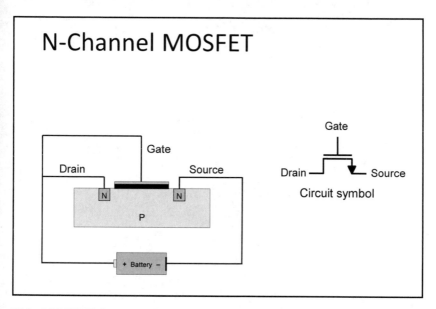

This MOSFET is an N-channel MOSFET because the channel becomes *negative* when the transistor is in the conducting state.

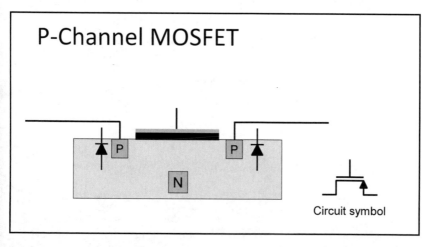

We can also make P-channel MOSFETs, in which we reverse the N- and P-type material.

Field-Effect Transistors: Junction FETs (JFETs)

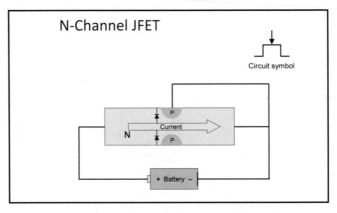

Like a MOSFET, the JFET is a device in which a voltage applied to the gate controls the flow of current through the transistor. Again, the transistor becomes a device that allows one circuit to control another.

Bipolar Junction Transistors (BJTs)

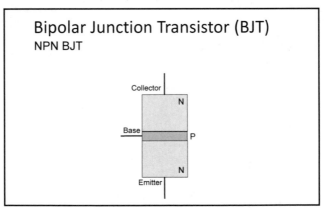

An NPN BJT consists of three regions: an upper N region (collector), a P region (base), and a lower N region (emitter). Notice that the P-type base region, separating the collector and the emitter, is very narrow; that's the key to transistor operation. Wires are connected to the three regions, so again, we have a three-terminal device.

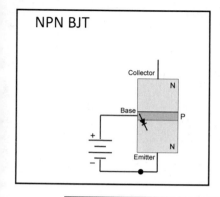

NPN BJT

If we connect a battery between the emitter and base, with its positive terminal at the base, we have a forward-biased PN junction between the base and the emitter; in that configuration, current will flow. The current consists of electrons flowing opposite from the direction of the current because electrons are negative.

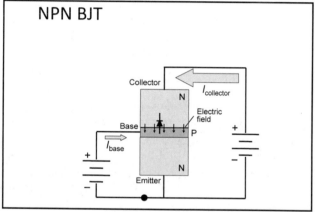

NPN BJT

If we connect another battery with its positive terminal to the collector and negative to the emitter, we reverse bias the collector-base junction. Remember that a reverse-biased junction has a strong electric field; here, that field points downward. Remember, too, that electrons, being negative, move opposite the direction of the field. Now, the current in the base-emitter circuit is continually injecting electrons into the base, where they would normally continue out the base wire and into the positive terminal of the battery. But the base is thin, so it's possible for electrons entering the base to "feel" the strong electric field at the collector-base junction, in which case they'll be whisked into the collector and can flow around the collector-emitter circuit. If the base is very thin, most of the electrons entering the base will end up in the collector circuit, and as a result, the collector current will be much greater than the base current—typically, on the order of 100 times larger.

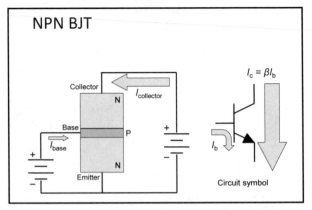

In this way, the BJT multiplies currents. It takes a small base current, which we can control through the base circuit, and produces a much larger but proportional collector current. The ratio of collector to base current is the transistor's *current gain*, and it's given the symbol β (lowercase Greek *beta*). For typical BJTs, β is generally in the range from about 50 to about 200 or so.

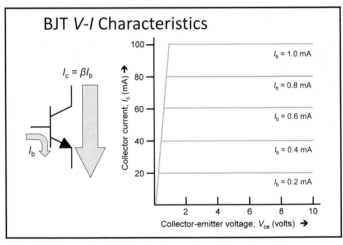

We can characterize a BJT in the same way we characterized other electronic devices, namely, by drawing *V-I* curves. But the transistor has a third terminal, and the collector current depends on the base current. So the BJT has not one but infinitely many characteristic curves, one for each possible level of base current.

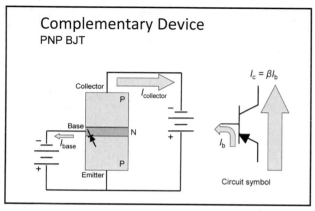

Complementary Device
PNP BJT

Just as with the FETs, there is a complementary device to the NPN BJT, namely, a PNP BJT transistor. Its base is N-type material, while its collector and emitter are P-type. It works in the same way as the NPN, but with all voltages and currents reversed.

Application: BJT Switch

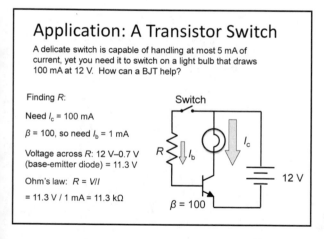

Application: A Transistor Switch

A delicate switch is capable of handling at most 5 mA of current, yet you need it to switch on a light bulb that draws 100 mA at 12 V. How can a BJT help?

Finding R:

Need I_c = 100 mA

β = 100, so need I_b = 1 mA

Voltage across R: 12 V–0.7 V (base-emitter diode) = 11.3 V

Ohm's law: $R = V/I$

= 11.3 V / 1 mA = 11.3 kΩ

β = 100

Let's assume that we have a manual switch that is very delicate, capable of handling, at most, 5 mA of current before it burns out. But we need it to switch on a 12-V light bulb that draws 100 mA— 20 times as much as the switch can handle. In this case, we can use a BJT as a simple switch. We have to choose a base-circuit resistance that will ensure the transistor turns fully on and passes the needed 100 mA with no more than 5 mA of base current. The diagram above shows the appropriate calculation.

Suggested Reading

Introductory

Brindley, *Starting Electronics*, 4th ed., chapter 8.

Lowe, *Electronics All-in-One for Dummies*, book II, chapter 6 through p. 302; pp. 306–311.

Platt, *Make: Electronics*, chapter 2, experiment 10.

Shamieh and McComb, *Electronics for Dummies*, 2nd ed., chapter 6, p. 126 through end.

Advanced

Horowitz and Hill, *The Art of Electronics*, 3rd ed., chapter 2 through section 2.2.1.

Scherz and Monk, *Practical Electronics for Inventors*, 3rd ed., chapter 4, section 4.3 through p. 437.

Project

Automatic Nightlight

A photosensor outputs 0 V until the ambient light decreases below a certain level, after which it outputs 5 V. You want to use this signal to turn on a light bulb that draws 250 mA at 24 V. Design a transistor switch to accomplish this and simulate it. Assign your light bulb the appropriate resistance. Verify that your circuit works.

Questions to Consider

1. What makes the transistor one of the most important inventions of the 20th century?

2. What does it mean that an FET is *voltage controlled*, while a BJT is *current controlled*?

*3. Determine the maximum permissible of the resistance (R) in the transistor switch described in this lecture when the following changes are made: (a) transistor β increases to 150; (b) battery voltage increases to 24 V; (c) the light bulb draws 500 mA when fully on.

Transistors and How They Work
Lecture 7—Transcript

Welcome back. This lecture is going to follow on our introduction to semiconductors and how they work, and look now at a more complicated device in the diode, namely the transistor. The transistor is arguably the most important invention, or certainly one of the most important inventions, of the 20th century, invented in the 1950s and came into full scale use in electronic circuits in the 60s and beyond, and quickly replaced those clunky old vacuum tubes that were used.

Think back to my first lecture when I talked about what electronics was really about. Electronics was about allowing one circuit to control another, and I introduced the several devices that we've used to do that—the vacuum tube and then the transistor—and I talked about how we put transistors on integrated circuits. Transistors are at the heart of all modern electronics, and what they do is allow one circuit to control another. In this lecture, I'm going to give you an introduction to how transistors work. Again, in keeping with the spirit of this course, it's kind of a zeroth order introduction. I'm not going to go into all the details; I'm not going to describe some of the subtleties of transistors. I'm going to ignore the fact that they have some resistance; that they have some capacitance; that they have some other properties that make them less than the ideal transistors I'm going to talk about here. If you were an electrical engineer designing real important circuits, you'd take all those things into account. But even with this basic introduction to how transistors work, we'll be able to build and understand rather simple circuits. In fact, by the end of just this lecture, we'll build what's probably the simplest non-trivial application, namely using a transistor as a switch. Later we'll move on to use transistors as amplifying devices. Then we'll go back, in the later part of the course, when we get to digital electronics to use them as switches; so a brief introduction to the several different kinds of transistors.

Before we get there, I have a conceptual diagram that suggests roughly what any kind of controlling device is like. I have this device here, and you'll notice it has three connections to it. That's the minimum number of connections you could make to have one circuit control another, and that's what the transistor has. Because there are only three connections, the two circuits you're controlling—the controlling circuit and the controlled circuit—need to have at least one point in common, and in this diagram that point is at the bottom of that arbitrary device I have that's going to let one circuit control another. That device is now going to become for us the transistor.

There are two main kinds of transistors. There are transistors called field effect transistors, FETs, and they're actually a little bit easier to understand so I'll introduce those first, although we won't use them very much until the later part of the course and I'll reintroduce them when we get there. Then there are bipolar junction transistors, which are gradually getting obsoleted by field effect transistors, although they're still widely used and they're particularly used in amplifier applications, audio amplifiers, that sort of thing. We'll spend a good bit of time on BJTs, although increasingly electronic devices use field effect transistors rather than BJTs, and that's particularly true of digital electronics.

Let's plunge in and begin to look at some simple descriptions of how transistors work. Before we get there, let me just remind you: What we're talking about are the change from vacuum tubes to these little tiny transistors, and then the ability to put many, many, many transistors on a single chip. Before we actually look at how transistors work, let me remind you that we can really cram today billions of these transistors on a single circuit, on a single chip, like a microprocessor at the heart of a computer. Here's a diagram that shows a microprocessor chip, much like the one I just pointed to. I enlarged the photograph of what the thing looks like. It looks like just a mess, but that mess is individual components built onto one piece of silicon, mostly transistors, but some resistors and other components as well. On the far right is a micrograph showing the nano scale size of the individual transistors that make up that integrated circuit. This particular one is field effect transistors, and so let's begin by talking about field effect transistors.

There are actually two kinds of field effect transistors. There are metal oxide semiconductor field effect transistors, also called MOSFET, widely used in today's semiconductor electronics digital circuits. Your cell phone is full of them, your digital TV is full of them, your computer is full of MOSFETs. There are also junction FETs, junction field effect transistors, or JFETs, and we'll look quickly at both kinds of field effect transistors to see how they work. Again, remember these are voltage controlled devices. They're a little easier to understand than the current controlled bipolar junction transistors, and you'll see in a minute what that distinction is.

Here's the basic block that makes up a MOSFET, a metal oxide semiconductor field effect transistor. It's a block of p-type—this particular one is a block of p-type semiconductor—and embedded in it are two little inserts of n-type material. Remember from the previous lecture, I emphasized that the p-n junction is at the heart of almost all semiconductor electronic devices. Here we see two—not one, but two—p-n junctions in this field effect transistor. I show a battery connected with its positive terminal on the right. Because the positive terminal is connected to the n-type region of the right-hand p-n junction—the junction being where that insert of n-type material joins the surrounding p-type material—that junction is reverse biased and no current can flow. The other junction, on the other hand, with the minus connected to the n-type material would be forward biased, but that doesn't do us any good because the right-hand junction is reverse biased and so no current can flow through this device, so what good is it? We could try reversing the battery. If we reverse the battery, it's the same picture. Now the left-hand junction is reverse biased and the right-hand junction is forward biased and still no current can flow, so what good is the MOSFET?

The MOSFET is a lot of good because I haven't shown you the rest of it. Remember, a transistor is a three-terminal device. It has three places it connects to the outside world, and I haven't showed you the third place. To get to the third place we grow—and grow is the right word; this is how it happens in the semiconductor manufacturing plants—layers of material one by one to make these complicated circuits. We grow an insulating layer, and it turns out (wonderful feature of silicon) that silicon dioxide—silicon, the second most abundant material on Earth, and oxygen, the most abundant material in Earth's crust—join together to make silicon dioxide; you know it as glass or quartz, and it's a very, very good electrical insulator. Just defusing a little oxygen and silicon into here makes a beautiful insulating layer. That's the "oxide" part in "metal oxide semiconductor"; we've already seen the "semiconductor" part down below. Then we're going to put a thin coating of metal on top of that oxide, and that thin coat of metal is called the gate.

It's completely insulated from the semiconductor by this insulating layer. That's going to be important in describing one of the virtues of MOSFET transistors, metal oxide semiconductor field effect transistors, because they have a very, very, very high resistance between the metal gate, which is where you put in the weak signal that's going to do the controlling, and the rest of the circuit. They draw very, very little current from the circuit they're connected to, and that's a virtue in many applications, as you'll see.

Here's our semiconductor material now. We are full of holes, mostly because it's p-type, but there are a few electrons in there. Remember, there are always a few so-called minority charge carriers that got bopped out of the crystal structure by random thermal motion. There are a few electrons in there, and I've shown some of them, probably more than there should be in proportion to the hole. But the p-type material is mostly holes, but with a few electrons. Now I'm going to apply a positive voltage to that insulated gate. I'm going to connect a positive volt and connect it to the positive end of the battery, and it's going to acquire a positive charge. That positive charge can't flow anywhere because there's that insulating material, but what's going to happen is that positive charge attracts negative charge, and there are a few negative charges in the p-type material. Not very many, but they'll be attracted into that region below the gate. That's the region between the two n-type intrusions into the p-type material. All of a sudden we have a situation in which we have a significant number of negative charge carriers in the region—it's called the channel—between those two intrusions of n-type material. Effectively what's happened is that whole channel, including the two intrusions, has become n-type. Now there are no junctions; we just have a piece of semiconductor. Semiconductors are conductors—maybe not great ones, but they're conductors—and that transistor can now conduct. By putting a positive charge on the gate of that transistor, we've turned the transistor on. We've enhanced the conductivity of that channel.

This particular field effect transistor, this particular MOSFET; not only are there are two kinds of FETs, there are two kinds of MOSFETs. There are n-channel MOSFETs and p-channel MOSFETs, and this is an n-channel MOSFET because the channel becomes n when you put positive charge on the gate. If you think about it, you can control the amount of positive charge on the gate; I could've put a different voltage on there or something. I can control that amount of charge and that will control the number of electrons that are drawn into the channel, and that will control the conductivity of the channel. You can almost think of this field effect transistor as a voltage controlled resistor. Equivalently, you could think of the current that flows in that circuit as being controlled by the voltage you put on the gate.

There's the metal oxide semiconductor field effect transistor. This particular one is an n-channel conductor. We can turn it on and off by putting charge on or off the gate, or we can vary the current that flows through it, varying its resistance by varying the amount of charge we put on the gate. That's how a MOSFET works, and it's fairly easy to understand in this context. Current is flowing through the MOSFET now as we've got that charge on the gate.

The symbol for a MOSFET—actually, there's unfortunately not a completely standardized way of drawing MOSFETs—this is the simplest symbol and it's the one I'm going to use. It shows two parts, the two places where those intrusions are. They're called the drain and the source for slightly obscure reasons we won't go into. The important electrode is the gate. You'll notice the little arrow going outward on the right-hand side of that symbol; that tells us this is an n-channel MOSFET because it shows that the junction between the p-type material and the n-type intrusion is going in the direction shown. Unfortunately, the symbol is upside down in some ways relative to the diagram, but if I turned the symbol upside down then the gate would be on the wrong side, so that's just the way it has to be. That was an n-channel MOSFET.

We can also make a p-channel MOSFET. Its symbol looks the same, except that little arrow goes the other way, and it's just the same except we've reversed the n- and the p-type material. One of the wonderful aspects of semiconductors is for every kind of transistor, we have a complement which uses p where the other uses n and vice versa. There are p-channel metal oxide semiconductor field effect transistors, and here's one of them.

Finally, the last type of field effect transistor we'll talk about—one we won't use much, but I will use some amplifier circuits that start with one of these, so I wanted you to see them—is an n-channel junction field effect transistor. What you see is a chunk of n-type material. I'd like you to imagine that this material is cylindrical, and that that intrusion you see of p-type material is a band that wraps all around this cylindrical structure. It doesn't have to be cylindrical, but that's the easiest way to think about it. That bottom piece of p is, in fact, connected to the top piece of p because they form a donut-like intrusion around the n-type material. We have two junctions here, two diodes basically; they're actually part of one diode because we have this p-type material going all the way around. There's the circuit symbol for a junction field effect transistor. It doesn't have quite that insulating quality that the MOSFET did, because there's a junction and junctions pass a small amount of current in the reverse direction. Now let's imagine connecting a battery across this n-channel and current can flow. No problem, because the semiconductor is itself a conductor. Not a very good conductor, but it's a conductor.

What happens if we apply a negative voltage to the p-type material? That causes the material to become, those junctions to become, reverse biased. Remember, that grows the depletion region. Now there's a depleted region around those junctions that's gotten bigger, and the depleted region is depleted of charge carriers. No charge can flow, and we've turned this JFET off. The JFET becomes also an on/off switch, or by varying the amount of that depletion it becomes something that controls the effective resistance of the channel.

With both our field effect transistors, we have devices in which applying a voltage to one electrode, the gate electrode, results in variation in the current flow in the second circuit. The transistors become devices that allow one circuit to control another. Those are the two field effect transistors. We have metal oxide semiconductor field effect transistors, we have junction field effect transistors, and we have them in both kinds because we can make alternate p and n; we can reverse the order of those two.

Field effect transistors, somewhat OK to understand; they're not trivial. But now we want to look at the bipolar junction transistor, and that's a lot more difficult to understand. I'm going to go through it carefully, and I'm going to use the big screen here because I'd like to be able to point out what's going on and do some animations. Here we're going to talk about the bipolar junction transistor, BJT. This is NPN BJT because it consists of three regions: an n region, a p region, and an n region. Their names are "collector" for the upper n region, "base" for that p-type region, and notice that p-type region is really narrow. That's crucially important to operate; really thin. It's separating the collector and the emitter, the bottommost n-type region, but it's really, really thin. That's going to be the key to transistor operation. We have wires connected to each of those; so again, we have a three-terminal device that we can connect to the outside world.

What we're going to do is connect positive end of a battery to the base. We'd also have some resistors to keep the current from being too big, but I won't show those. Here we have, then, a forward biased p-n junction between the base and the emitter, and so in that configuration current will flow. The current consists of electrons flowing around backwards from the direction that the current goes because electrons are negative. If I just connected the circuit up like that, I'd have a current flow through that forward biased base emitter junction. Easy enough.

Now let's do some more circuitry. Let's take a bigger battery, perhaps, and connect its positive end to the collector and it's negative end to the emitter so it becomes the controlled circuit. Here's the controlling circuit, here's the controlled circuit, and let's see how that works. Remember that there's an electric field established by that diffusion of carriers—charge carriers, holes and electrons—across the junction. The electric field points from the n toward the p because the n has become a little bit positive because holes have diffused into it, and the p has become a little bit negative because electrons have diffused into it. There's this electric field in there.

What happens? If you were an electron and you came to around and into this n-type region, you'd want to keep going happily around like that. But because that base region is so thin, you might kind of overshoot the mark and get into the region of that electric field. If you do, that electric field will grab you—you're a free electron, you're negative, so you want to go opposite the direction of the electric field—and you'll be whisked up into the collector. Even though the collector to base junction is reverse biased—we've got the battery connected the wrong way to make that junction conduct— nevertheless, if we inject or emit (and that's why it's called the emitter) free electrons into that base, some of them will be caught in that electric field and they'll be whisked across and they'll make a current. If you make that base region thin, then most of the electrons that get in there because of the current in the base circuit will, in fact, be whisked into the collector and they'll make a current, and a much bigger current, in the collector circuit.

That's reverse biased collector junction, and here's what it looks like. There it goes; and now most of the electrons are going around in the collector circuit and we have a current flowing in the collector again, opposite the direction of the electrons, from the positive terminal of the battery. It's bigger than the base current; it's typically a lot bigger than the base current. It's typically on the order of 100 to 200 times the base current, maybe 50 times, but somewhere on the order of 100.

Here's the circuit symbol for the bipolar junction transistor. The base is this middle piece, the collector is the piece with no arrow on it, and the emitter is the piece with the arrow. The base current flows in the base and out through the emitter. By the way, that arrow tells you the direction of the base emitter diode. The diode from the base of the emitter points that way if you do it with a diode symbol. We have a base current, IB, which we establish—I sub B for base—with some external circuit. As a result, there's a big collector current, much bigger, and it's bigger by a factor that's called beta—the Greek symbol beta, the Greek letter beta—and beta is typically on the order of, as I said, 100 or so.

What the bipolar junction transistor does is it multiplies currents; it takes currents and makes them bigger. It takes a small base current, which you can control by whatever you put on the base circuit, and it turns it into a much bigger collector current. That beta is called the current gain. That's what characterizes the transistor, and I'll be using beta a lot, and I'll often assume it's about 100, although it need not be exactly 100. That's the workings of the bipolar junction transistor, and now we'll move on to figure out what we can do with bipolar junction transistors and look a little bit more at how they work.

Let's first begin trying to characterize a BJT in the same way we characterized other electronic devices: namely, by drawing their current voltage curves. I'm just going to draw this in one quadrant, positive for both, because you can only make a BJT work by connecting it the right way. With an NPN BJT, you want positive to the collector, so we only have positive voltages and positive currents here. The problem is we've got a device here that has a third terminal, and what the collector current is depends on that base current. For every base current, I have a characteristic curve. Here's what the characteristic curve looks like, approximately, for a transistor; and actually, this isn't a bad approximation. This is for 0.6 milliamps of base current, and I'm assuming a beta of 100 in this case; so 0.6 times 100 is 60, and we have a collector current of 60 milliamps. That says unless you're getting very, very low collector to emitter voltages, you're going to have 60 milliamps of collector current flowing no matter what. You want to make the base current a little smaller, make it 0.4 milliamps, you'll get 40 milliamps of collector current. You make it smaller still, 0.2 milliamps, you'll get 20 milliamps; 0.8 you'll get 80; and 1 milliamp, you'll get 100 milliamp. Those are the BJT characteristic curves, and there are many of them because the bipolar junction transistor has not one but many characteristic curves—in fact, infinitely many—for each possible level of base current that flows in the device.

I described in detail the workings of the NPN bipolar junction transistor. Just as with the field effect transistors, there's a complementary device, a PNP transistor. Its base is n-type material; its collector and emitter are p-type material. Works the same way but everything's reversed: The batteries are reversed; the currents go in reverse. We'll come across a very clever amplifier circuit that exploits this complementarity between two kinds of transistors.

That's an introduction—a whirlwind introduction—to lots of different kinds of transistors; but basically the two kinds: the voltage controlled field effect transistors and the current controlled bipolar junction transistors.

I want to end by talking about a particularly simple application of the bipolar junction transistor as a switch. Here's the application. We're going to make a transistor switch, and I'm going to assume something slightly artificial. I've got some actual manual switch; it's very delicate. It's capable of handling at most 5 milliamps of current before it burns out; it gets damaged. But I need it to switch on a light bulb that draws 100 milliamps, 20 times as much as this switch can handle, and it's a 12 volt lightbulb. How can a bipolar junction transistor help us with that? It can help us by building a simple transistor switch circuit. We're going to start with our bipolar junction transistor shown there, and I'm going to assume it has a beta of about 100. I'm going to connect the lightbulb as a resistance in the collector circuit of that transistor. I'm going to connect a battery or power supply; it's 12 volts because it's a 12 volt lightbulb. Then I want to figure out how I can get a 5 milliamp current, or maybe even less, to make that lightbulb light. This transistor has a beta of 100. If I passed a base current of only 1 milliamp, that would give us a collector current—beta times that is the current gain—of 100 milliamps, and we light our lightbulb completely satisfactorily. We have to figure out how to do that.

Here's our switch, and I'm going to connect the switch back to the 12 volt power supply. I'm going to use the switch to control the current that flows through a resistor and then through the base emitter junction. We want to figure out how big to make that resistor, so let's find that resistance. We need 100 milliamps of collector current; that's what it takes to light that lightbulb. If we get that, we'll have 12 volts across the lightbulb and essentially 0 volts across the transistor. Beta is 100, so we need a base current of only 1 milliamp. The voltage across the resistor is almost the 12 volts. Not quite, because remember—and we're going to use this to design carefully here— that there's about a 0.7 volt drop across a forward biased diode junction, which is what the base emitter junction is, so the base emitter junction has about 0.7 volts. The base emitter diode takes away 0.7 volts; we have 11.3 volts across that resistor R. Ohm's law says R is the V/I; 11.3 volts over 1 milliamp is 11.3 kilo ohms. If we have a resistor of that much, we'll get 1 milliamp, and I'm going to use a slightly smaller resistor in a circuit I'm actually going to build. Let's actually go and look at a real version of that circuit with just a few minor modifications.

Here we are using the circuit board arrangement that we're going to be using—it's sometimes called a bread board, by the way—and over in this region I've wired a version of that simple transistor switch I just described. Here's the transistor. You'll notice it's a transistor with a kind of metal case around it; that's a slightly beefier transistor that can handle bigger currents and dissipate a little more power than some of the tiny transistors. I chose that because this lightbulb draws significant current and I wanted a beefier transistor so it wouldn't heat up. My lightbulb is in the collector leg of the circuit. There's a little resistor, which is the resistor going to the base. This circuit is very much like the one I just described for you. What we'd like to do to understand this circuit is to understand what the currents are through the collector part of the circuit and through the base part of the circuit because that will show us that this transistor is, in fact, doing this current amplification, which is its thing to do. This current gain beta will actually be able to measure it.

Remember from the third lecture, I talked about electronic instruments and how to make measurements. Making current measurements was a little trickier than making voltage measurements, because to make current measurements you have to break the circuit, place the ammeter in series with whatever component you want to measure the current through, and go from there. I want to know the current through the lightbulb, which is also in our diagram. It's a current through the collector of the transistor. I also want to know the current through that little resistor, which goes to the base emitter junction because that's the base current. I need to measure both those currents. I'm going to come out of the power supply. On these boards, this yellow terminal is a positive power supply that can go from almost 0 up to about almost 20 volts, and it's adjustable with this knob. I've got it adjusted to about 12 volts because that's what our bulb wants. We've also got a fixed 5 volt power supply, and I'm actually going to use that for the base current. I'm, in fact, going to my delicate switch I'm talking about, it's going to be simply me plugging this wire into there. I've tried to color code things to avoid confusion: The red wires represent current in the collector circuit and the black wires are carrying the current in the base circuit, black for base.

Let's look at what happens; let's trace the wiring. Out of the positive 12 volt supply we come with current flowing into the input of the upper meter. This is the ammeter, set to measure milliamps, DC milliamps. We go through ammeter, which remember is basically like a piece of wire, like a short circuit, 0 resistance effectively if it's a good ammeter. We go out through the red wire and into the top of the lightbulb, through the collector, through the whole transistor, and to ground. That's where that loop of circuit goes. The other loop is going to start with the 5 volt power supply once I connect it. It's going to flow through the lower ammeter. The black wire's B for base, black for base; it's going to flow out of the common wire there. It's going to flow into this resistor through the base emitter junction, and to ground both of these power supplies or reference to ground. That means when I say there's 12 volts here that means 12 volts with respect to ground. There's ground available all over this place, including at that black post; and any of the places that are marked here in black are ground, and we're connected back to ground through here.

I'm going to now demonstrate this device. I'm simply going to turn the power on. The power's on, but nothing's happening; there's no current in either meter. But now I'm going to close my little delicate switch and we're going to look at what the current meters read. First of all, we close the switch and the lightbulb lights; so our transistor switch is working, because that switch wasn't switching the current to the lightbulb, it was switching the current to the base of the transistor, not to the lightbulb. Let's take a look at what we see here. Here's the upper one; that's the red curve, and that's the collector, and we see 0.223 amps, 223 milliamps, flowing there. The meter has auto-adjusted its scaling so it's in amps, so 223 milliamps is flowing there. Look down on the lower meter, it says 2.80. But look carefully, it's 2.80 milliamps. There are 223 milliamps in the collector circuit and only 2.80 milliamps in the base circuit. If you think about how those are related, 223 and 2.80, they're differed by a factor of not quite 100 because if I multiplied 2.80 by 100 I'd get 280; I've only got 223. This tells me I've got a transistor whose beta is on the order of, but a little bit less than, 100.

Let's wrap up. We've seen a whole variety of transistors. We've come to understand how those different transistors work. We've developed the transistor switch and seen how to design it. If you're interested in doing the project for this lecture, your project will be very similar. It'll consist of designing a switch that's supposed to switch on a night light automatically, and you'll have to go through the same design procedure I did for this simple transistor switch.

Transistors as Amplifiers
Lecture 8

I n the preceding lecture, we learned about several kinds of transistors, including MOSFETs, JFETs, and BJTs. One of the earliest applications for transistors was as amplifiers, devices that make weak signals stronger. We're all familiar with amplifiers in the audio realm; a radio has an audio amplifier, as does a stereo system and a cell phone. In this lecture, we'll develop a simple circuit for an audio amplifier. Then, in the next lecture, we'll combine that with some other circuits to make a complete audio amplifier system. Key topics in this lecture are as follows:

- The common-emitter BJT amplifier
- Biasing

- Load line analysis
- Distortion
- Better biasing.

Common-Emitter Amplifier

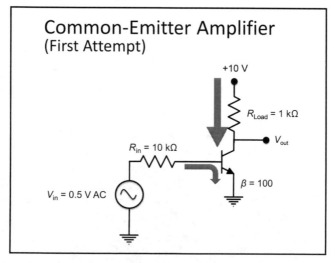

A transistor is a three-terminal device that that allows one circuit to control another. Because it has only three terminals, one of those terminals must be shared by both circuits. We can build a transistor amplifier in which the collector is common, the base is common, or the emitter is common. The common emitter (CE) is probably the most often-used configuration. The diagram here shows a first attempt at a CE amplifier—but this circuit won't work because the transistor's base-emitter junction becomes reverse biased when the AC input signal V_{in} goes negative.

Biasing

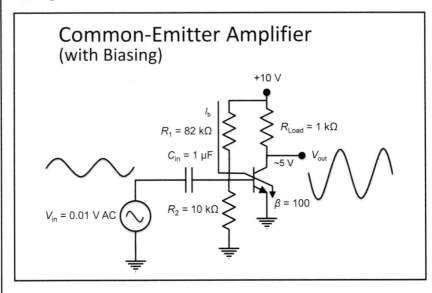

A functioning CE amplifier requires circuitry for *biasing*, which provides a steady base current even when there's no input. Biasing is generally chosen so that the output, taken at the transistor's collector, is about half the supply voltage when there's no input voltage. With biasing, the base-emitter junction is always conducting, and the output swings lower and higher as the input swings higher and lower. The output swing will be greater than the input swing, making this circuit a *voltage amplifier*.

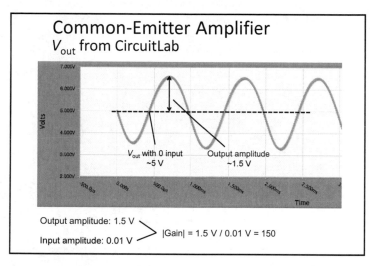

Common-Emitter Amplifier
V_{out} from CircuitLab

V_{out} with 0 input
~5 V

Output amplitude
~1.5 V

Output amplitude: 1.5 V
Input amplitude: 0.01 V
|Gain| = 1.5 V / 0.01 V = 150

A probe in CircuitLab shows the collector voltage of the transistor in a properly biased CE amplifier. This is the voltage labeled V_{out}, here a sine wave swinging 1.5 V either side of 5 V. We can calculate the gain of this amplifier as follows: The output amplitude (the peak of the output swing) is 1.5 V; the input amplitude is 0.01 V. The gain, then, is: 1.5/0.01 = 150.

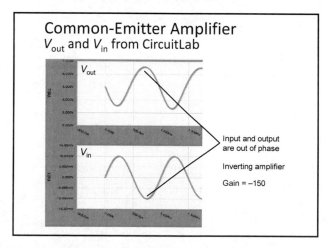

Common-Emitter Amplifier
V_{out} and V_{in} from CircuitLab

V_{out}

V_{in}

Input and output are out of phase

Inverting amplifier

Gain = –150

Because the input and output are out of phase, we have an inverting amplifier, which means the gain is actually –150.

Load Line Analysis and Distortion

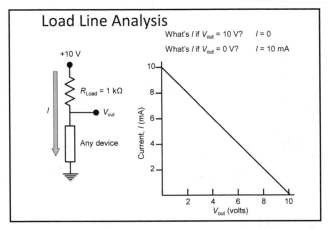

Load Line Analysis

What's I if V_{out} = 10 V? $I = 0$

What's I if V_{out} = 0 V? $I = 10$ mA

+10 V

R_{Load} = 1 kΩ

I

V_{out}

Any device

Current, I (mA)

V_{out} (volts)

To understand how this amplifier works and learn its limitations, we conduct *load line analysis*. Imagine an arbitrary device in series with a resistor (called the *load resistor*) and a battery or power supply. We can draw a plot of V_{out} — the voltage across the arbitrary device—versus I, the current through the device. If the output were 5 V, we would have 5 mA flowing through the load resistors and through the arbitrary device. In general, V_{out} and I must lie on a line that goes from the maximum current (10 mA here, when V_{out} = 0) to the maximum voltage (10 V here, when $I = 0$). This line is the *load line*.

We now replace the arbitrary device with the collector-to-emitter portion of a transistor. We choose a biasing circuit whose steady base current puts the collector voltage at about half of the power supply voltage; this establishes the *operating point*, as shown here.

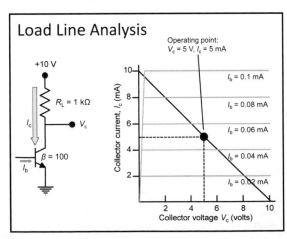

Load Line Analysis

Operating point:
V_c = 5 V, I_c = 5 mA

+10 V

R_L = 1 kΩ

I_c

V_c

$β$ = 100

I_b

Collector current, I_c (mA)

Collector voltage V_c (volts)

I_b = 0.1 mA

I_b = 0.08 mA

I_b = 0.06 mA

I_b = 0.04 mA

I_b = 0.02 mA

152

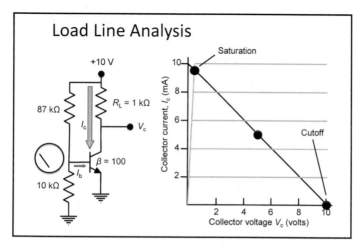

Load Line Analysis

When we apply an input signal, the base current varies and, therefore, so does the collector current. But the output voltage must remain on the load line. So the voltage swings down and up the load line as the input voltage goes up and down. But the voltage swing is limited by two special points: cutoff, where there's no current flowing through the transistor and there's maximum possible voltage at the output, and saturation, where there's the most possible current and very little voltage. No matter what happens at the input, the output can't go beyond these points. If it's pushed beyond those limits, the amplifier will exhibit distortion, and if it's an audio amplifier, you'll hear that distortion.

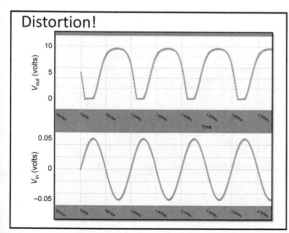

Distortion of V_{out} shows clearly in these CircuitLab plots of output voltage (top) and input voltage (bottom).

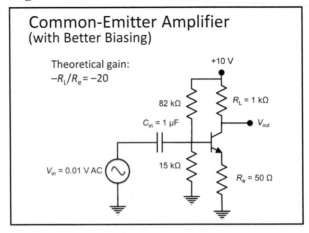

A better biasing scheme adds a small (here, 50 Ω) resistor in the emitter lead. Adding this resistor gives us a fixed gain that is approximately independent of the transistor's own characteristics. In fact, that gain becomes the resistance of the load divided by the resistance of the emitter resistor, in this case: $1000/50 = 20$.

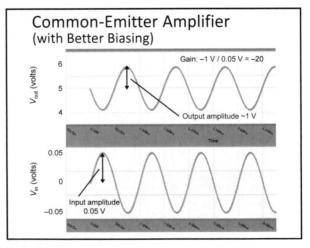

The output from CircuitLab for this CE amplifier shows that it has, in fact, a gain of -20.

Suggested Reading

Introductory

Lowe, *Electronics All-in-One for Dummies*, book II, chapter 6, pp. 302–306.

Shamieh and McComb, *Electronics for Dummies*, 2nd ed., chapter 6, pp. 135–137.

Advanced

Horowitz and Hill, *The Art of Electronics*, 3rd ed., chapter 2, section 2.2.6.

Scherz and Monk, *Practical Electronics for Inventors*, 3rd ed., chapter 4, section 4.3, pp. 440–442.

Project

Audio Preamplifier

Simulate a CE amplifier using a 2N2222 transistor, power supply voltage V_{cc} = 15 V, and a gain of –10. Use trial and error to bias with an operating point within 0.5 V of $V_{cc}/2$. Verify the gain and operating point. Determine the input voltage range for which distortion isn't visually obvious.

Questions to Consider

*1. What should be the voltage gain of the amplifier shown below?

2. Why is it necessary to bias the transistor in a CE amplifier?

*3. What circuit quantities determine the load line's (a) slope and (b) intersection with the voltage axis?

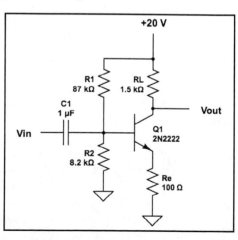

Transistors as Amplifiers
Lecture 8—Transcript

Welcome back. Last time you learned about transistors, all kinds of transistors. You learned about MOSFETs, and JFETs, and BJTs, or bipolar junction transistors. One of the earliest applications for transistors—and indeed of the vacuum tubes that preceded them—was to be amplifiers; devices that took weak signals and made them stronger. You're particularly familiar with amplifiers in the audio realm, so I'm going to emphasize audio amplifiers in the next couple of lectures. But then we'll generalize the amplifier concept considerably beyond that.

We want to think about applications for audio amplifiers. Briefly, your radio has an audio amplifier, your stereo system has an audio amplifier, your cell phone has an audio amplifier. Anything that produces sound that you can hear coming out of electronic circuit, I guarantee you has an audio amplifier on it. We'd like to develop a simple circuit for an audio amplifier in this lecture, and then in the next lecture we're going to combine that with some other circuits and make a complete audio amplifier system. Let's switch to our big screen and take a stab at designing a simple audio amplifier. The amplifier we're going to start with is called a common emitter amplifier. Before I go into the amplifier itself, let me explain the meaning of that term "common emitter."

You'll recall that a transistor is a three-terminal device, and yet it's a device that allows one circuit to control another. But because it has only three terminals, one of those terminals has to be shared by both circuits. You can build a transistor amplifier in which the collector is common, the base is common, or the emitter is common. Perhaps the common emitter is the most common configuration, but the others are used as well, and we'll see some of them in the next lecture.

We have a common emitter amplifier. In this circuit the emitter is grounded. Ground is a common point; it's the negative terminal of the power supply. I should also add that I'm adopting here another shorthand notation: I mark plus 10 volts at the top of that resistor. That means that point is connected to a 10 volt power supply. I'm no longer going show separately a battery, or a power source, or something else. It's assumed. That 10 volts is understood to be taken with respect to ground, so there's 10 volts between here and here. By the way, if you're doing the projects and the simulations, you can explicitly put a battery or a DC power source into your circuits. You can also, at least in CircuitLab, simply put a node name here and label it plus 10 V or minus 10 V and it'll assume that there's a 10 volt or whatever voltage you want power supply there. That's one of many common sort of shortcuts we're going to be using. We no longer show all the wires; we no longer show the voltage sources; and so on.

What I have so far is the collector and emitter of the transistor. I have my resistor here; it happens to be 1 kiloohm. I call that the load resistor because that's the resistor in which the power produced by this amplifier is being dissipated; in this case, just turned into heat. On the other hand, if I wanted an output from the amplifier, an output voltage, I'd take it right there at the collector or the junction between the collector of the transistor and that load resistor.

This is our first try. There's the output circuitry, if you will. Now let's add an input. I put up a half volt AC power source; that's actually a pretty big input to an amplifier. Amplifiers usually start with voltages in the millivolts, or tens or hundreds of millivolts. But that's 500 millivolts; that's a half a volt. That's pretty big input voltage. I have an input resistor of 10 kiloohms. Remember, we can't put a voltage source right across the base emitter junction or we'd burn it out because it's basically a short circuit when it's forward biased. This is our first try. Will it work? No. It won't work, and here's why: Remember that that base emitter junction is a diode. A diode doesn't really start conducting in the forward direction until there's about 6/10 or 7/10 of a volt or more across it. As that AC swings up to half a volt, 5/10 of a volt, it's not enough to turn that diode on. Even if that diode were perfect and didn't have that extra 6/10 or 7/10 of a volt drop across it, this still wouldn't work—it would work a little better, but it still wouldn't work— because it would only allow current to flow when the input went positive. When the input went negative, that's still a diode and it would be shut off. This circuit doesn't work. We have to get a little bit more sophisticated about our common emitter amplifier.

Here's the idea of the amplification. When current does come in here, if we could make that happen by putting a big enough voltage in, the idea is base current would flow, and larger collector current flows because we know that the transistor is basically a current amplifier. Here's how it goes: If the input voltage increases, and I have an AC voltage there so it's going to go up and down, the base current increases. The base current remains small. We've got a big resistor there and small voltage, not much current; but the base current will increase as the voltage goes up. What that means is, because the transistor has this big current gain beta, the collector current will also increase and the collector current is large. Now there's a larger current flowing through that load resistor. The voltage across that load resistor by Ohm's law is the current times the resistance. If the current goes up, the voltage across that resistor increases. That means the output voltage, right here, goes down. There's a bigger voltage drop across that resistor. That decreases the collector voltage, which is the same as the output voltage.

As a result, as the input voltage swings up and down, the output voltage will also swing down and up rather than up and down, and that's going to be important in a minute. The bigger the current flowing in, the bigger the collector current, and the lower the output voltage goes. You increase the input and the output goes down. You get larger voltage swings at the output, and that's what we mean by amplification. In principle, if it weren't for this problem with the base emitter diode either not being able to turn on or not being able to turn on at least when the input goes negative, we'd have an amplifier. But, as I say, because of that issue with the diode, this amplifier isn't going to work, so we've got to get a little bit more sophisticated.

Here's the more sophisticated way to build this amplifier: It's with circuitry called biasing. If this were an electrical engineering course, we'd probably have several lectures on how to bias a transistor. I remember in my own electrical engineering course in college, we actually had to spend a lot of time biasing a transistor, and it was one of the harder things we figured out how to do. It involved thinking a little more sophisticatedly about the transistor and its own internal resistances and capacitance, all kinds of stuff. We're going to skip all that. We're going to use trial and error to bias our transistors when we need to do that.

I haven't shown any input yet, but what I have shown is an extra connection through a rather large resistor, 82 kilo ohms to the base, and then there's this extra resistor from the base to ground. That doesn't make a lot of sense unless you recognize that there's also some resistance in the base emitter junction. I'm assuming the transistor has a beta of 100; I've still got that 1 kilo ohm load resistance. Here's the idea: There's a current—I'm going to label it IB because it's the base current—and it's going to flow down through this resistor and through the base emitter junction. It's going to provide a level of base current, even when there's no input. If there's base current, there's also collector current, and therefore this voltage at the output or at the top of the collector is smaller than the 10 volts, but it's not as low as the 0 volts.

If I've picked these resistors carefully—and that's the tricky part of biasing a transistor so I'm not going to go into details, and you could do this in your project by trial and error—if I pick those carefully, I'll get this output voltage when there's no input to be roughly half the power supply voltage. It's sitting halfway between the 10 volts at the top and the 0 volts at the ground. That's a bias transistor, and it's interesting because now we have base current already, whether or not there's an input. We also have a slight disadvantage of that: The transistor and these resistors are all dissipating power, so the circuit does use up power even when nothing's happening.

Now let's add an input. In comes the input. Now I've reduced the input voltage to a more typical input. That's 0.01 volts AC; that's 10 millivolts; that's typical of what you might get. I'm going to couple it in through a capacitor, because if I didn't do that, the direct current flows associated with these resistors would get messed up by whatever else was connected out there. But because I've got a capacitor and a capacitor blocks direct current, DC cannot get through a capacitor. It's like an infinite resistance to direct current; to 0 frequency. Then all we're going to see, the capacitor's going to let through the changes associated with the AC, but it won't let through any DC voltage that's associated with this source.

Now I have a situation in which if I put a small sinusoidally varying waveform in at the beginning, the current through the base emitter junction will also vary slightly in response to that, and that will cause a much bigger variation in the collector current. That will cause, as we saw before in our basic idea, this output voltage at the collector also to vary. It will vary much more, and there we have amplification. Also notice, as I said before, when the input voltage goes up, the base current goes up, the collector current goes up, and the collector voltage, which is the output voltage, goes down. When that one goes up, this one goes down.

That's a common emitter amplifier with biasing, and that's a decent start at an amplifier.

Here's a sinusoidal output. This is a probe in CircuitLab looking at the collector of that transistor, the voltage I've labeled V-out, and it's a sine wave. If we look at it in a little bit more detail, you'll notice it's swinging about, just about 5 volts. That means I've biased the transistor successfully. It sits normally between 0 and 10 volts, its power supply voltage, and it sits roughly halfway in between, and that's going to be important for reasons you'll see shortly. There's the output from CircuitLab. Not only do we have it sitting at about 5 volts, we also have a variation of about 1.5 volts. It's going from 5 up to about 6.5, and down to about 5 less 1.5, about 3.5. It's swinging 1.5 volts either side of where it would sit if there weren't any input, at roughly 5 volts.

Remember that I had a 0.01 volt AC input signal. It was going up 0.01 volts and down 0.01 volts below 0, and so on. We can use that to calculate the gain of this amplifier. How much does it amplify? How big is this output voltage compared to the input voltage that we started with? The output amplitude is 1.5 volts; that's the peak of the output. The input amplitude was 0.01 volts. What that means is the gain of this amplifier is 150—1.5 divided by 1/100 gives us 150—so this amplifier takes its small input voltage, which is varying some small amount, and it amplifies that variation by a factor of 150. That's a good amplification factor; that's impressive.

You'll notice that I've put absolute value signs—the vertical lines around the word gain—because this picture, this number, this 150 doesn't really quite describe the entire gain. We can see what the gain really is—that is, what its sign is also—if we look at the input and the output waveforms. Again, I've asked CircuitLab now to look at both the input and the output, and they

160

aren't on the same scales. The input variation is much, much smaller, but I've blown up so they look about the same. But they are different in amplitude by a factor of 150. Because they're out of phase—that is, when the input goes high the output goes low, and when the input goes low the output goes high—it's an inverting amplifier. It's turned the signal upside down, and so the gain we're going to call minus 150. That's a common emitter amplifier with gain as stimulated in CircuitLab, and we'll do a real amplifier shortly. So we have an inverting amplifier with a gain of minus 150 there.

I'm going to pause now and do what's one of the more sophisticated things I'm going to do in the whole course probably, and particularly in this part of the course, and that's look at an idea that electrical engineers talk about called load line analysis, because we're going to have to get a little more sophisticated about this amplifier if we're to understand how it really works and what its limitations are. I'd like you to imagine a very abstract situation in which I take a power source—in this case it's our same 10 volt power supply, our load resistor of 1,000 ohms—and then I'm going to put between the bottom of that load resistor and ground any device. It could be another resistor, it could be an inductor, it could be a diode, it could be a capac; it could be anything. Later it'll become the transistor. I'm going to ask the following question: What is the current, if the output voltage happened to be 10 volts? If V-out there on the right at the junction between the resistor and the arbitrary device, if that were 10 volts, what would be the current through that resistor, and therefore, through the device? The answer is: It would be 0. Why is that? Because both sides of the resistor would be at the same voltage; there would be no voltage across the resistor; Ohm's law tells us there would, therefore, be no current through the resistor. I would be 0 if the out were 10 volts. I'm not saying the out is 10 volts, I'm saying if that were the case. Maybe the device allows that, maybe it doesn't; we'll get to that. I would be 0 in that case.

What's I if V-out happened to be 0? That would certainly be the case if the device were just a piece of wire, for example. In that case, the entire 10 volts would appear across that load resistor. It's a 1,000 ohm load resistor, so with 1 volt across 1 kiloohm we'd get a current of 10 milliamps; 10 volts across 1,000 ohms across 1 kilo ohm gives us 10 milliamps.

That's the possibility; the two extreme possibilities in this case. We could have an output of 10 volts, in which case the current would be 0; we could have an output of 0, in which case the current would be 10 milliamps; and we could have anything in between. You could easily convince yourself, for example, if V-out were 5 volts, then we'd have 5 milliamps flowing through the load resistors and also on through whatever that device was. In fact, we can describe all this graphically by drawing a plot of V, the output voltage there, the voltage across that unknown device or that arbitrary device versus the current through it. One of our points was 10 volts and no current, and another point was no volts and 10 milliamps. Another we just thought about without calculating out was 5 volts and 5 milliamps. You can quickly convince yourself that all the possible cases lie on that line. When I have the circuit set up like this, the output voltage and the current through the whole string there of the load resistor and the arbitrary device must be on that line somewhere.

The other thing we need to think about is we have some arbitrary device down there. That device also constrains the voltage and current, and the constraint is given by the VI curves we've been plotting before. If it was a resistor, it would be a diagonal line up from the origin there at the corner, but it's a transistor in this case. I'm now going to replace that arbitrary device by the collector to emitter part of the transistor; we'll worry about the base later. Here are the characteristic curves that we developed before for a bipolar junction transistor for different values of the base current. There they are. So what this is telling us is the output of this amplifier—the current through the load resistance and the output voltage—must lie on that line, and as we vary the base current, the actual value of V-out and the corresponding current, slide up and down that line.

There are two special points we need to identify: We need to identify, for example, a point roughly in the middle; there are actually three special points I want to talk about. That's the point I biased my amplifier to be at; it's at that middle point. That's, by the way, called the operating point or also the quiescent point. If I've biased my transistor carefully, that point lies roughly in the middle of that load line; that diagonal line called the load line.

As I vary the base current, where we are on that load line is going to swing back and forth. The two extremes we can swing to are the point called cut off where there's no current flowing through the device—in this case the transistor—and there's the maximum possible voltage at the output, and the other extreme is called saturation where there's the most possible current flowing in that circuit, and that's set by the value of the power supply and basically the load resistor, with a little bit of fudginess for that steep curve at the start of the transistor curves.

We have to live somewhere on that load line, and our amplifier output has to lie on that load line. If we try to push the amplifier further than that, things aren't going to work out very well, and what's going to happen in particular is we're going to get what's called distortion. We're going to get sound if it's an audio amplifier that just doesn't sound very good. Let's take a look at how that would work. In fact, let's not only take look, let's take a listen.

What I have over here now is, in fact, an amplifier very similar to the one I just described and very similar to the one I had built in CircuitLab, which I just talked about. It's built down here in this corner and you can basically see most of it. Let me describe what I have here: There's the input coupling capacitor, there's the transistor, there are the two little resistors that form the biasing network in the base circuit, and there's the load resistor. That's all; that's my audio amplifier. I've got a lot of other wires going on here, and let me explain them to you, but they're not all that significant. Here's my function generator. It's set up to produce a 440 Hertz tone or sine wave. That's A above middle C for you musicians, and it'll sound like that. That's set to be peak to peak voltage of 100 millivolts, so it's going to swing up 50 millivolts and down 50 millivolts.

In fact, you can already see that on the big screen; that's what we're seeing. There's that variation; the yellow is the input. Here comes the wire out of the function generator. It's going into a connector here, down into the circuit through this yellow wire, and into the input of my amplifier, going in through that capacitor, that purplish-brownish thing. There's a T connector here, and another cable is going out, the yellow cable, and into channel one of the oscilloscope; channel one being the one that displays in yellow. What's at the other end of this? At the other end of this there's a red wire coming out of that output point. That is, it's connected directly to the collector of the transistor and also to the bottom of the load resistor. The top of the load resistor is connected to my 10 volt power supply, which is connected to this yellow thing, and I've adjusted the voltage so it's 10 volts. That red wire goes out and it goes into this connector whose green cable takes it into channel two of the oscilloscope, the green channel. In addition, this connector takes it into the voltmeter, which is now reading almost 0; it's reading a few millivolts, perhaps.

Now I'm going to turn this circuit on, and what do we see? Take a look on the voltmeter, first of all. The voltmeter is a DC voltmeter. It's reading about 5.9 volts, 6 volts, something like that. I didn't quite hit the biasing right. I would rather it be at 5 volts, so it would be right in the middle of that load line, but we're a little closer to 6 volts. That's OK; it doesn't matter; not crucial, although it does mean we'll have a little bit less dynamic range in our amplifier, a little less range of input. It's running at about 6 volts almost, and it's called the quiescent point or the operating point. That's where the amplifier sits when there's no input.

Now there's input right now, but this is a DC voltmeter, so it's not noticing that fluctuation associated with that input. There's our operating point. Then we look on the oscilloscope—and you can see that best on the big screen— and you can see that there's an output voltage. Let's just take a little look at that output voltage. You might say, "The output voltage isn't nearly as big as the input voltage," but that's not the case. Take a look. The output voltage is at 5 volts per division. That means one of these divisions is about 5 volts. The output voltage is swinging up about 2 1/2 volts, and then down about maybe a little less than that. The 0 point of it's maybe somewhere along here, so maybe it's swinging up about 2 volts and down about 2 volts.

You'll notice something else interesting: These little marks here show where the 0 level is of voltage. I can adjust those with a knob, where they appear on the oscilloscope. You'll notice for the input voltage, which is a sine wave swinging up and down above and below 0, that's right in the middle of the sine wave. But you'll notice for the output, it's not; it's way above that by about 5 volts. Lo and behold, 5 volts. There's our 5–6 volt operating point voltage at the output of the amplifier. We see the operating point on here, too, and we see this fluctuation by roughly 2 volts.

We've got an output amplitude of roughly 2 volts; we've got an input scale at 200 millivolts per division. What are we doing? We're going from here, up maybe, I don't know, half a division maybe. That's about 100 millivolts. We've got 100 millivolts and 2 volts. What do you have to multiply 100 millivolts—that's 1/10 of a volt—by to get 2 volts? You've got to multiply it by, let's see: If you multiplied 100 millivolts or 1/10 of a volt by 10, I'd get 1 volt; so I've got to multiply by 20 to get 2 volts. This amplifier has a gain of about 20. This wave form is actually 20 times bigger than this one, and if I put them on the same scale on the oscilloscope, the green one would be way off scale. We have an amplifier. It's amplifying, and it's producing a gain of about 20. Furthermore, if you look carefully, you can see that where the input is up, the output is down; so they're out of phase. That's actually a gain of minus 20.

Now let's see what happens if we turn up the input voltage. I'm going to adjust the level of the input voltage by 10 millivolts on a step. I'm going to raise it up, raise it up, and you can see as I do that the output voltage goes up, and up, and up, and up, and up, and up, and the input keeps going up. Look what's happening: The output voltage is beginning to be no longer a sine wave; it's beginning to flatten out at the top. That's the highest voltage we can have; that corresponds to the point called cut off on that load line analysis. What's happening here is we've reached the maximum possible voltage that can be produced in that circuit. The transistor doesn't really make the voltage or the power, it just controls the power coming from the battery, and it's a 10 volt battery or 10 volt power supply and so we can't go anymore. That waveform is distorted; this is no longer off high fidelity amplifier. If we go up higher you can see that distortion worse and worse, and it's beginning to distort at the bottom now, too, because we're running into saturation as well. Because my operating point was 6 volts, it didn't happen quite symmetrically. If it had been 5 volts, we would have hit saturation cut off at almost the same point.

There's a bad situation. Why is it bad? Let me have you hear why it's bad. Let's bring it back down to about 100 millivolts, where we started, and let me just do one other thing. This little amplifier I built is not a power amplifier, it's not good enough to drive a loudspeaker itself, so I've connected here a loudspeaker, and I'm coupling into it through a capacitor. It's coming out this red cable to an amplified loudspeaker. It's got its own amplifier, but we want to hear what this little amplifier's doing, and that's what we'll hear. I think I probably want to go over here so you can hear the sound coming from that loudspeaker, and it sounds pretty good; nice steady tone. Now I'm going to turn up the gain on that input, and I think I'll just hold this right near my microphone so you can hear it. Up we go, bigger and bigger. It just gets louder and louder but still sounds pretty good. But just about the point where you could see visibly the distortion on the trace from the oscilloscope, you also heard that raucous sound. That's because we no longer have a pure sine wave. In fact, if you mathematically analyze that, we'd see that that had to be a mix of frequencies and there's kind of a dissonance there that makes it sound really lousy.

That's the limitation of an amplifier as imposed by that load line: We can't push the input too far or the output will go so far that we swing between saturation and cut off, and then the output distorts.

We're going to look at a better biased amplifier. Here's a picture of a slightly better biasing scheme. What you'll notice about this scheme is there's one more resistor and it's in the emitter lead, and it's a rather small resistor; it's only 50 ohms. What's that one for? That resistor does a number of things. First of all, it protects the transistor against what's called thermal runaway: the transistor getting hot, more electrons are liberated and holes, more power is dissipated, it gets hotter. Thermal runaway can occur and destroy the transistor. If that starts to happen, the voltage drop builds up across that 50 ohm resistor and helps prevent that. That's one possible purpose for that. There's a little bit less distortion because of this. There's a little bit of feedback involved; we'll talk about that later.

But probably the most important reason for putting this extra resistor is when you buy a bunch of transistors, their beta, their current gain, isn't really very well fixed; it might be 50 for one of them and 150 for another one from the same bin, the same type. What putting this load resistor, this emitter resistor, here does is to give you a fixed gain that's independent, approximately, of the transistor's own characteristics. In fact, that gain becomes the resistance of the load—in this case, 1,000 ohms—over the resistance of that emitter resistor, 50 ohms; and if you take 1,000/50 you get 20. In fact, this is the circuit I built here. It has that extra resistor, and that extra resistor is giving us a gain of only 20 instead of the 150 that we calculated before. By the way, if we wanted a little more gain, we could put a capacitor across that 50 ohm resistor. The 50 ohm resistor would still do its job as far as direct guard is concerned, but the capacitor would act as a lower "resistance" to AC, and we could increase the gain that way. That's a common way to increase the gain, but I didn't bother to do that on this particular circuit. There's our amplifier with better biasing. I'm going to simply add an input and couple it in through a capacitor, and then I have a complete common emitter amplifier.

We'll end by looking at what the output of CircuitLab looks like for this common emitter amplifier. There are the input and output voltages. If you look at the input and output voltage swings and do the calculation, you'll find that this amplifier has a gain of just about 20; actually, minus 20 again because when the amplifier is up, when the input is up, the output is down.

That's a real audio amplifier, one stage. We'll take that and combine it with other circuits to make a complete audio amplifier system. In the meantime, if you'd like to design and build your own audio amplifier with your simulation package, you can do that and that's what the project is for this lecture.

Building an Audio Amplifier
Lecture 9

In this lecture, we'll continue our ongoing exploration of audio amplifier circuits. A complete audio amplifier requires several stages of voltage amplification, followed by a power output stage to drive a loudspeaker. As we put together a complete audio amplifier in this lecture, we'll learn about two new amplifier circuits: the emitter follower and the complementary symmetry configuration. We'll cover the following key topics:

- Multistage amplifiers
- Coupling common-emitter stages for voltage gain
- The emitter follower
- Complementary symmetry
- Eliminating distortion.

Multistage Amplifiers

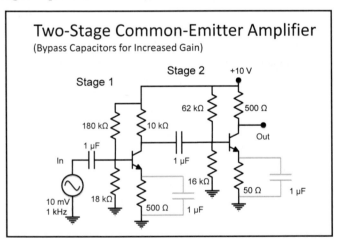

Both the stage 1 and stage 2 CE amplifiers shown here are voltage amplifiers. Their primary purpose is to increase the input voltage. Because the output of the first stage swings about its operating point rather than about 0 V, we need to couple the two stages with a capacitor that blocks DC but lets the varying AC signal through. We've also added bypass capacitors around the emitter resistors. The AC signal "sees" these capacitors as lower-resistance paths to ground than the emitter resistors, and this recaptures some of the gain we lost by adding those resistors.

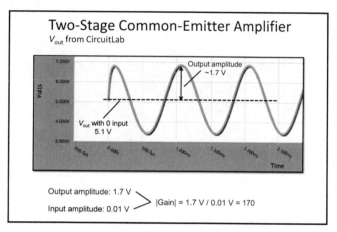

The output from a simulation of this amplifier shows that the voltage is swinging roughly around 5 V; that's the operating point. The output amplitude is 1.7 V; the gain is 170.

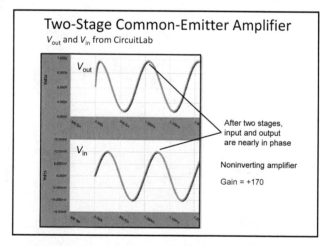

We can see that this is a noninverting amplifier because the signal went through one CE amplifier, which inverts the input signal (here, a sine wave).

When the input goes high, the output goes low. That low output is then fed into the second stage, and when the output of the first stage goes low, the output of the second stage goes high, giving us a positive gain overall.

Volume Control

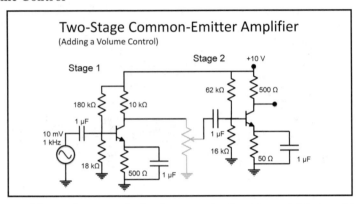

We can add a volume control using a *potentiometer*, which is a kind of variable resistor that acts as a variable voltage divider. Here, we've put the volume control between the two amplifier stages. Turning the knob moves the variable contact (the arrow on the potentiometer symbol), tapping off anywhere from zero to all of the first-stage output and supplying the corresponding voltage as input to the second stage.

The Emitter Follower

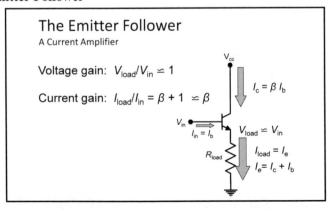

An emitter follower is a current amplifier. It's also called a common-collector configuration because the collector of the transistor is connected directly to the power supply, which along with ground, is another common point in the circuit.

Complementary Symmetry
To reach the final stage of our amplifier to drive a speaker, we can make use of the complementarity of NPN and PNP BJTs by developing a complementary-symmetry (CS) circuit.

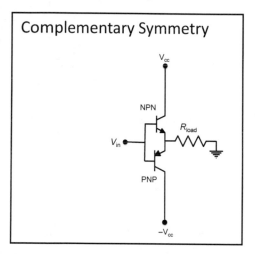

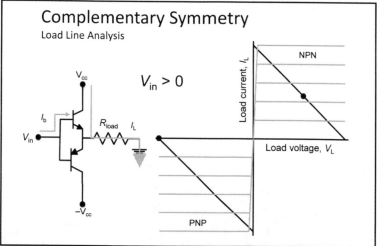

Here, a positive input voltage turns on the NPN transistor (the upper transistor in the diagram), which amplifies the positive-going half of the AC cycle. A negative voltage turns on the PNP (lower) transistor. Together, the

171

two transistors amplify the entire cycle. Each transistor is biased at its cutoff and moves into the active region of the load line only when it's conducting.

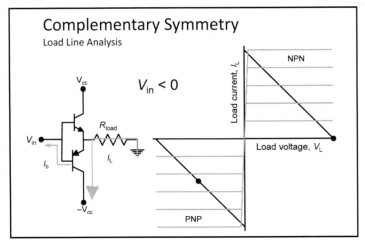

This device will amplify when the input signal goes positive by turning on the NPN transistor and when the input signal goes negative by turning on the PNP transistor. When there's no input, neither transistor is on, and we're not dissipating power.

Eliminating Distortion

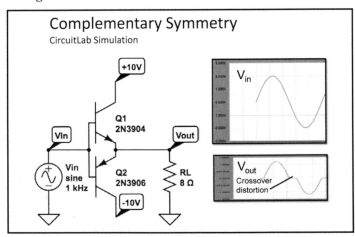

We still have a problem that causes minor distortion in the output voltage and, thus, the sound from our amplifier. This glitch occurs because turning on the NPN transistor doesn't require just a positive voltage; it requires a voltage going above the 0.7 V needed to turn on the base-emitter junction. And turning on the PNP transistor requires the input going below that same 0.7 V. *Crossover distortion* is caused by the transistors turning off during the interval when the input voltage is between +0.7 V and −0.7 V.

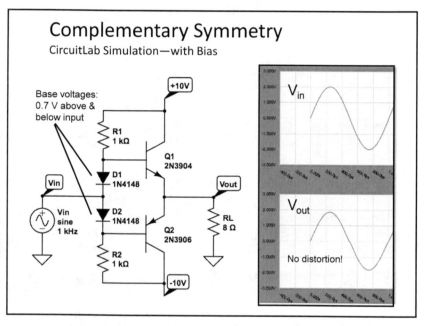

An easy trick for eliminating crossover distortion is to simply add a couple of diodes and resistors in a string from +10 V down to −10 V. What the diodes do, effectively, is bias the base of the NPN transistor at +0.7 V, so it's just on the verge of turning on. The lower diode biases the base of the bottom transistor, the PNP transistor, at −0.7 V, also just on the verge of turning on. If the input signal swings even the slightest bit positive, the upper transistor turns on. If the input swings even the slightest bit negative, the lower transistor turns on, and there's no crossover distortion.

Introductory

None of the introductory books treats the transistor amplifier circuits introduced in this lecture. For practical amplifier construction, you would start with integrated-circuit amplifiers, and you wouldn't have to know how their internal circuitry works.

Advanced

Horowitz and Hill, *The Art of Electronics*, 3rd ed., chapter 2, sections 2.2.1 and 2.4.

Scherz and Monk, *Practical Electronics for Inventors*, 3rd ed., chapter 4, section 4.3, pp. 439–440.

Project

Gain and Power

Simulate the combination of a two-stage CE amplifier with a diode-biased CS output stage to make a complete audio amplifier. Use the two-stage CE amplifier from earlier in this lecture and the CS amplifier. Couple the CE and CS amplifiers with a 1-µF capacitor. For input, use a 0.025-V, 1-kHz sine wave. Find the voltage gain and calculate the power delivered to R_L.

Questions to Consider

1. Why is it necessary to use a capacitor between two stages of a multistage audio amplifier?

2. Why is a CS output stage in an audio amplifier more energy efficient than a single-transistor output stage?

Lecture 9: Building an Audio Amplifier

Building an Audio Amplifier
Lecture 9—Transcript

Welcome back to our ongoing exploration of audio amplifier circuits. Let's begin by looking at the circuit we developed last time, which was a single transistor common emitter amplifier.

I'm showing it now and I'm calling it stage one, because we're going to develop here a two-stage common emitter amplifier. The amplifier I have here is almost identical to the one I showed you last time. If you look at it carefully you'll see the only difference; it's configured exactly the same way, the only difference is the resistors have larger values. That's because this is the very first stage of an amplifier that's going to end up at the far end powering a loudspeaker, but that's to come by the end of the lecture. This particular stage doesn't need to handle much power, and so we don't need very large currents flowing in it. There's stage one of what's going to be a two-stage common emitter amplifier.

Stage one. The idea that there's stage one suggests there's stage two. Stage two looks identical in configuration almost. I don't have an input here yet. Notice the resistor values are all considerably lower. They're lower because there are bigger currents flowing in this stage. Both of these common emitter amplifiers are what we'd call voltage amplifiers. Their primary purpose is to take an input voltage and make it bigger. In the process they also make the power available bigger, but their primary function is to amplify voltages; get them up so they're out of the level where they might be subject to electrical noise, get them big enough to feed into a real power amplifier that can deliver power to a loudspeaker. It's a two-stage amplifier at this point. Each of them is connected to a 10 volt power supply. We could connect them to exactly the same power supply, although if we were really sophisticated we might isolate the power supply slightly so that a heavy current flow in the stage two amplifier doesn't lower the voltage to the stage one amplifier; a lot of sophisticated things if you were designing a real amplifier that we aren't doing here.

The question is: How are we going to join them? I can't just take a wire from the output of the stage one amplifier to the input of the stage two amplifier, and the reason I can't do that, remember, we have this operating point established. For both these amplifiers, if I've picked the resistors right, it ought to be about 5 volts. That output ought to be sitting at 5 volts when there's no input, and it ought to swing maybe 1 or 2 volts above 5 volts, and 1 or 2 volts below 5 volts. More than that and we're likely to have distortion. We want to capture those swings in the output of stage one, but we don't want to capture the 5 volt steady operating point level. What device will let us do that? What device will let us do that is the device that blocks direct current, doesn't let DC through, doesn't let zero frequency through, but let's variations through. That is, of course, a capacitor.

We're going to couple of these two circuits with a capacitor. I put a 1 micro farad capacitor, which is good enough for audio work probably in this situation. Maybe it'd have to be a little bigger if we wanted to get to really bass frequencies because they're changing rather slowly. But there's now our coupled amplifier, and I've joined the two power supply connections to one power supply. In this case, probably not the best practice, but it will do for us. They're now coupled, and I have a two-stage common emitter amplifier. Let's look at what this amplifier does and let's add one more modification to this amplifier. I mentioned in the previous lecture that those resistors from the emitters to ground have multiple functions. They help stabilize the transistor against thermal runaway and things like that, but they also establish a gain that's independent of the beta of the transistor to some extent. They mean you can take arbitrary transistors off the shelf and put them in basically the same circuit and they'll work the same. But as we saw last time, we'd started with a gain that was of 150 or something. When we put that resistor in, that gain dropped to about 20, so we kind of lost gain. You can get some of that gain back. You can gain back some of that gain by putting capacitors because, again, the capacitors are like the resistors for the AC, for the audio frequency voltage variations. They'll act as lower resistances, and since the gain is the load resistor over that emitter resistor, there will be a higher gain for those time varying signals that we're interested in. This helps boost the gain. That's why those capacitors are there.

Let's now see how this device works. Again, I'm not going to wire up the whole thing on my board, but I wired it up in CircuitLab. By the way, if you'd like to pause right now and do this in CircuitLab if you're planning to do the project, or CircuitLab, or do circuits, if you're planning to do the project for this particular lecture, this is going to be the first part of it, but you're going to have to do more. If you want to explore the circuit right now, you could press pause, do that, and come back and we'll look at the output.

Here's the output from my CircuitLab simulation of the two-stage common emitter amplifier. You can see that the voltage is swinging about roughly 5 volts. That's my operating point; it came out 5.1 volts in this case. Close enough to the middle, between the 0 and the 10. My two-stage common emitter amplifier, the swing is from the 0 level up to about 1.7 volts above it, so we're going from about 5.1 volts to 6.8 volts. But the important thing is that swing; that's the amplitude of the output variations. We've got an output amplitude of 1.7 volts, we've got an input amplitude, I set that input voltage to 0.01 volts peak, and so we've got a gain—and I'm putting the absolute value sign around it because we've got to think about the sign— whose magnitude is 170. The input voltage variations going through this two-stage common emitter amplifier have been boosted by this factor of 170. That's about what we had with our first one-stage amplifier, but we didn't have that emitter resistor in there that stabilizes things and makes the gain basically independent of the individual characteristics of the transistor. With two transistors we've got back that high gain, but we have a much more stable circuit and it doesn't depend on what particular individual transistor we put in it. There's our two-stage amplifier.

I cautioned you last time we've got to take into account the sign of the gain. Is the gain positive or is the gain negative? Is it a non-inverting amplifier or is it an inverting amplifier? Let's look now at the input and the output graphs from CircuitLab; we'd see the same things on an oscilloscope if we had a real circuit here. This now is a non-inverting amplifier. Why is it non-inverting? Because we started and went through one common emitter amplifier, which we know, from the previous lecture, inverts; turns the sine wave upside-down. When the input goes high, the output goes low. Now that output that's going low is fed into the second stage. When the output of the first stage goes low, the output of the second stage goes high, and that's when the input of the first stage went high. It's a double negation: It's minus 1 times minus 1; it's two negatives make a positive. This has a positive gain, and in this case the gain is 170. So we built a two-stage common emitter amplifier here with a gain of 170.

This amplifier is missing something we'd probably like to have, and that's a volume control. We might also want to have tone controls, base and treble controls. We already know how to make those. We'd make those with filters, and we could insert those filters at various points in the circuit. But it'd be awfully nice to have a volume control. By the way, what we have here is what would probably in the old days be called a pre-amplifier. You might've actually, in a stereo system of the 1960s or 70s or even 80s, bought a separate pre-amplifier and then a power amplifier, and the volume control probably would've been in the pre-amplifier. How do we add a volume control?

Let me show you a device that we're going to use to add a volume control. Here I have a device that's called a potentiometer. Don't worry, you're not going to see a big clunky thing like this with a ceramic base and all this heavy duty stuff in an audio amplifier. This is an ancient, and very heavy, and very power-handling capability potentiometer. What's a potentiometer? A potentiometer is another kind of resistor. What kind of resistor is it? It's basically a resistor, and what we have here, in fact, is coils of wire wound around a piece of ceramic. They start here, and they go all around here. The resistance in this particular case just happens to be 100 ohms, so between this terminal and this terminal is just a 100 ohm resistor. But what we also have is a middle terminal, and that middle terminal connects through this mechanism to this adjustable slider. That slider can move, and so it contacts different parts of this resistance. Right now it's in the middle, and so the resistance from here to here is 50 ohms, half of the total resistance, or from here to here. The resistance from here to here is 50 ohms. What I've got is a voltage divider, which if I put a voltage across these two outer terminals, I'll get half that voltage at the middle. On the other hand, if I put, say, plus 10 volts here and 0 volts here, now I'll have 5 volts in the middle. If I crank all the way over here so it's connected to the 10, I'll have 10 volts. If I crank all the way down to here, I'll have 0 volts. What this potentiometer allows me to do: It's a variable voltage divider. It allows me to divide the voltage as I choose by turning, what you'd see as the user, this knob, which turns that variable piece.

It's a little bit more than a variable resistor. A variable resistor would have two terminals and a knob, and if you turn the knob, you'd change the resistance. This does something more than that: It taps off a fraction of the voltage that's applied across the whole thing. By the way, if you wanted to have just a plain variable resistor, you could make it by taking the resistance across either pair of terminals with the middle, this pair or this pair, and you'd get the resistance between the first point and the middle point, and that would vary as you turn this slider.

That's a potentiometer, and one of its many uses is be a volume control in an audio amplifier. In the old days, a big powerful one like this might've been used for something like controlling the brightness of a lightbulb. But we know that's a horribly inefficient thing to do nowadays because most of the power would be dissipated as heat in this resistor, and we have far better electronic ways to handle that problem. I mentioned that that's not what you'd find for a volume control in an amplifier. A volume control probably in an older amplifier or one of those tube radios I showed you might look like this. It's a similar thing; you just can't see it because the variable or the resistance is hidden behind this metal case, but you do see the three terminals: the two that mark the end of the resistance, and the one in the middle.

What we have, if we want to add a volume control to our amplifier, is a potentiometer, and I've placed the potentiometer in this case at the output of the first amplifier and before the input to the second stage. I've got to be a little bit careful in doing this. I'm introducing a new resistor into the circuit that wasn't there, and that's going to mess up the biasing and the operating point of the stage one amplifier unless that potentiometer has a much, much bigger resistance than that 10 kilo ohm resistor that I've got as the load resistance in my stage one. I might need 100 kilo ohms, 200 kilo ohms; that's getting awfully big. If I were redoing this amplifier, I might actually put the volume control at the output of the second stage, because then the potentiometer resistance would only need to be much bigger than the 500 ohm load resistance in that case. But it doesn't really matter; the idea is here. As I slide that slider up and down—and you can see the symbol for the potentiometer; it's a resistor, and then it's got this little pointy arrow which represents the slider that you can imagine moving up and down along that resistance—if I move it all the way up, I get the most volume I can have; I get the full output of stage one. If I move it all the way down to the ground, I get zero, and even though stage one is producing an output, there's no input to stage two. That's our volume control.

So far, we have a nice two-stage common emitter voltage amplifier. We've increased the gain by putting those capacitors across the emitter resistors. We've added a volume control. Pretty good. The problem, though, with this and many other amplifiers is in this configuration, even if there's no input to these amplifiers, there's still current flowing through their load resistors and through the transistors because we're trying to hold that operating point about halfway between; in this case, the 10 volt supply voltage and the 0 volts of ground. That means we're dissipating power in the load resistors, and we're dissipating power in the transistors, and we're wasting energy.

At the early stages of an amplifier, it's not a problem because it's not a lot of power. But when you build a 200 watt audio amplifier, you'd really like it when the music is quiet for that thing not to be drawing much power out of the power line. If you simply continued with bigger and bigger power versions of this common emitter amplifier, you'd be dissipating hundreds of watts, even when the thing was completely quiet. You don't want that. That's a crazy waste of power, contributing to climate change, doing all kinds of bad stuff that you don't want to do, wasting your electric bill, heating up the amplifier and making the components wear; all kinds of bad things. We need to develop a power amplifier that can handle lots of power and drive a loudspeaker, for an audio amplifier, but that won't dissipate power when things are quiet. That circuit is called an emitter follower. Not quite; the emitter follower's going to be part of that circuit.

The emitter follower is a current amplifier. It's also called a common collector configuration, because you'll notice that the collector of the transistor is connected directly to the power supply, which, along with ground, is another common point in the circuit shared by everything. This is a common collector—not a common emitter, but a common collector—amplifier, also called an emitter follower. Why is it called an emitter follower? Because we take the output at emitter, and the output basically follows the input. What happens is we have a voltage gain that's basically 1; it's actually a little bit less than 1 if we're dealing with small voltages because of that 0.7 volt drop across the base to emitter junction. This amplifier doesn't have—I don't want to say it has no voltage gain—0 voltage gain, it has a voltage gain of 1. It doesn't gain you anything in voltage. It passes the input voltage through as an output voltage.

But what it does do is amplify current. There comes a small base current in. Down comes a much bigger collector current, beta times the base current, beta being the current gain of the transistor. Both of those combine and go through the emitter resistor, and the output voltage is approximately the input voltage minus the 0.7 volts at that junction. But there's a lot more current, so this is a current amplifier, and because it's amplified current there's also a lot more power available to pass through that load resistance than there would've been coming into the input. It's a power amplifier. So were our voltage amplifiers, but this one is amplifying power by amplifying current. You could replace that load resistor by a loudspeaker if you wanted to.

There's the emitter follower. A current amplifier, a power output stage sort of, for big power amplifiers. But so far it's still not all that good. Let's just take a quick look at it. The current gain of this thing is I through the load. The load is now in the emitter lead, so it's carrying both the emitter current and the base current, so it's beta plus 1. Beta, for the current coming through the collector (remember, beta's about 100 or so); 1 for the extra current coming in through the base. Beta plus 1 is close enough to beta because beta's on the order of 100. Who cares whether it's 100, or 101, or 99, or 150, or 151; it doesn't matter. It's essentially, approximately, beta.

The problem with this amplifier is it, too, when we're in the quiescent state has to be biased or that transistor will turn off when the input voltage goes negative. We, therefore, are going to have to bias this one just like we did the other amplifier or in some similar scheme, and we're going to be dissipating a lot of power when things are quiet. That now matters because that resistor, if it's a loudspeaker, typically has 8 ohms of resistance. It's very low; lots of current is going to flow; there's going to be significant voltage, the power supply voltage for a big power amplifier, it might be 30 or 40 volts, maybe even more. A lot of power; we don't like that. We're wasting energy; we're heating up that load resistor; we don't need to. What can we do that would make this better?

Remember when we talked about semiconductors, I mentioned that we have—of course, the beauty of semiconductors is we have these two kinds of semiconductors—p-doped and n-doped, p-type and n-type. They're complementary. One has positive charge carriers, holes; the other has negative charge carriers, electrons. We've been dealing almost exclusively since then with NPN bipolar junction transistors. But I pointed out when I introduced transistors that we could equally well make a PNP bipolar junction transistor that's complementary. What we're going to do now is develop a complete output amplifier stage, the final stage that's going to drive the speaker, and we're going to make use of that complementarity of the two transistors. We're going to develop a circuit called complementary symmetry.

I'm going to show you the simplest rendition of complementary symmetry. I should point out that if you look at circuits for real amplifiers, there are, in fact, asymmetries between the workings of PNP and NPN transistors, and they have to do with the different ability of holes and electrons to move and carry current. Sometimes you'll see circuits that you'll scratch your head about; it looks like they have two NPN transistors, but they're actually simulating a PNP transistor with an NPN. The circuit I'm going to show you is going to be truly complementary and use one NPN transistor and one PNP transistor. How does this work?

Here's our basic emitter follower amplifier. I'm first going to take that load resistance and move it so it goes out to the right. It's the same circuit. Remember, you can draw a circuit any way you want as long as you've got the connections right. That's the same circuit I just showed you, it's just got the load resistor going off to the right. Now I'm going to bring in a PNP transistor, and I'm going to connect its collector not to ground, but to a negative supply voltage. Maybe I've got plus 10 volts at the top of the collector of the NPN transistor, and I've got minus 10 volts at the collector of the PNP transistor. You'll notice I've connected their two bases together. Just to emphasize that word symmetry, I'm going to redraw it with the input connection to the middle. Now this thing is completely symmetric. Draw a horizontal line through it, especially if I'd run that ground connection sideways, we'd have perfect symmetry. We do have perfect symmetry electrically. We've got an NPN transistor in the top, a PNP transistor on the bottom; positive power supply at the top, negative power supply, again, relative to ground. The collector of the lower transistor is 10 volts, say, below ground; the one at the top is 10 volts above ground if we're talking 10 volts; there's 20 volts between the plus supply and the minus supply. How are we going to get an input into this thing?

Let's take a look at what we've got with our load line analysis. There's the load line analysis we did in the previous lecture for an NPN transistor. We have this load voltage, we have the load current. They have to lie somewhere along that diagonal line, the load line. They also have to lie somewhere along those characteristic curves, the transistor curves. I pointed out when I first drew these for a transistor, transistors only work in one direction, so we don't draw all four quadrants. But now we have two transistors. There's cut off saturation for the NPN transistor, but we also have the PNP transistor with its own load line and its own saturation and cutoff.

What's the idea behind this complementary symmetry amplifier? Here's how it works. Suppose V-in is 0. If V-in is 0, then both transistors are at cut off. There's no current flowing anywhere, and the out is 0. But what if V-in is greater than 0? If V-in is greater than 0, we get some current flowing in through the input lead, through the base emitter junction because that's going to be forward biased, out through the load resistor, and to ground if V-in is 0. That means the upper transistor, the NPN transistor, is conducting, it's working, it's transisting, it's being a transistor. We get a bigger collector current, and the collector current flows through the load resistance and to ground, and we've got amplification. But that's only going to work if V is greater than 0. In fact, this is what was wrong. This looks like the common emitter amplifier I started with as our first try and I said it doesn't work. It didn't work because if the input went negative, the base emitter diode was shut off and the thing didn't pass any current, nothing happened.

But now we have that second transistor. What happens if V-in is less than 0? If V-in is less than 0, everything goes the opposite way. We get a small current flowing in from ground, if you will, through the load resistor, and in through the base emitter junction that's forward biased now going to the left. In other words, if we put it negative V-in, we draw a little bit of base current like that and that causes a much bigger collector current, and, again, the collector current goes the other way because it's PNP transistor. Now we get current flowing from right to left, a big current flowing from right to left, in the other transistor.

The beauty of this thing is it'll amplify when the input signal goes positive by turning on the NPN transistor, and it'll amplify when the input signal goes negative by turning on the PNP transistor. That's why it's complementary. We have these two complementary transistors and they complement each other. One of them amplifies the positive going half of the cycle, and the other amplifies the negative going half of the cycle; beautiful. The beauty of it is when there's no input, when the input is 0, neither transistor is on, we're dissipating no power. We only draw power when, crash, the cymbals crash, and a lot of current has to flow. When the music is quiet, no power. Great, it solved their power problem.

However, it's got another subtle problem, and here's the subtle problem. I did a simulation of this circuit in CircuitLab to show you. It's got an 8 ohm load. I've got a V-out labeled; I've got 10 volts there at the NPN transistor, minus 10 at the bottom. I've got an input sine wave coming in. I'm showing you pictures of V-in and V-out. They're not identical for two reasons: One is V-out is actually a little bit less than V-in. There are a number of factors there; these transistors you're using in these simulations are in some sense like real transistors. They've got some aspects that we've been ignoring; that's part of the reason for that. The other part of the reason is there's that 7/10 of a voltage drop across the diode. That 7/10 of a volt drop across the diode leads to the other problem. Look at that V-out graph, the lower graph. You see we get a nice rising sine wave, but then we get this little flat part. Uh oh. Then we go down to the negative half as the lower transistor, the PNP transistor, kicks in and starts amplifying. But there's that little short time when it's 0. Why? Why don't we want that?

Let me tell you first why we don't want that, and then I'll tell you why it occurs. Here's why we don't want that: I'm going to play you a sound that's basically two seconds of what the output of this amplifier sounds like, followed by two seconds of; actually, it's going to start the other way, it's going to be two seconds of pure 440 Hertz A above middle C tone, as it should be, as the input would sound, for example, in this case. Then I'm going to play two seconds of the output, and it's going to repeat that. You'll be able to hear the difference between what the input signal sounds like and what the output signal sounds like. Take a listen. I think you'll agree that the distortion is plainly evident in that case. This isn't high fidelity. High fidelity means faithfully reproduce. Fidelity: Faithfully reproduce the input. We aren't faithfully reproducing the input, and it's because of that little glitch that occurs.

Why does that glitch occur? It occurs because what does it take to turn on the NPN transistor? It doesn't take a voltage going positive, it takes a voltage going above that 0.7 volts that it takes to turn on the diode. What does it take to turn on the PNP transistor? It doesn't just take the input going negative, it takes the input going below that 0.7 volts when the PNP transistor will turn on and current will flow from right to left through it; problem. That little glitch is caused by the transistors turning off during that interval when the input voltage is between plus 0.7 volts and minus 0.7 volts. Neither transistor conducts, and the output, therefore, is 0. That's called crossover distortion. It's occurring when we cross over from the upper transistor conducting to the lower transistor conducting. By the way, you'll notice I've got a 2N3904— that's the designation of a transistor—and a 2N3906. There are transistors that are made to be basically matched, complementary pairs, one NPN, one PNP. The 2N3904 and 3906 are such a pair.

What do we do about this crossover distortion? How do we get rid of the crossover distortion? There's an easy trick. Our trick for getting rid of the crossover distortion is to simply put a couple of diodes, running in a string with resistors in series with them nice and symmetrically, from the plus 10 volts down to the minus 10 volts. There's going to be current coming down through that string of resistor and diode. Both diodes are forward biased. Current is flowing down through that string, and both diodes, therefore, have 0.7 volts across them. The input is coming to the gap between the two diodes, and what the diodes effectively do is they take the base of the upper transistor, the NPN transistor, and they make it 0.7 volts positive. It's just on the verge of turning on. The lower diode takes the base of the bottom transistor, the PNP one, and it makes it 0.7 volts below ground and, therefore, just on the verge of turning on that one. Now if the input signal swings even the slightest bit positive, bingo, the upper transistor's on. If the input swings even the slightest bit negative, the lower transistor goes on and we have no crossover distortion and the graphs of the input and output show that these things are identical. What a great and elegant solution to that problem relating to ultimately that internal electric field of the p-n junction.

I could stop there but I'm not going to; I want to say one other thing. I'm going to now give you a mantra, which I'm going to be repeating from now till the end of course. It's a very strange mantra for someone who wants you to learn electronics and build electronics. The mantra is; well, let me tell you the mantra: The mantra is don't build, buy; or buy, don't build. Nobody today would go out and build an audio amplifier from scratch, unless you want to do it; it's kind of a learning process. You'd go out and buy an integrated circuit audio amplifier. I'm going to just show you, in the last few minutes, what one sounds like.

What I have over here—let me turn on the oscilloscope on big screen so you can see it—is a so-called LM386 integrated circuit audio amplifier. It's got 10 transistors in it, seven resistors, and a couple of diodes, not surprisingly. It has a kind of complementary symmetry output; although, as I mentioned, it fakes it with NPN transistors. It's all on that little chip, everything. I'm coupling it out to a loudspeaker that's built into my little bread board here, and we're going to listen to the output of that loudspeaker once I turn it on. There's the music. Once again, the music is my sister's hammered dulcimer playing. If you want to know what the music looks like, if you've ever wondered what music looks like, there it is on the oscilloscope screen. That's the input, so when I turn off the amplifier, it doesn't turn off. That's a cute little amplifier. It costs a few cents, maybe a few tens of cents, to buy one of those chips. It can produce up to a watt of power. Not a huge amount, but significant for many applications.

If that's not beefy enough for you, go to this one. Don't build, buy it. This is an LM3886 integrated circuit audio amplifier and it can produce 68 watts continuously, 134 watts in peaks, and it costs about $10. It's used in high end televisions, stereo systems, and others, and it's basically a complete audio amplifier. You hang a few more components, resistors, and capacitors on the outside of both of these to give them the gain you want and some other features but basically, they're audio amplifiers all in themselves but they work on the principles we've just learned here.

That's the end of this lecture. If you want to do the project, you're going to build your own complete audio amplifier.

The Ideal Amplifier
Lecture 10

In the preceding two lectures, we learned about transistor amplifiers that are designed specifically to amplify in the range of audio frequencies. But it's important to note that amplifiers have other functions besides those related to audio, including instruments for scientific, medical, or consumer use; electronic thermometers are an example of the latter. In this lecture, we'll look more theoretically at the characteristics we would ideally like to have in an amplifier. Then, we'll see how surprisingly close we can come to an ideal amplifier at a remarkably low expense. Key topics to be covered include:

- An ideal amplifier
- DC amplification: the differential amplifier

- An integrated-circuit operational amplifier
- The op-amp as a comparator
- Application: battery tester.

An Ideal Amplifier

Desired Amplifier Characteristics

- High gain: V_{out} = Gain × V_{in}
  - Or gain easily controlled to desired level
  - Ideal: infinite
    - Tempered with negative feedback
- Low output resistance
  - Source or sink substantial current
  - Ideal: zero
- High input resistance
  - Draw minimum current from input source
  - Ideal: infinite

V_{in} ─▷─ V_{out}

Among the qualities we'd like to have in an amplifier are high gain, low output resistance, and high input resistance. Ideally, we'd like to have infinite gain, zero output resistance, and infinite input resistance.

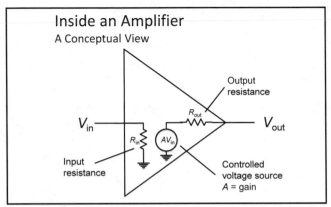

Inside an Amplifier
A Conceptual View

V_{in} — R_{in} — R_{out} — V_{out}

Output resistance

Input resistance

Controlled voltage source A = gain

AV_{in}

A conceptual view of an amplifier shows the following: The input comes in and controls the output, giving an output voltage that is the gain (A) multiplied by the input voltage. If there's current at the output, either coming from the amplifier or going into it, that current must go through the output resistance. A real amplifier approximates the goals of an ideal one to the extent that the input resistance is very high, the output resistance is very low, and the gain is very large—much larger than 1.

Desired Amplifier Characteristics

- High gain: V_{out} = Gain × V_{in}
- Low output resistance
- High input resistance
- Wide frequency response
  - Depends on application
  - Audio: 20 Hz–20 kHz
  - Instrumentation: DC to high frequencies

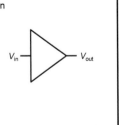

In addition to high gain, low output resistance, and high input resistance, we usually want an amplifier to have a wide and flat frequency response, meaning it amplifies a broad range of frequencies with the same gain across the whole range. That is, for the range of frequencies we're interested in amplifying, we would like the amplifier not to amplify one frequency more than another.

DC Amplification: The Differential Amplifier

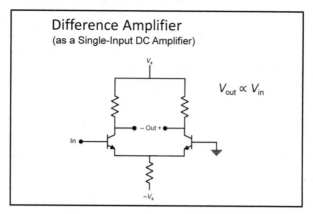

Difference Amplifier
(as a Single-Input DC Amplifier)

$V_{out} \propto V_{in}$

Our two-stage common-emitter preamplifier won't work for DC. Capacitors are open circuits to DC; they respond only to variations in voltage. To address this problem, we need a *difference amplifier*, which has two inputs. In the implementation of a difference amplifier shown here, we have two NPN transistors with separate load resistors and a common emitter resistor. The difference amplifier gives us an output proportional to the difference between its two inputs.

An Integrated-Circuit Operational Amplifier

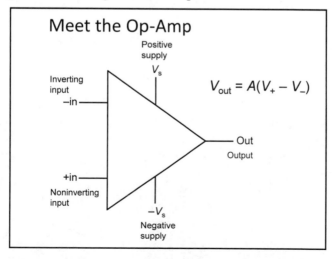

Meet the Op-Amp

Positive supply
V_s

Inverting input
−in

Noninverting input
+in

Negative supply
$-V_s$

Out
Output

$V_{out} = A(V_+ - V_-)$

The *operational amplifier* (*op-amp*) is a difference amplifier, usually built on a single, compact integrated circuit, that approaches the ideal amplifier. It has two inputs, both with high input resistance, a single low-resistance output, high gain, and a broad frequency response that includes DC (zero frequency). Originally, op-amps were designed to do mathematical operations and build analog computers; they're still used for many applications in analog electronics.

An op-amp has two power supply connections, one to a positive power supply, typically +15 V, and one to a negative power supply, typically –15 V. It has a noninverting input, also called the *plus input*; an inverting input, also called the *minus input*; and one output.

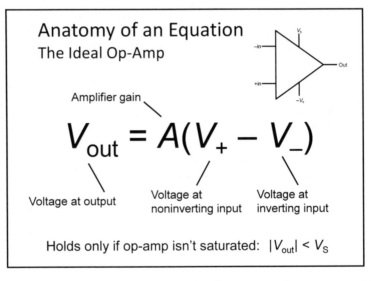

As we can see from the expression for the output of an op-amp, it's a difference amplifier: $V_{out} = A(V_+ - V_-)$. It amplifies the difference between V_+ and V_- by the factor A, which is the amplifier's gain. The ideal op-amp has a very large A.

Op-Amp	Gain	R_{in}	I_{in}	Bandwidth	Cost	Comments
TL081	200,000	$10^{12}\,\Omega$	40 mA	3 MHz	$0.22	CircuitLab default
DC	1,000,000					DoCircuits generic
411	200,000	$10^{12}\,\Omega$	20 mA	4 MHz	$0.70	Horowitz & Hill default
741	200,000	2 MΩ	25 mA	1 MHz	$0.22	Workhorse, but becoming obsolete
L165	10,000	0.5 MΩ	2 A		$1.00	Power op-amp

Notes: All values are "typical" and may vary. Quantity pricing is shown; individual pricing may be several times higher.

The table shows specifications for five op-amps.

The Op-Amp as a Comparator

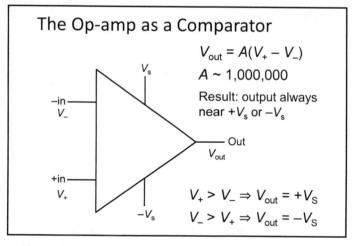

The Op-amp as a Comparator

$$V_{out} = A(V_+ - V_-)$$

$$A \sim 1,000,000$$

Result: output always near $+V_s$ or $-V_s$

$$V_+ > V_- \Rightarrow V_{out} = +V_S$$
$$V_- > V_+ \Rightarrow V_{out} = -V_S$$

An op-amp in the comparator configuration simply compares $V+$ and $V-$ to find which one is larger. Its output goes to (or near) either the positive or negative supply voltage. It only gives information about which input is larger but doesn't tell any more about the specific input voltages or their difference—only whether that difference is positive or negative.

Application: Battery Tester

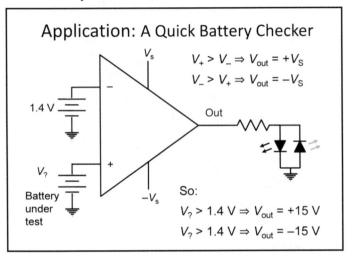

Application: A Quick Battery Checker

$V_+ > V_- \Rightarrow V_{out} = +V_S$

$V_- > V_+ \Rightarrow V_{out} = -V_S$

1.4 V

$V_?$

Battery under test

Out

V_s

$-V_s$

So:

$V_? > 1.4\ V \Rightarrow V_{out} = +15\ V$

$V_? > 1.4\ V \Rightarrow V_{out} = -15\ V$

One application of a comparator is the simple battery checker shown here, which compares the voltage of the battery test against a standard voltage, here chosen as 1.4 V.

Suggested Reading

Introductory

Brindley, *Starting Electronics*, 4th ed., chapter 9 through p. 146.

Lowe, *Electronics All-in-One for Dummies*, book III, chapter 1; chapter 3 through p. 382; pp. 387–390.

Shamieh and McComb, *Electronics for Dummies*, 2nd ed., chapter 7, pp. 157–158.

Advanced

Horowitz and Hill, *The Art of Electronics*, 3rd ed., chapter 4, section 4.1, stressing the introduction to op-amps.

Scherz and Monk, *Practical Electronics for Inventors*, 3rd ed., chapter 8, section 8.3.

Projects

Differential Amplifier Simulation

Simulate the two-transistor differential amplifier shown in this lecture. Use ±15-V supplies and 56 kΩ for all three resistors. Put a voltmeter between the two output terminals and check to have it display its voltage. Find the differential voltage gain:

- Attach voltage sources to both inputs (transistor bases)
- Set both initially to 0
- Run DC simulation and record output voltage shown on meter
- Raise one input by 5 mV, drop the other by 5 mV, and record output
- Continue until you have about six measurements
- Make a graph of output versus differential input and use it to find the gain.

Low-Battery Warning Circuit

A plug-in smoke detector has a 4.5-V backup battery (three 1.5-V AA batteries in series). It has built-in ±15-V DC power supplies that normally operate its circuitry unless there's a power failure and the battery has to take over. No other power supplies are available.

- Design a low-battery warning circuit that will light an LED if the battery voltage drops below 4 V. The LED should draw 15 mA of current when it's lit. No need to build this one; just sketch a circuit.
- Hint: Consider using an op-amp and either a voltage divider or a zener diode.

1. Why can't the audio amplifier circuits of Lecture 9 be used to amplify DC or slowly varying signals?

2. The ideal amplifier should have a *high* input resistance but a *low* output resistance. Why the difference?

3. With any reasonable voltage (more than a few microvolts) across the inputs of an op-amp, the output is sure to be at one of its two limiting values. Why?

The Ideal Amplifier
Lecture 10—Transcript

Welcome back. In the previous two lectures, you got an earful of audio amplifiers, transistor amplifiers, which are designed specifically to amplify in the range of audio frequencies. Today I want to go a little bit more general, and I want to look more theoretically at what we'd really like to have in an amplifier. We're going to open up to a world of wonderful little amplifiers that can do almost anything for us in the world of analog electronics.

Before I get there, let me point out that audio isn't the only thing we want to amplify. For example, you go into your doctor and you're given an EKG. The doctor or the nurse puts electrodes all over your body. Those electrodes are measuring tiny voltages, very small millivolt voltages, produced by your heart muscle and available at your skin. Those voltages have to be amplified so they can be recorded, displayed, whatever. Maybe you have an electronic weather station at home. It's got a temperature sensor in it; it's got a barometric pressure sensor; it's got various sensors. They produce typically very low level voltages. In order to display them, to memorize the maximum and minimum temperature or whatever else your instrument does, you need to amplify those voltages. At Middlebury College we've got a new 143 kilowatt array of solar trackers. My students and I are studying them to see how efficient they are. We have a special instrument that measures the intensity of sunlight, it's called a pyranometer. It has in it a little sensor that produces just a few millivolts. In fact, in bright direct sunlight, it produces only 10 millivolts, 1/100 of a volt, and that needs to be amplified.

We need amplifiers that aren't just audio amplifiers, and so in this lecture we're going to look more at what we'd really like to have in an amplifier, and then we're going to see how remarkably close, for how remarkably little money we can come to that ideal amplifier.

Let me begin by suggesting what we'd like to have in an amplifier. What an amplifier does is to amplify; so we'd like it to have gain, and we'd probably like it to have pretty high gain. Remember, we built an audio amplifier with a gain of only 20, but we put some capacitors, got the gain up to 170; so we want kind of high gain. Either that or we want a gain that we can control.

I want an amplifier with a gain of 30; I want to be able to control it and make sure the gain is 30. I'm going to tell you something remarkable that you won't really believe fully until we've gone through the next couple of lectures: The ideal gain we'd like to have is infinite. What? Yup, infinite. We're going to temper that gain and make it whatever we want with a magic called negative feedback, but that comes in a few lectures. Take my word for it: We'd like the ideal amplifier to have infinite gain. We're not going to achieve that, but maybe we can get a gain of a million, or 10 million, or 100,000, or maybe 1,000, anyway. We want a big gain, because we can get the amplifier to give us any gain we want smaller than that. Maybe that's a little bit surprising to hear, but that's we want: infinite gain, or a very high gain.

We'd also like the amplifier to have low output resistance. What do I mean by that? An amplifier is basically like a battery in the sense that it's a source of voltage. Remember, a battery has an internal resistance. If you put too low a load resistance across the battery, you draw so much current that the battery terminal voltage drops below what it's supposed to be. We had a 6 volt battery, we put a lightbulb across it, we only got 5 volts at the terminals. We don't want that to happen with an amplifier. An amplifier—like a battery, like any other source of voltage—has an internal resistance. In the case of an amplifier, it's called the output resistance because it's the resistance that you would see looking into the amplifier from its output. If the output resistance is low, then the amplifier can deliver a lot of current; it means it can deliver a lot of power. Or, and I say this carefully, it can not only source current, deliver current coming out, it can sync current coming in because typically our amplifiers output voltages swing positive and negative. So we think of them as sourcing current, pushing current out, or syncing current, pulling current in. If it's got a low output resistance, it can sync or source lots of current. Ideally, that output resistance should be, of all things, 0. It isn't, but the lower it is, the better the amplifier, meaning the more current and, therefore, the more power the amplifier can deliver.

We'd like something else: We'd like high input resistance, because an amplifier is a little bit like a voltmeter. It senses an input voltage and then does something to it; in the case of an amplifier, amplifies it. We'd like to make sure that the voltage at the input of the amplifier when the amplifier is connected is, in fact, the voltage you want to do something to. If the amplifier has too low an input resistance, too low a resistance between its input and ground, it'll draw lots of current from the source whose voltage you're trying to measure, and that may alter the voltage you get from that source. We want high gain, ideally infinite; low output resistance, ideally 0; and high input resistance to draw minimum current from the external source that we're connecting to the input; and ideally that should be infinite.

If I tried to conceptualize an amplifier; here's kind of a conceptual view of an amplifier: That big triangle represents not a transistor or other single component, it represents a whole circuit that's been designed to be an amplifier. This is a conceptual view; this isn't really what's inside an amplifier. On the left we see the input to the amplifier. The input isn't really a resistor connected from the input to ground, there's all kinds of circuitry, but it presents an effective resistance at the input, and that's the input resistance. Then we have a controlled voltage source. I'm going to use the letter "A" to stand for the gain. Whatever the input voltage is, this controlled voltage source makes the voltage equal to A times V-in. That's the gain, the amplification. I show these completely disconnected; that can't be. What I'm showing here conceptually stands for a whole bunch of circuitry that involves a lot of transistors, mostly, and other components, like the amplifiers we designed in the last lecture that took an input voltage and produced a bigger output voltage and had some gain factor. But conceptually, we look at this way. Then between the controlled voltage source and the output itself is that output resistance.

There's a conceptual view of the insides of an amplifier. Again, not what's really inside, but conceptually what's inside. In comes the input, the input controls the output, the output produces a voltage that's A times the input voltage where A is that gain, and sends out current if you ask for current, but it has to go through that output resistance.

What would the ideal amplifier look like? I suggested already what the ideal amplifier looks like: The resistance at the input is infinite. It's as if the input wasn't connected to anything. That can't be, but we can get pretty close. Then we have this controlled voltage source with a gain of A, and A ought to be infinite, and then we ought to have 0 output resistance. That's conceptually what an ideal amplifier ought to look like, and a real amplifier approximates that goal of an ideal one to the extent that the input resistance is very high, the output resistance is very low, and the gain is very large; much, much, much larger than 1.

There's something else we want in an amplifier. We'd like an amplifier to have high gain, low output resistance, high input resistance, and typically we'd want it to have a wide frequency response and a flat frequency response. What I mean by that is for the range of frequencies we're interested in amplifying, we'd like it not to amplify one frequency more than another. If it did, suppose it amplified the range of 400–500 Hertz, more than other ranges, and every time we played A above middle C at 440 Hertz that note would sound louder in the output of the amplifier than it was supposed to, than it did in the music that we were inputting. That's why we want a flat frequency response. I've shown a curve here that might be the typical frequency response for an audio amplifier. Our ears can hear from about 20 Hertz, 20 cycles per second, up to—if you're not too elderly—about 20,000 cycles per second; you lose high frequencies as you age. But 20 Hertz to 20 kilohertz is about the audible frequency range. If you're a bat, you've got to hear higher; you need a different audio amplifier. But an audio amplifier ought to be flat over that frequency range. The amplifiers we've been talking about with those capacitors in them and stuff are designed to cover that range fairly well.

The amplifiers we're going to be dealing with in the rest of the analog part of this course are amplifiers we might use in instruments, like measuring the output of my sunlight measuring instrument, or in the amplifiers in an EKG machine, or something like that. They typically need to go from DC, they need to be able to take a millivolt input at DC, and if the amplifier has a gain of 1,000, they need to be able to produce from that millivolt, that thousandths of a volt, an output of 1 volt. We'd like the ideal amplifier to go from DC to frequencies that are very high. How high? It depends on the instrument. That oscilloscope I've been using in this course, it has lots of amplifiers in it, and those amplifiers can take frequencies from DC—they can amplify a steady level; you saw that when I connected a 6 volt battery to it—and they can actually go up to the gigahertz range, billions of cycles per second in that oscilloscope. That's one of the things that make it a very high-end oscilloscope.

We'd like a wide frequency response; let's look at the issues that come up if we want a wide frequency response. Here's our two stage common emitter pre-amplifier coupled with the input coming in through a capacitor and the stage one output going to the stage two input through a capacitor. This one won't cut it. It's no good for DC. Capacitors are open circuits to DC; they only respond to variations in voltage. If we want an amplifier that goes all the way down to direct current, this one isn't going to do it, and we've got to do something better to make a DC amplifier. When you hear someone say "DC amplifier," it doesn't just amplify DC, but it will amplify DC as well as signals that are varying at a range of other frequencies. You plug a battery, a 1.5 volt battery or something, into your audio amplifier—don't do it because you might damage it—but plug a small millivolt level signal into your audio amplifier and it's not going to come out the other end, because there are capacitors in there, and the capacitors block the DC. The amplifiers we've seen so far are just not going to cut it as DC amplifiers.

Here's what we're going to do instead: We're going to build a circuit called a difference amplifier, and this difference amplifier is going to have two inputs. I've drawn one side of the difference amplifier here. It's got an NPN transistor; it's got a resistor from the power supply voltage to the collector; it's got another resistor in the emitter lead. Then I'm going to add to this an identical piece. Symmetrically, exactly the same: Same transistor, same resistor, and a common resistor in the two emitters. I'm not going to go through the complicated mathematics of this, but I'm going to try to motivate you to understand what this circuit does. By the way, you might have to have a little variable resistance in there to compensate for the slight differences in the transistors, but I'm going to assume those two resistors are exactly identical and the transistors are exactly identical. Let's ask what happens.

We have several inputs here: We have an input one on the left and an input two on the right. We're going to take the output as the voltage between the two collectors, and let's figure out what happens. Suppose V1 and V2 are equal. I don't care what they are—within limits; they better not exceed the power supply voltage or the negative supply voltage—but suppose they're equal. If they're equal, then equal currents flow through those two collector resistors; the voltage drops across those two collector resistors are the same; so no matter what those two input voltages are, as long as they're equal the collector voltages of the two transistors are the same. Therefore, V-out on the left (V-out minus I'll call it) and V-out on the right (V-out plus I'll call it) are, in fact, the same. Therefore, the output voltage—which I'm going to define as the voltage between those two collectors, taking it to be positive if it's more positive on the right, and negative if it's more negative on the right, more positive on the left—will be 0. This is, in some sense, finding the fact that the two input voltages are the same, and it's giving us a 0 output if they're the same. It's a difference amplifier.

What if V1 is greater than V2, and therefore V2 is less than V1? If V1 is greater than V2, we're going to get a bigger base current coming in on the left, and that means we're going to get a bigger collector current on the left than we have on the right. Since V, the drop across the resistor, is I times the resistance, we're going to have a bigger drop across the left-hand resistor. Therefore, the voltage at the minus out, the voltage at the collector of the left-hand transistor, is going to be lower because there's a bigger current flowing. Therefore, there's more of a voltage drop across that left-hand resistor and the voltage at the bottom end of it, at the collector of the transistor, is lower than on the right-hand transistor. Therefore, V-out minus is less than V-out plus, and V-out isn't equal to 0. In fact, it's bigger than 0, because the voltage on the right is bigger than the voltage on the left. You can easily convince yourself that the output voltage is, in fact, proportional to V1 minus V2. That's the output voltage, V1 minus V2. In this case, we have V2 less than V1, and we get an output voltage, which by our definition is positive.

This is a difference amplifier. The output taken across those two terminals is, in principle, proportional to—and the resistors determine the proportionality—the difference of those two voltages. It doesn't care what their actual values are. I can have 2 volts at one, and 1 volt at two, and I'll get the same output as if I have 3 volts at one and 2 volts at two, because all it depends on is the difference. That's ideal; ideally, that's what happens. It's more complicated than that, but in principle a difference amplifier just gives you an output proportional to the difference between its two inputs.

If you wanted to make a DC amplifier, you could do it like this: You could ground one of the inputs—that's silly, aren't we wasting that whole transistor—and then just go in through the other one, and the other input will then tell us V-out being proportional to the difference between V-in and 0, between V-in and ground, and we'll get an output proportional to the input. That will be true even if there's DC because there's no capacitors involved in coupling this. You might say, "That's sort of a waste of that right-hand transistor," but it isn't. What happens is several things can happen: If the circuit heats up, the transistor characteristics change a little bit. But if we have a difference amplifier, both transistors change the same way, and to a first approximation the output isn't affected. This is actually a nice circuit for precision amplifiers that measure DC voltages carefully.

That's great; we know how to make this difference amplifier. Now we're going to take this difference amplifier and use it as the input stage of what's called an operational amplifier, or op-amp. The next few lectures are going to be on op-amps, and they're wonderful devices. I have to tell you, I'm very enthusiastic about op-amps, so I hope you'll share some of my enthusiasm.

The concept of an op-amp is a little compact amplifier, differential amplifier or difference amplifier. It's got two inputs, it's got an output, it's got a high gain, it's a low output resistance, it's got a high input resistance, it's got a broad frequency response, and it can do DC. All those things in a little tiny package, cheap. We're going to see how we build that. That's the op-amp concept. It's called operational amplifier because they were originally designed to do mathematical operations, and we'll see how some of that works. That function is fairly much obsoleted; we used to actually build analog computers out of them. But they're still useful for a great many applications in analog electronics.

Let me begin by talking about some operational amplifiers. Here's a very common operational amplifier; it's actually one that dates back to the 1970s. It's a single integrated circuit—I'll show you one in a minute—built on a single chip. It's got a dozen or so transistors, and a few resistors, and some other things in it, and we're not going to go through the details of its circuit. But I want to point out roughly what it has. At the left you see, circled, a part of the circuit containing a number of transistors, one, two, three, four five, six, seven. The upper two transistors are the important ones. You can recognize that those are essentially in this difference amplifier configuration, so we have an inverting input and a non-inverting input; a plus input and a minus input as they're called. You'll see often in integrated circuits that there are transistors used in place of resistors. It's cheap and easy to make transistors on integrated circuits. It's easy to get more sophisticated control than with a resistor. That's why you see a lot more transistors, and I'm not going to go through what all of them do. But that left-hand block is basically the input stage. Then, down at the lower right, there's a stage that provides the voltage gain, the voltage amplification. Then on the right, you should recognize a complementary symmetry output stage with a PNP transistor and an NPN transistor.

The 741 operational amplifier—again, one that's been around since the 1970s or so—does it all for us, and I'll actually be using 741s because they're very easy to work with. If you work in CircuitLab, for example, you'll find that their default operational amplifier is something called the TL081, and its circuit looks a little different but quite similar. It actually has junction field effect transistors at the input to make its input resistance even higher. We work with that one also.

Whichever circuit we have, we're going to ignore individual transistors from now on. You'll be glad to know we're not going to deal with all the details. We're going to shrink that whole thing down into a triangular structure, and the triangle represents all that circuitry that's inside. There's what an operational amplifier looks like, sort of symbolically. Let me just tell you a little bit more about the op-amp before I show you a few of them.

Meet the op-amp: The op-amp has two power supply connections, one to a positive power supply, typically plus 15 volts; one to a negative power supply, typically minus 15 volts, but it could be something different. It has a non-inverting input called the plus input; it has an inverting input called the minus input; and it has an output. Here's what it does: Its output is its gain times V plus minus V minus. It's a difference amplifier. It takes the difference between V plus and V minus, and it amplifies it by some factor, A, which better be big. That's what an operational amplifier does.

The ideal op-amp has a very big A, and V-out is AV plus minus V minus. Here's sort of an anatomy of that equation, because we're going to use it, not a ton of times actually, a few times so we get the hang of op-amps and then we won't need it anymore. V-out, the voltage at the output, is the difference between the voltage at the non-inverting input, the voltage at the inverting input, multiplied by the amplifier gain. That's our picture of the ideal op-amp. It's just this triangular symbol. I should warn you that I'll often draw that symbol, and other people who do circuits will often draw that symbol, without the power supply connections shown. They're implied. In your circuit simulators, you can get op-amps that either have explicit power supplies where you have to connect the voltage source, or they're assumed. In the real world, they can't be assumed; you really have to connect power to an op-amp. One of the first problems my students have when they work with op-amps, sometimes they forget to connect the power because the circuit doesn't show it. But it's got to be there. So often, we won't show those two power connections coming off the top and the bottom; we'll just show the plus and minus input and the output. Watch out, sometimes we'll put the minus input on the top. I'll usually but not always do that. Sometimes the minus input will be on the bottom, and some of your circuit simulators will come up with the minus input on the bottom and some on the top, and you just have to watch out for it. There's the ideal op-amp.

I have to say something more about this: This output will be equal to A times the difference of the inputs only if that value is less than about the supply voltage. You can't possibly get more than 15 volts positive out of an amplifier with a positive 15 volt supply, and you can't get more than 15 volts negative out of an amplifier with a negative 15 volt supply. For some amplifiers, you can actually swing all the way up to those limits. For others, like the 741, you can probably get up to a maximum of about 12 volts out and minus 12 volts, so somewhere in between there. Beyond that, the amplifier is said to be saturated and it's sitting at its maximum possible voltage, which is either at or close to the power supply voltage.

Let's look at some op-amps that are out there in the real world. First, I'll give you a little sense of the specifications available, and then we're going to actually look at some real op amps. I've got a list here of five op-amps, some of which you'll be using if you're doing the project, some of which you may not use. The TL081, which I mentioned is CircuitLab's default—that's in the comments column there—has a typical gain of around 200,000. Remember I said we wanted a big gain, a lot bigger than 1; well, 200,000 is a lot bigger than 1. We'd like a big input resistance. The input resistance for this amplifier, because it's got field effect transistors at the input, is about 10 to the 12 ohms, a tera-ohm. That's huge; it can supply a maximum of about 40 milliamps at the output. That's another way of telling you that it has a modest output resistance. It's got a bandwidth of about 3 megahertz, which means it can amplify from DC up to about 3 million cycles per second, and it costs about $0.22 if you buy them in quantity.

Dew circuits, if you're using dew circuits for your simulations, have a sort of generic op-amp, although you can specify a particular op-amp and it'll give the real specs for that. But if you just bring up their generic op-amp, it's got a gain of a million, and that's all we're told about it.

A very common op-amp, and the default in the book *The Art of Electronics* by Horowitz and Hill, which is on the reading list for this course, is a 411 op-amp. It has a gain of typically 200,000. They just specify the minimum. They can go up higher than that because an infinite gain would be fine. So 200,000 might be 2 million; it turns out not to matter a bit. Again, an input of about a tera-ohm, 10^{12} ohms; it can sync about a source or sync about 20 milliamps and goes up to 4 megahertz and costs about $0.70. These costs are pretty variable. This old workhorse, getting obsolete but still great, the 741; a gain, again, of 200,000 or more. Typically, the input resistance is 2 megaohms. It's a lot lower because it has bipolar junction transistors. It can source or sync 25 milliamps and goes up to about a megahertz, and they cost a few tens of cents also. Finally, one other one I want to show you because I'll be using it in some circuits is a 165. It has a gain of only about 10,000 and an input resistance pretty low of only about half a megaohm, 500,000 ohms, but it can source or sync two amps. It's a power op-amp, costs about $1. We'll be using those in some neat circuits I'll be showing you in a couple of lectures. Those are some op-amp specifications.

Let's look at a couple of op-amps, or actually a few more: I have some op-amps over here. There's a 741 down there; it just looks like a little square thing with eight pins. They all look pretty much the same; there's not much to see here. But that happens to be a 741 op-amp, and it's actually connected in a circuit that I'm going to demonstrate in a minute. Up here is a 747. It's really just two 741 op-amps on the same package, on the same integrated circuit. That's a 747. Up here is the 411 that I just mentioned, which is almost identical to the 741. It's got a little broader frequency range, but it has a much higher input resistance, and there are cases where we'll want to use that. Finally, here's the 759 power op-amp. It's even got a heat sync, and we've attached a wire to that heat sync. It's one of the contacts for the op-amp; it has that heat sync because it can get pretty hot because it can dissipate several watts of power. Those are what op-amps look like. They're not very exciting. It's much more exciting to see that triangle with the plus input and the minus input and the output. If you're into electronics, that really excites your imagination.

I've mentioned that these op-amps ought to have infinite gain, and I want to say a little bit more about that. What possible good could that be? In the next few lectures, we'll see, as I said, how to use the magic of negative feedback to tame that gain, to make the op-amp have any gain you want, or indeed to do various other things. But you can even do something with the infinite gain, and here's why: Because any of these op-amps—even the ones with the smallest gains, even the smallest input voltage and the smallest voltages we can get down to very realistically, without them being swamped by noise—are in the millivolt range. If you have a gain of 200,000 and you put in a millivolt, a thousandths of a volt, times 200,000 it's going to try to give you 200 volts out, and you just can't do that because you've got plus minus 15 volt power supplies. What happens in practice for real op-amps: If you take a real op-amp, connect the power to it, and don't connect anything to the inputs, or connect something to the inputs but don't do this negative feedback magic I guarantee you the output will either be at plus limit (typically around plus 15 volts; maybe plus 12 or something like that) or minus limit; it'll be saturated either way. That in itself can actually be useful. There's no way you can get the input voltage between those two so small—unless you use the magic of negative feedback—to get down within the range between the maximum and minimum voltages. An op-amp left to its own devices goes to one of those two limits. By the way, I have to caution you, the op-amps in your circuit simulators are a little more perfect than real op-amps, and sometimes they will converge on 0, which is something that a real op-amp won't do.

I have one example of this to show you. This is the op-amp as what's called a comparator. It simply compares the two voltages and asks which one is bigger, and it only gives you information about that. Imagine I have an op-amp with a gain of a million. The output is always going to be near plus the supply voltage or minus the supply voltage. If V plus is bigger, it's going to try to go, let's see, we've got a gain of a million. Let's say, whatever's at the input, we're going to try to get plus a lot of voltage if V plus is bigger, and we're going to try to get minus a lot of voltage if the minus is bigger. We're going to either be at plus 15 if V plus is bigger or minus 15 if V minus is bigger. That's why it's called a comparator. The only thing the output knows about is which of the inputs is bigger. It doesn't tell you any more information than that.

Why would you want that? Well, here's an example: Maybe you want to test batteries quickly. Here I have a circuit for an op-amp; it could be a 741, could be another one. I've got it wired up. This time I'm showing the power supply connections, which I won't always show. I'm going to connect a volt—and I'm making this to test 1.5 volt batteries, which is a common battery voltage—so I'm going to connect 1.4 volts to the minus input, and I'm going to connect the battery I'm trying to test to the plus input. At the output I've got a resistor and a couple LEDs and they're back to back. If the output is positive, the green LED will be forward bias, current will flow through it to ground. If the output is negative, the red LED will be lit and current will flow from ground up through it and into the amplifier. If V plus is greater than V minus, V-out is positive, limit on 15 volts. If V plus is less than V minus, V-out is negative and will light the red LED. If V minus is greater than V plus, it will light the other LED under the other condition. In this case, with 1.4 volts at V minus, if V plus, the battery voltage, is greater than 1.4 volts, we'll get plus 15 and we'll light the green LED; and if it's the opposite we'll get minus 15 and we'll light the red LED.

Let's see how the thing works. Here I have the circuit. I've got my function generator, which also can produce DC voltages. It's set to produce 1.4 volts, and you can see the 1.4 volts on that meter. Here's the circuit. There's the

741 op-amp. There are the two LEDs. If I were more sophisticated I'd have it be off until I connected something, but, as I said, these op-amps go to one limit or the other and right now it just happens to be going to the negative situation. Here, I've got a little piece of wire that's connected to ground— that's the ground for my battery under test—and here's a battery I'm going to test. I'm going to put the ground in contact with the ground. Here's the plus input of the op-amp, and I'm going to connect that to the battery. It's green, and that battery is good. That's all I know about it: It's got a voltage of more than 1.4 volts. Here's another battery. Connect it, and the red stays lit. That one's got a voltage of less than 1.4 volts; that's all I know. Finally, here's a third battery. Looks like that one is also good. Three batteries tested with an op-amp in so-called open loop configuration, no feedback. All it can do is go to one output or the other, but that's useful in giving us a quick check on the battery. Let's just see if it did its job correctly. Here we have the battery that tested bad, and lo and behold it measures in at 1.34 volts. That's below the 1.4 volts that we set kind of arbitrarily as our standard for dividing between a good and bad battery.

There we've seen an application of an op-amp with its almost infinite gain, producing either plus about 15 volts at the output or minus and only able to distinguish the inputs and outputs, which one is bigger than the other. It's a comparator. It's actually a useful circuit, and now we want to move on and see how we contain that infinite gain by making circuits that do whatever we want them to do.

But let me just wrap up by saying we've developed this concept of the operational amplifier; a nearly ideal amplifier with huge gain, low output resistance, high input resistance, cheap, available in a small, easy to use complete package with all its transistors in it. A wonderful device that we're going to have a lot more to do with.

That's all for this lecture, but if you want to move on to the project you can build yourself a difference amplifier. That's the last circuit you'll be building with individual discrete transistors, and then you can move on to work with an op-amp as a comparator. From now on, our analog circuits will all use op-amps rather than individual transistors.

Feedback Magic
Lecture 11

T he term *feedback* refers to a change in a system that follows as a result of some earlier change. There are two kinds of feedback: negative, in which the additional change has the effect of diminishing the original change, and positive, in which the additional change enhances the original change. Negative feedback is stabilizing; if something goes one way, the feedback brings it back the other way. Positive feedback is destabilizing; it can lead to runaway effects. In this lecture, we'll look at some intriguing demonstrations involving negative feedback. Key ideas include the following:

- The feedback concept
- Electromechanical feedback:
  a servomechanism

- Optical feedback:
  the intelligent light bulb
- Thermal feedback:
  a constant-temperature system.

The Feedback Concept

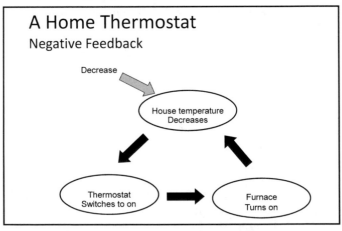

A Home Thermostat
Negative Feedback

Decrease

House temperature
Decreases

Thermostat
Switches to on

Furnace
Turns on

Your home thermostat is a negative feedback system based on a temperature-sensitive switch. If your house temperature decreases below the level you've set, the switch turns on. That turns on the furnace, which produces heat, and the house temperature increases. If it gets too hot, the switch turns off, turning off the furnace and dropping the house temperature.

Electromechanical Feedback: A Servomechanism

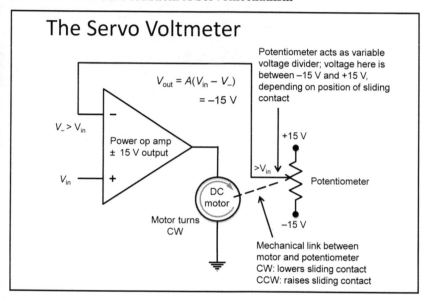

The Servo Voltmeter

$V_{out} = A(V_{in} - V_-)$

$= -15$ V

Potentiometer acts as variable voltage divider; voltage here is between −15 V and +15 V, depending on position of sliding contact

$V_- > V_{in}$

Power op amp ± 15 V output

+15 V

$>V_{in}$

V_{in}

+

Potentiometer

DC motor

Motor turns CW

−15 V

Mechanical link between motor and potentiometer CW: lowers sliding contact CCW: raises sliding contact

The demonstration servo voltmeter consists of a single op-amp with its inverting input connected to a potentiometer—a variable voltage divider. At the output, there's a DC electric motor that runs on ±15 V, turning one way if the voltage is positive and the other way if it's negative. The feedback loop consists of a mechanical connection between the shaft of the potentiometer and the motor shaft.

A variable input voltage is supplied to the noninverting input of the op-amp. If this input voltage is greater than the voltage at the inverting input, the motor turns in such a way that it moves the movable tap on the potentiometer toward higher voltages, thus increasing the voltage at the inverting input. If the voltage at the inverting input exceeds the voltage at the noninverting input, then the motor goes the other way, lowering the voltage at the noninverting input. Thus, the position of the potentiometer's movable contact corresponds to the voltage at the noninverting input. An indicating needle is connected to the motor-potentiometer shaft, indicating the input voltage. The mechanical feedback causes the voltage at the inverting input to "follow" the voltage applied at the noninverting input. The system is called a *servomechanism* because the position of the indicating needle slavishly follows the position of your hand as you turn the knob on the input voltage. Similar systems are used for accurate control of heavy machinery or in prosthetic limbs.

Optical Feedback: The Intelligent Light Bulb

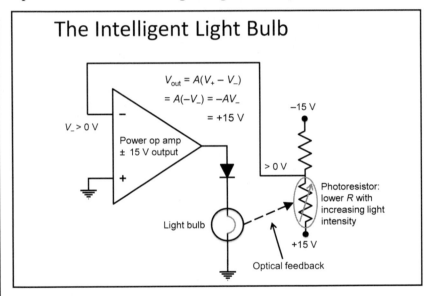

The Intelligent Light Bulb

$$V_{out} = A(V_+ - V_-)$$
$$= A(-V_-) = -AV_-$$
$$= +15 \text{ V}$$

$V_- > 0$ V

Power op amp
± 15 V output

−15 V

> 0 V

Photoresistor:
lower R with
increasing light
intensity

Light bulb

+15 V

Optical feedback

We can morph the servo voltmeter into another circuit with the same power op-amp, but instead of the potentiometer, this circuit has a resistor and a photoresistor in series across the ±15-V power supplies. The photoresistor's resistance depends on the intensity of light falling on it, decreasing with increasing intensity. Optical feedback occurs because the light bulb at the op-amp output shines on the photoresistor. The system is analogous to the mechanical-feedback servo voltmeter, except here, the feedback acts to keep the light intensity at the photoresistor constant—even if it's moved closer to or further from the light bulb. This circuit might be used as a sensor in an energy-efficient office building to sense and adjust levels of ambient lighting.

Another optical feedback system is "funny face," a circuit with two photoresistors in a "face" mounted on the shaft of a motor connected to the output of an op-amp. Feedback causes the face to turn so that it follows a bright light.

Thermal Feedback: A Constant-Temperature System

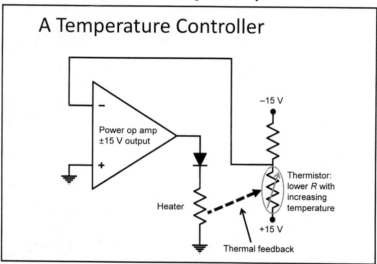

A Temperature Controller

Power op amp ±15 V output

−15 V

+15 V

Heater

Thermistor: lower R with increasing temperature

Thermal feedback

If we replace the photoresistor with a *thermistor*—a thermal resistor whose resistance drops with increasing temperature—then we create a temperature-control feedback system.

Suggested Reading

Introductory
None of the introductory books covers feedback at the level of this lecture.

Advanced
Horowitz and Hill, *The Art of Electronics*, 3rd ed., chapter 4, section 4.1, stressing feedback.

Reverse Engineering

Design a plausible circuit for "funny face." Hint: You need only the following components:

- Power op-amp
- Two photoresistors
- DC motor, reversible with polarity change
- Power supply for op-amp (assumed).

Questions to Consider

1. Give several examples of negative feedback, not necessarily electronic.

*2. Identify the nature of the feedback loop in the following devices that were discussed in this lecture: servo voltmeter, intelligent light bulb, and temperature controller. Is the feedback electronic, thermal, optical, or mechanical?

Feedback Magic
Lecture 11—Transcript

Welcome to the first of two lectures on what I like to call the magic of feedback.

You're familiar with the term feedback: "Give me some feedback." "How am I doing?" What's feedback? Feedback in electronics and in lots of other areas is a change in a system that follows as a result of some earlier change; so something happens, and as a result of that thing that happens, something else happens. There are two kinds of feedback: There's negative feedback, in which the additional change has the effect of diminishing the original change. Negative feedback is stabilizing. If something goes one way, the feedback brings it back the other way. Then there's positive feedback. In positive feedback, the additional change that happened as a result of the first change enhances the original change. Positive feedback is destabilizing. It can lead to runaway effects. I want to give you some examples of feedback from realms other than electronics before we get into talking about electronic feedback. Then we'll look at some really intriguing demonstrations involving negative feedback in electronic circuits.

Let's look at, for example, grade feedback. This is one you're familiar with if you've had kids in school, or been through school yourself, or like me, are a teacher. What happens? Somebody gets a low grade, and that affects their grade in the course they're taking. What does that do? As we go around this loop here sort of counterclockwise, that affects their motivation. What do they do? They study harder, and so the next test gets a better grade and their class grade goes up. Here's a feedback effect in which the low grade caused a secondary effect—motivation, studying more, and getting a better test—and then the class grade goes back up. Negative feedback: The negative feedback, the secondary change that followed the first change, mitigates to some extent the first change; it can't reverse it, it turns out, but it mitigates it. That's a common example of feedback: feedback associated with grades.

Let me give you another feedback that we worry about a lot in this age of climate change. This is called ice-albedo feedback. You hear a lot about this in connection with the melting of ice in the Arctic. Here I have in the top box

Earth's temperature. Suppose for some reason—we're not going to specify what—Earth's temperature goes up a little bit. If Earth's temperature goes up a little bit, then going around this loop counterclockwise, the ice cover decreases because some ice melts. When the ice melts, especially ice that's floating on the oceans, the ice itself is quite reflective of sunlight but the ocean surface is quite dark and absorbs sunlight. What happens when the ice melts is the reflectivity—it's also called the albedo—decreases, and as a result there's a greater increase in the amount of solar energy absorbed and that, of course, raises Earth's temperature. This is a destabilizing positive feedback; and it's this destabilizing runaway nature of this kind of feedback that gets us particularly worried about them.

There's a feedback in sort of the psychology of students, a feedback in the climate system; let's look at a few more feedbacks. Here's a feedback in your house: You have a thermostat in your house. I like to think of a thermostat as one of the simplest feedback devices, and also one of the simplest devices that, if I can use this word, exhibits a little tiny modicum of intelligence. What happens with a home thermostat? Suppose for some reason the temperature in your house decreases; somebody opens a window, or it simply gets colder outside. What happens? The thermostat, which has a temperature sensor in it—and I'll show you one in just a moment—switches on. That turns on the furnace, and the furnace produces heat, and the house temperature increases. That's a negative feedback. If the house temperature starts to go down, the thermostat kicks in, and produces this negative feedback loop that warms things up again.

Let me show you an example of a household thermostat. I have one here. This is an old fashioned household thermostat. You may still have this kind in your house; but if you do—at least if it operates the way this one does— you're not supposed to anymore because what this thermostat has inside it is a small metal coil, and that metal coil is made of two different metals that expand differently with temperatures, so the coil tightens or untightens depending on whether the temperature is going up or down. Resting on that coil is a small glass tube containing two metal pieces that aren't connected to each other and a drop of liquid mercury, which is why these thermostats are now no longer supposed to be used. In fact, I know my recycling center will give me $5 if I turn in one of these; so we're trying to get the mercury out of circulation. Nevertheless, as the spring winds or unwinds, or I set the temperature to different values, that drop of mercury can go the other way as the tube tilts, and it closes a switch, and that's what turns on the furnace and causes the system to heat up. The thermostat is wired in such a way that if the house temperature drops, the coil turns in such a way that that switch, switched by that little drop of mercury, turns on and turns on the furnace. Then, of course, as it gets too hot, the furnace turns off again. The thermostat senses now it's too hot, and the furnace turns the thermostat off. There's another example.

Let me give you just a couple of other amusing examples. Here at the upper left is a picture from my garden in Vermont. In Vermont we don't have a very long growing season, although it's gotten a lot longer in the last few decades because of global warming. But still, I want to grow lettuce in March, so I have this cold frame I built. Inside this cold frame, you can see some lettuce growing, but you can also see that arrow. The arrow is pointing to a cylinder that's connected to the lid of the cold frame, and the cylinder has a fluid in it that expands when it warms up. If it gets too hot inside that cold frame, which you don't want to have happen—lettuce doesn't like it too hot—that cylinder expands and it opens the lid of the cold frame, and the cold frame cools down. If it gets too cold in the cold frame, that fluid contracts and the cylinder pulls the lid back down again. There's another example of a negative feedback system increasing the growing season. Another negative feedback system that you use when you're driving is the cruise control in your car. The cruise control senses how fast the car is going. You've set it to go 60 miles an hour; if it senses it's going 61, it backs off on the throttle, drops down to 59, turns the throttle back on, and so forth and so on.

Negative feedback is with us. It's common throughout much of the technological world, and it's also common in the natural world.

You get the idea of how feedback works. Let's back off now and look a little more philosophically or a little more abstractly at feedback. What do you need for a feedback system? First you need a sensor that senses the quantity that we're dealing with. In the grade feedback that sensor is the brain. In the ice-albedo feedback that sensor is the ice undergoing melting. In the thermostat that sensor is the temperature-dependent switch. Then you need a mechanism for acting back on the system to either mitigate or enhance the change, depending on whether it's negative or positive feedback. In the grade feedback that mechanism is motivation. In the ice-albedo feedback that mechanism is the absorption of solar energy. In the thermostat that mechanism is the production of heat by the furnace.

I'd argue that operational amplifiers, in fact, make wonderful facilitators for feedback systems. In the rest of this lecture, I'm going to show you three examples where we use operational amplifiers to facilitate a feedback system. They're not entirely electronic feedback; they use the feedback loops in these things. It's called a feedback loop, that thing that takes you from the effect back to the cause and mitigates or enhances it. The feedback loops in these are definitely not electronic, and in the next lecture we'll get to purely electronic feedback systems. I'm going to begin with some demonstrations of some of these. They're rather crazy, funky demonstrations; let's move over to our demo table and take a look at the first one.

Here I have a crazy-looking thing. It's a voltmeter. It says "volts," "less volts," "more volts"; less volts to the left, more volts to the right. What does this thing consist of? It consists of a single operational amplifier, a power op-amp; this one's the 759 that I showed you last time that can handle a lot of power, it can put out a lot of current. The op-amp is connected, and I'm going to show you the connections in a little bit, to this potentiometer. I introduced the potentiometer in connection with the volume control in an audio amplifier, but it's the same thing. It's that variable voltage divider. At the back—maybe if I turn you can see it a little better—there's an electric motor. It's a simple DC electric motor. It runs on 12 volts, plus or minus 12 volts; and if you put plus 12 volts it turns one way, and if you put minus 12 volts it turns the other way. It can go in either direction. The feedback loop consists of a connection between the shaft of the potentiometer that sets how much voltage you're dividing and the output shaft of the motor. Those two are connected, and that's the feedback.

Now I'm going to demonstrate the operation of this thing, and then we'll talk about how it works. I'm going to turn it on, and you can see the needle starts to jiggle a little bit. I have an input voltage to this system—the system is a voltmeter—and I'm getting the input voltage by using another potentiometer, the one that's built into my board here. I've got plus 15 volts at one end of it and minus 15 volts at the other, so if I turn the knob I can dial any voltage between minus 15 volts and plus 15 volts, and that's what's going into the input of my voltmeter over there. Let's turn the voltage down a little bit and the needle follows; it goes down. Down it goes, down it goes. You'll notice the needle's wiggling a lot; we'll talk about that. It's wiggling back and forth; not a lot, but a little bit. Turn the voltage up, up goes the needle. Turn it up a lot, the needle goes up until it finds the right position to correspond to that input voltage. If I wanted to, I could've calibrated this thing and I could've made a nice voltmeter that was accurately calibrated.

You might think about this and say, "I could do more with this than just make it be a voltmeter." I could, for example, make that motor a really hefty one, and this could be not just turning an indicator needle but it could turn some heavy piece of machinery and follow exactly what I do over here with this little knob. My puny hands turning this knob are making that big needle go back and forth, and that needle, instead of a needle it could be the bucket of a huge power shovel; or as you'll see in a minute, it could be an artificial limb moving. It could be all kinds of things. This is called a servo mechanism, "servo" meaning "slave." This system is a slave to this simple little knob here. The way the slaving works is by the magic of negative feedback, and the feedback is the connection between that potentiometer and that meter.

Let's look at the circuit behind that servo voltmeter I just demonstrated. We start with a power op-amp; that 759 op-amp can deliver currents up to hundreds of milliamps at its output. It delivers plus or minus 15 volts roughly, maybe more like 12, but it delivers the kind of voltage that motor wants. Remember, the output of the op-amp when there's no feedback—and there isn't any feedback yet—is always at one limit or the other. If it's a one limit the motor goes clockwise, if it's at the other limit the motor's going to go counterclockwise, the motor depending on the polarity of the voltage it gets. The bottom end of the motor is at 0 volts, the top is at either plus or minus 15 volts, and depending on which it is it goes one way or the other. That's sort of the business end of the amplifier.

We have this potentiometer. One end of it's connected to 15 volts; the other end's connected to minus 15 volts. That potentiometer is acting like a variable voltage divider. The voltage here swings anywhere between plus 15 or minus 15 depending on where that sliding contact that's adjusted by turning the potentiometer shaft is. That potentiometer is connected back to the inverting input of the op-amp, the negative input. The input voltage that I'm trying to measure with my voltmeter—the voltage I was getting at by turning that little knob on my so-called bread board—that's connected to the plus input. How does it work?

Suppose the voltage here, tapped off by the potentiometer, happens to be greater than that input voltage; suppose that just happens to be the case. Voltage at V-in, voltage here, which is the same as the voltage at the minus input, is bigger than the voltage at the plus input. Remember how an op-amp works. V-minus is greater than V-in. An op-amp produces an output voltage, which is that huge amplification factor times V plus minus V-minus, but V-plus is my input voltage, so I'm calling it V-in. That, in this case, will be minus 15 volts because remember, an op-amp just goes to one limit or the other unless it has really good negative feedback. That's going to go to minus 15 volts. What's it going to do? That's going to turn the motor, in this case, counterclockwise. There's the mechanical link—that's the feedback—between the motor and the potentiometer. I've got it set up so counterclockwise lowers the sliding contact and clockwise raises it. In this case, the motor turns counterclockwise and the sliding contact goes down. That moves this voltage lower; it moves it more toward minus 15 volts. Even though that voltage was greater than the input voltage, now it's less. We have a bigger voltage on the positive input, and consequently we get an output voltage that's positive, and the motor turns the other way. What's going to keep happening in this relatively primitive circuit is the motor's going to be either one side—the output voltage here of the potentiometer is going to be on one side or the other—of the input voltage that we've selected. If it's on one side it will go to the other side, and as soon as it gets to the other side it will go back. That's why the needle's sitting there jiggling just a little bit. But it does a pretty good job of following that input voltage. If I track the input voltage up, it goes up. If I track the input voltage too fast, it takes a while to get to the right place, but eventually it gets there and then it starts that bobbling back and forth, just as this circuit describes.

There's an example of a very simple servo mechanism: just an amplifier, a DC motor, and a potentiometer, and that all important feedback link.

Here's a picture of a practical application. This is a thought-controlled robotic arm that physiologists and biomedical engineers are working on. The sensor actually looks at what the nerve impulses in this man's arm are doing in terms of what he wants the hand to do, and through a servo mechanism similar to this but a little more sophisticated, a robotic hand follows exactly the motions of his hand exactly the way my needle on my voltmeter followed the motions of my hand as I turned that dial. That's another example of a feedback circuit.

Let's look at one other feedback circuit. Before we do, I'm going to take this servo voltmeter and I'm going to morph it into another circuit. I'm going to do that because I'd like you to see that all these feedback circuits I'm demonstrating are very similar. It's now morphed into something that looks a little bit different. It's the same power op-amp, it's got almost the potentiometer, but instead it's got a resistor and a photoresistor—something I introduced in the very first lecture that has a resistance that varies with the amount of light on it—and I've got a lightbulb, and I've got a diode. We'll talk about all those things, but before we do, I'd like to demonstrate what that circuit looks like.

I'm going to go over here. I'm going to disconnect the servo voltmeter so we won't have it going, and I have to move the lightbulb. Here we are now set up to show you a circuit that I call the intelligent lightbulb. Why is it intelligent? Here's my photoresistor. This is the resistor whose resistance depends on the amount of light falling on it, and you'll recall that the resistance decreases with the amount of light. If you look at the circuit on the big monitor with the photoresistor down there at the bottom, if the resistance decreases, that's going to pull that middle point between the photoresistor and the other resistor toward plus 15 volts. That's going to put a big voltage into the negative terminal, the inverting terminal, of the op-amp. The positive is grounded, it's at 0 volts; so if there's a big voltage at the negative input, the output of the op-amp is going to go negative, and that's going to turn that lightbulb off. Right now, the photoresistor is seeing the lights in the studio, and what it's done is swung toward the plus 15 volts at that junction point, which is connected back to the minus input of the op-amp. There's a big voltage going in there, and consequently the op-amp voltage output is negative, and that diode is blocking any current flowing through the lightbulb.

We're going to not worry about the studio lights; instead, we're going to worry about that light. Now I'm going to point the photoresistor at that light. Right now, the situation is that we are getting some light on the photoresistor. As I move the photoresistor toward the lightbulb, you'll see that the light goes dimmer and dimmer and dimmer. Then I pull it back and it goes brighter and brighter and brighter. Why is this an intelligent lightbulb? It's intelligent because if I put my eye right next to that photosensor and I move my eye in right next to it, the light looks the same intensity to me the whole time. What's this circuit doing? It's a very simple negative feedback circuit, and what it's actually doing is trying to keep the level of the light falling on that photoresistor the same, regardless of what else is happening. How is it doing that? If the level of the light gets too big, the photo resistance goes down, we go toward plus 15 volts at the minus input of the op-amp, and we turn the lightbulb off. If it gets too dim, we go toward the minus 15 volts. That means we have a lower voltage on the negative input than the 0, the ground at the positive; that turns the power op-amp's output to plus 15. Because of the way the diode is, that turns the lightbulb on. It's just like the voltmeter wobbling back and forth. What the servo voltmeter does is produce a level of light that depends on how much light the photodetectors sees. There's a circuit that I call, again, the intelligent lightbulb.

I've argued that this is a circuit that tries to maintain a constant level of light on this photoresistor. Another way of putting that is this circuit tries to maintain the voltage at this point at a fixed value. What fixed value? It tries to maintain it at 0 because the positive input is at 0, because as soon as this level gets below 0, the lightbulb turns on. That brings the photoresistor back toward plus 15 volts, and brings this point above 0. That turns the lightbulb off, because there's now a bigger voltage at the minus input; we get minus 15 out. Nothing can go through the diode in that condition. We're going to switch back and forth, bobbling back and forth, just the way the servo voltmeter did between plus and minus 15 volts at the output, and the lightbulb being alternately on and off. We're going to do that in just such a way that keeps the photoresistance constant, because that's what keeps this voltage at 0 volts. That's going to mean that the intensity of light falling on the photoresistor stays the same. If I pull the lightbulb away, the intensity of light falls off with the distance from its source. The lightbulb is going to have to get brighter to maintain those conditions, and that's exactly what we saw happening. Now let's take a more sophisticated look at that by connecting an oscilloscope to this point, the output of the amplifier, and to this point, the junction of the two resistors, or equivalently the minus input, the inverting input, of the amplifier. Let's look at the oscilloscope.

What we have on the oscilloscope screen now is yellow trace that represents the output of the amplifier, and that's at a 5 volt per division scale. Both the yellow trace and the green trace have their zeroes right across the middle of the screen. If you count, we're up about 2 1/2 divisions; that's about 12 1/2 volts. That's about the maximum that comes out of this particular op-amp. The lightbulb is fully on, as it better be when we have 12 1/2 volts coming out of that op-amp. Notice that the green, which is that junction between the two resistors, or equivalently the inverting input of the op-amp, is at minus 15 volts. This is a situation when the op-amp is said to be out of control; the negative feedback isn't working yet.

But now watch what happens as I bring the photoresistor near the lightbulb. Now we see the output of the op-amp swinging back and forth between plus and minus roughly 12 1/2 volts. Look what happened: As soon as it came in control, look what happened to that junction point; the minus input, the inverting input of the op-amp. It went smack dab to 0 and just stays there. As I move the lightbulb in and the op-amp gets dimmer, the lightbulb turns on less and less of the time. We see only one of those pulses as I go further out. We see more and more of them and they get wider and wider. That's the op-amp doing this on/off, on/off, on/off to the lightbulb, and it's doing it in such a way to keep the average output of the lightbulb at just the right level to keep the resistance of that photoresistor the same.

I do want to show you one other thing before we move on, and that is what happens if I turn up the gain on the green channel of the oscilloscope. I'm going to turn that up, and the first thing that happens is the trace disappears altogether. Gone; we don't even see it because it's gone way off the bottom of the screen because it's at minus 15 volts and now I'm at 100 millivolts per division. But watch what happens as I bring the op-amp back into control. Here we go: Take the photosensor, bring it near the lightbulb, and now we see a very slight variation in the voltage at the minus input. Ideally, it would be exactly 0 but, in fact, it's varying very slightly; and it's those slight variations above and below 0—you can see it swinging above and below the middle of the screen—that are driving the op-amp with its enormous gain to swing equivalently between plus and minus 15 volts. This is all happening in such a way that we maintain the voltage at the minus input at almost 0 and we maintain the light intensity at whatever value it is. It stays the same as far as the photoresistor here is concerned.

Now you can imagine some practical uses for this kind of circuit. One practical use might be as some kind of sensor in an energy efficient building that senses what the level of ambient lighting is and turns the lights up or down accordingly to keep the amount of light falling on your desk the same; mixing sunlight with artificial light, and automatically doing that, turning down the artificial lights, to keep the total light the same. That would be a practical application. Here's another practical application of this circuit: We can take this same circuit and morph it very simply into a similar thing in which instead of having a lightbulb, we've got a heater. Instead of a photoresistor, we've got a thermistor, which is a thermal resistor whose resistance drops with increasing temperature. We have thermal feedback, and that device could be a temperature controller.

We have many, many applications for these simple negative feedback circuits. I want to end with a kind of fun one that's the last one, which also has some practical applications, but it's going to be really silly as I demonstrate it to you. I'm going to set up the equipment a little bit differently and I'll show you my last feedback demonstration.

Here we are with a really silly looking demonstration, which I call funny face. Many of my students are amused by funny face, and a few of them are freaked out by funny face. I'm not going to tell you much about how funny face works, except to say that he works very much like the other feedback demonstrations we've seen. If you want to do the project for this lecture, the project consists of reverse engineering funny face and figuring out how he works. But here's funny face, and he's got a couple of eyes, and he's mounted on a motor. This fancy arrangement here is nothing but some electrical contacts that feed electricity into this region without preventing it from rotating. We used to have wires but they'd get all wrapped up on themselves. Funny face has a friend. Here's his friend, and he really likes her; she's his friend. He likes his friend so much that he'll follow the friend wherever she is. We're in a studio with lots of bright lights, so the friend doesn't work too well. Instead of the friend, I'm going to use this bright LED flashlight; LEDs, remember, a kind of diode. I'm going to turn on the power to funny face. I have funny face plugged into our bread board; we're only using it for the power supply in there, so I'm not showing you the rest of it. It has the same 759 op-amp you've seen before; the power op-amp that can drive these motors. We're going to turn funny face on; and there goes funny face, bye-bye. Let's see what funny face does when I bring a friend nearby. Funny face really likes that friend. That's funny face. You may have noticed funny face bobbling about kind of indecisively. Funny face isn't very well designed for doing what it's supposed to do. We can do better.

Funny face is silly; it's a fun little demonstration. We often set it up in the hall at Middlebury College and people can come by and wave the friend around and watch it go. But funny face is sort of a metaphor for some much more serious circuits that do similar things. I just want to end by showing you one of those circuits.

Here's my student, Misha Gershall, who did a senior thesis with me recently. Behind him is an apparatus he built that is designed to track the sun, and it's designed to improve the efficiency of sun-tracking solar collectors. Misha's tracker has somewhat more sophisticated electronics, but the idea is still the same. It has two motors so it can turn in two mutually perpendicular directions. So instead of just rotating in one direction like funny face, it can point anywhere in the sky. The thing on the right—that black thing sticking up—is comparable to funny face's two eyes, except it's got four eyes, which is what allows it to move into two mutually perpendicular directions. I'm not going to say any more about Misha's project right now, except to say that it's a more sophisticated version of the same feedback system here we have in funny face. We'll get back to Misha's project in the final lecture of this course, in which we'll see a whole day's time lapse movie of this thing tracking the sun across the sky.

That's a quick introduction to the magic of feedback; in particular, negative feedback that tries to mitigate a change and keeps a system in a stable location: the servo voltmeter at a location appropriate to the input voltage; the intelligent lightbulb with an intensity of light that's appropriate to the settings of the two resistors that make it up; and funny face with a position that's appropriate to maximizing the amount of light that falls on funny face.

If you'd like to do the project for this one—and I recommend it; this isn't one you're going to be building with a simulator—your job is to reverse engineer funny face; that is, to design a plausible circuit for funny face. What do I mean by reverse engineer? I mean take something that you don't have the diagram of how it works and figure it out. You'll actually find, if you buy certain pieces of software or hardware, there are sometimes rules that say, "If you use this, you aren't allowed to reverse engineer it." But I'm going to ask you reverse engineer funny face, and I'm going to tell you, you really only need the following components: one of these power op-amps that we've talked about; two photoresistors rather than one; and a DC motor that's reversible, depending on which polarity it gets. Of course, put it in parentheses, because you always need a power supply for your op-amp. You can reverse engineer funny face and see if you can figure out how to make that simple but quite impressive circuit that maximizes the amount of light falling on it.

Electronic Feedback
Lecture 12

The preceding lecture showed a number of demonstrations involving negative feedback: mechanical, optical, and thermal. In all those cases, the circuits had op-amps whose outputs swung between two limits, +15 V and –15 V. But they did so in a way—because of negative feedback—that kept the voltage at the negative input close to the value it had at the positive input. In this lecture, we'll look at circuits in which the feedback mechanism is electronic. That gives rise to a remarkable versatility in op-amp circuits, including circuits that "tame" the op-amp's huge intrinsic gain. Important topics in this lecture include the following:

- Op-amp circuits with negative feedback
- The inverting amplifier: a detailed analysis

- Op-amp rules: simplifying op-amp circuit analysis.

Op-Amp Circuits with Negative Feedback

An inverting amplifier is a simple op-amp circuit that uses negative feedback. The noninverting (+) input of this op-amp is at ground, 0 V. We have an input resistor, R_{in}, and we'll connect the voltage we want to amplify to the point marked "In." We'll connect a feedback loop from the output of this amplifier back to the inverting (–) input. For this configuration, we want to know the gain: What's the ratio of V_{out} to V_{in}?

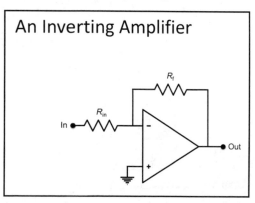

An Inverting Amplifier

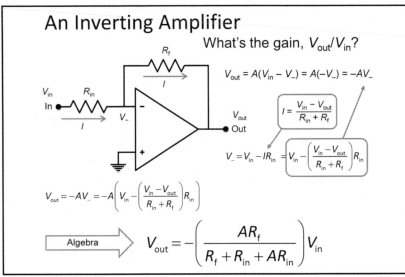

An Inverting Amplifier

What's the gain, V_{out}/V_{in}?

$$V_{out} = A(V_{in} - V_-) = A(-V_-) = -AV_-$$

$$I = \frac{V_{in} - V_{out}}{R_{in} + R_f}$$

$$V_- = V_{in} - IR_{in} = V_{in} - \left(\frac{V_{in} - V_{out}}{R_{in} + R_f}\right)R_{in}$$

$$V_{out} = -AV_- = -A\left(V_{in} - \left(\frac{V_{in} - V_{out}}{R_{in} + R_f}\right)R_{in}\right)$$

Algebra $\Rightarrow$

$$V_{out} = -\left(\frac{AR_f}{R_f + R_{in} + AR_{in}}\right)V_{in}$$

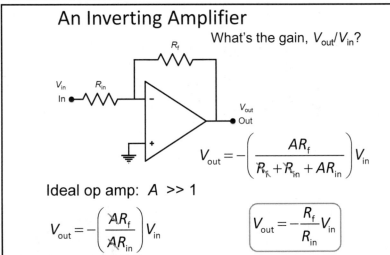

An Inverting Amplifier

What's the gain, V_{out}/V_{in}?

$$V_{out} = -\left(\frac{AR_f}{R_f + R_{in} + AR_{in}}\right)V_{in}$$

Ideal op amp: $A \gg 1$

$$V_{out} = -\left(\frac{\cancel{A}R_f}{\cancel{A}R_{in}}\right)V_{in}$$

$$V_{out} = -\frac{R_f}{R_{in}}V_{in}$$

To answer this question, we must walk through some heavy mathematics, but doing so will allow us to analyze other op-amp circuits almost trivially. Because of the fact that the amplifier's gain is huge ($A \gg 1$), we get the remarkably simple result $V_{out} = -R_f/R_{in}$. Negative feedback has "tamed" the huge gain and allowed us to make an amplifier with any gain we want! And the value of A doesn't matter at all—as long as it's big.

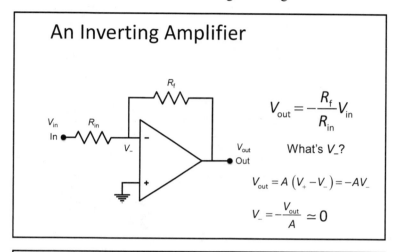

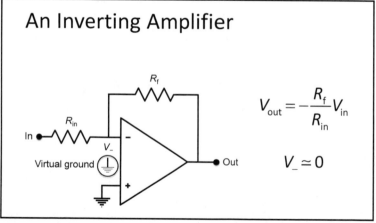

To find the voltage at the inverting input to the op-amp itself, we again go through some math; the result is $V_- \simeq 0$. Because of that, we say that V_- is at *virtual ground*. It's as if that point were connected to ground, or 0 V—but it isn't really. Rather, the negative feedback works actively to keep V_- at very nearly 0 V.

Op-Amp Rules: Simplifying Op-Amp Circuit Analysis

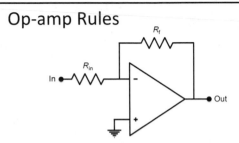

Op-amp Rules

1. No current flows into op-amp inputs
   - Huge input resistance
   - Enhanced by negative feedback
2. With negative feedback, $V_+ = V_-$
   - $V_- > V_+$ makes V_{out} negative; feedback pushes V_- down
   - $V_- < V_+$ makes V_{out} positive; feedback pushes V_- up

We can identify two simple rules for analyzing op-amps: (1) No current flows into the op-amp inputs, and (2) whenever negative feedback is in control, V_+ and V_- are almost identical. (For the inverting amplifier, V_+ was 0, so V_- became very nearly 0, as well.)

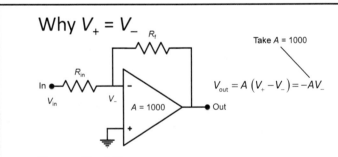

Why $V_+ = V_-$

Take $A = 1000$

$$V_{out} = A\left(V_+ - V_-\right) = -AV_-$$

- Suppose $V_- = 1$ V
- Then V_{out} "wants" to be -1000 V
- V_{out} heads toward negative supply voltage of -15 V
- Because of feedback resistor, V_- also heads downward
- As soon as V_- goes negative, V_{out} heads toward $+15$ V
- That drives V_- positive, and the cycle repeats
- Feedback is ~instantaneous, so V_- assumes a small average value

If V_- were greater than V_+, we get a large negative output, which, through the feedback resistor, would be felt at the inverting input. That decreases the inverting-input voltage until it's no longer greater than V_+. If V_- becomes less than V_+, the output goes positive, and that results in increasing V_-. This all happens almost instantaneously, with the result that V_+ and V_- remain essentially equal.

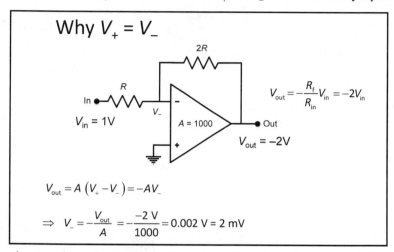

Why $V_+ = V_-$

$$V_{out} = -\frac{R_f}{R_{in}}V_{in} = -2V_{in}$$

$V_{in} = 1V$

$A = 1000$

$V_{out} = -2V$

$$V_{out} = A\left(V_+ - V_-\right) = -AV_-$$

$$\Rightarrow V_- = -\frac{V_{out}}{A} = -\frac{-2\,V}{1000} = 0.002\,V = 2\,mV$$

Again, a mathematical analysis gives us confirmation of op-amp rule 2: that V_+ and V_- are very nearly equal. The difference voltage $V_+ - V_-$ is, therefore, very small, and for that reason, we get an output voltage that is not at one of the limits but somewhere between those limits.

Suggested Reading

Introductory
Brindley, *Starting Electronics*, 4th ed., chapter 9, pp. 150–152.

Lowe, *Electronics All-in-One for Dummies*, book III, chapter 3, pp. 382–385.

Advanced
Horowitz and Hill, *The Art of Electronics*, 3rd ed., chapter 4, sections 4.1.3 and 4.2.1.

Scherz and Monk, *Practical Electronics for Inventors*, 3rd ed., chapter 8, section 8.4 through p. 642.

Delving into Math

The op-amp in the circuit shown has gain A. Note that the op-amp has been flipped, so the positive input is at the top. Analyze the circuit to find the exact relation between V_{out} and V_{in}. Show that when A is very large ($A \gg 1$), you get a simpler relation that doesn't depend on A. With $A = 1000$ and $V_{in} = 1$ V, find the "error signal," $V_+ - V_-$.

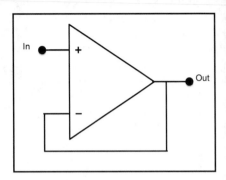

Gain-of-3 Inverting Amplifier

Design a gain-of-3 inverting amplifier, simulate it, and verify its gain both for DC and AC. How big is the "error signal"? Check the (time-dependent) voltage at the op-amp's inverting input.

Questions to Consider

1. Why is the circuit shown below called an *inverting* amplifier?

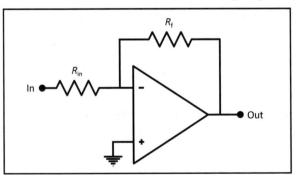

2. In the circuit above, identify the point called *virtual ground* and explain this term.

3. State the two op-amp rules.

Electronic Feedback
Lecture 12—Transcript

Welcome to the second of my two lectures on the magic of negative feedback. If you think back to some of the demonstrations I showed you in the previous lecture, there was a lot of negative feedback involved. There was optical feedback, there was mechanical feedback between a motor shaft and a potentiometer, there was possibility of thermal feedback, and so on. In all those cases, the circuits I showed you had op-amps whose outputs swung between its two limits, plus and minus 15 volts. But they did so in such a way, because of the negative feedback, which kept the voltage at the minus input close to the value it had at the plus input, which in some of the circuits was, in fact, 0.

We're going to look now at circuits in which the feedback mechanism is a lot faster. It's electronic feedback. It doesn't require a mechanical movement or a lightbulb to change temperature and brightness. It's going to be direct connections from the output of an amplifier back to the input, and we'll see that that gives rise to a remarkable versatility in using op-amps. What it's going to do for us most importantly is to tame that enormous op-amp gain—a thousand, a million, 10 million, whatever it is—that I said I didn't really matter, and you'll see why it doesn't matter, and you'll see why we can tame that down to make any gain we want.

This is going to be the most heavily mathematical lecture I have for you, but bear with me briefly. Let's begin by looking at a very simple op-amp circuit that employs negative feedback. We'll analyze the circuit here on our big screen. This circuit is called an inverting amplifier; you'll see why as we get through it.

We're going to start with the plus input of our op-amp at ground, 0 volts. We're going to have an input resistor, call it R-in, and we're going to connect whatever we want to amplify to that point marked "in." The all-important feedback, the negative feedback: We're going to take feedback from the output of the amplifier back to the negative input. That's why it's going to be negative feedback, because it's going to the inverting input of the amplifier.

We want to know what the gain is. What's V-out versus V-in? What's the ratio of V-out to V-in? That's what I'd like to show you.

Before I do that, I have to issue a math alert. My students at Middlebury College are used to this, especially in the upper class electromagnetism courses, where we'll go on for a day deriving an equation involving complicated vector calculus. It's not going to be that bad here. But is a math alert, and the beauty of a math alert—besides the ugliness of all the math—is that heavy math is coming, but it's going to bring us important insights. In this particular case, it's going to obviate any future math. Once we get through this bit of heavy math, we'll have a rule that will allow us to analyze circuits like this almost trivially; we won't have to go through this math again. So we're operating briefly under a math alert, and bear with me as I go through the math that finds exactly the answer to that question: What's the gain of this circuit. What's V-out versus V-in?

Let's remember what we know about op-amps. We know that intrinsically that triangular thing, the op-amp, produces an output that's its intrinsic gain A—which is supposed to be a big number, but right now I'm going to ignore that fact and later we'll talk about A being big—times the difference between the input voltage at the plus input versus the input voltage at the minus input. By the way, here's one of many instances in which I'm no longer going to show you the power connections to the op-amp. They're assumed; they're there. If you wire a real op-amp you better put them or the thing won't work. If you wire an op-amp in one of these circuit simulators, they may be assumed, depending on which op-amp you choose.

Here we are, and we're asking: What's this output voltage over that input voltage? What's the ratio? Because that's the gain of the amplifier. Here we go. In this particular case, the plus input is connected to ground, 0 volts. So V-plus is 0. V-out, which was A times V-plus minus V-minus becomes A times 0 minus V-minus, and that's just A times minus V-minus or minus A times V-minus. The output is minus A times whatever's at the input, but we don't know what's at the input. We only know what we put in over here; what we're calling the input voltage to the entire amplifier.

Let's think about what happens here. That input voltage is going to drive some current, I, through that resistance. We want to figure out what that current is, and we can do that using Ohm's law. Here's how we can do that: We know one thing about ideal amplifiers. They have essentially infinite resistance at their inputs. What that means is no current can flow into the input of the amplifier. That was one of our important characteristics of ideal amplifiers. No current goes into here. Now, it's not quite true; tiny bits of current go in. Depending on the input transistors it could be microamps, nanoamps, or picoamps; 10 to the minus 12 amps. It's small. We ignore it. That means any current that's flowing through R-in, the only place it can go is up through R-F. These resistors are truly in series, even though you see a little junction there, because that junction doesn't lead anywhere electrically; it leads to an infinite resistance or almost infinite resistance that we can ignore. That current is flowing through the series combination of R-in and R-F. What's the voltage across that combination? It's the difference between the input voltage and the output voltage. I, Ohm's law, is the voltage across that string of two resistors, V-in minus V-out, divided by the total resistance, which since they're in series is the sum R-in plus R-F.

That's the current that's flowing. It still doesn't answer our question, what's the gain, but we're getting there. That's the current; there's the current flowing. We've got it figured out.

We can figure out what V-minus is if we knew what that current was. We can figure out what the voltage here is because there's a drop across that input resistor, R-in, and that drop is given by the voltage across that input resistance times the current. Whatever V-in is, the voltage here's going to be less by an amount equal to the current times the resistance because that's the voltage dropped across that resistor. So V-minus becomes V-in minus I R-in; that's what V-minus is.

Now I'm going to take this somewhat complicated expression for I and I'm going to stick it in that expression for V-minus. Oof, grind through a little algebra; well not much algebra, I'm just sticking that expression right in there in place of the I. There we now have an expression for V-minus. What do I do with that? I put it in up here in the expression for V-out in terms of V-minus. We have this expression for V-out, but the reason it's a little bit complicated to solve is that V-out also appears on the right-hand side. We have to do some algebra to extract that. You may not want to do that algebra; you may want to do that algebra. If you'd like to, I invite you to pause, press your pause button, and work through that algebra. If you don't want to do it, I'm not going to do it for you on screen here; I'm simply going to give you the result. But I assure you, it takes nothing more than ninth grade algebra to achieve this result. V-out becomes minus—that's why it's called an inverting amplifier—this somewhat complicated expression involving the resistors in the circuit and the intrinsic gain of the op-amp itself times V-in. The gain, in fact, is everything that multiplies V-in. It's negative, it's an inverting amplifier, and there's its value.

That doesn't really tell us too much; that's not very enlightening. Let's go further with that. Here we are, our inverting amplifier. We found that this is the gain, and now remember that the ideal op-amp has I gain much greater than 1. I kept saying it doesn't really matter. Is it 1,000? Is it 100,000? Is it a million? Doesn't matter, as long as it's much greater than 1. Here's why: Because if the gain A is much greater than 1, then A times R-in is surely much bigger—a million times bigger, 1,000 times bigger, I don't care, but much bigger—than R-in itself; A times R-in. I can cross out R-in in the denominator there. It's negligible compared to A times R-in because A is so big.

I'm going to make another assumption, which is the assumption that this feedback resistor isn't huge. What do I mean by not huge? It's not so huge that it's comparable to A times R-in, but it still might be much bigger than R-in. It could be 100 times R-in; it could even be 1,000 times R-in if A was a million; and it would still be negligible compared to A R-in. I'm going to make the assumption that R-F is nowhere near as big as A times R-in, but it still may be, again, much bigger than R-in. A feedback resistor can be bigger than the input resistor, it just can't be A times bigger. So we're going to neglect the feedback resistor also.

Now we have this expression that's quite a bit simpler. In fact, it's really a lot simpler than it looks because now we have A in the numerator and A in the denominator, and we can cancel the A. There's the justification for the statement I've been making over and over and over again: that the intrinsic gain of the amplifier that might be 1,000, or 10,000, or 100,000, or a million, or whatever, doesn't matter. In this configuration with negative feedback it just doesn't matter. We get that V-out for this circuit is simply minus R-F over R-in. You want an amplifier with a gain of two, make R-F twice as big as R-in and you'll have an amplifier; it won't be 2, it'll be minus 2, because this is an inverting amplifier configuration. You can tell an op-amp is in an inverting configuration if the input to the whole circuit dumps in to the minus input of the op-amp. Guarantee you, that's an inverting configuration; the output comes out with the opposite side of the input. It doesn't mean the output comes out negative, by the way. A negative input produces a positive output, and a positive input produces a negative output.

Because this op-amp that we developed—and I showed you the circuit for it a few lectures ago—was developed and designed with no capacitors coupling through it, it will amplify DC. If I put 1 volt in and R-F is twice R-in, I'll get minus 2 volts out. If I put a sine wave in, I'll get a sine wave varying twice as much but going the other way, and so on. V-out is simply minus R-F over R-in, a remarkably simple result coming after all that heavy math.

Before we go on, let me point out we did the heavy math once. We're never going to have to do it again. You see an amplifier like this, you know how to make it like that. Not only that, you see another op-amp circuit, and the results of this heavy math are going to help you to analyze that circuit very simply without ever going through that heavy math again. We want to look a little bit further at the circuit to see what happens.

We know what the gain is; it's minus R-F over R-in. Let me ask another question: What's the voltage here? What's the voltage at the inverting input to the op-amp itself? What's that voltage? V-out, as we know always for the op-amp, is A V-plus minus V-minus. In this case, because V-plus is at ground, right there, then this is simply minus A V-minus, as we wrote before; or equivalently, V-minus is minus V-out over A. This is a typical op-amp. Its output swings between, say, plus and minus 15 volts. A is 1,000, a million, whatever, 10,000, 100,000; whatever it is, A is huge, and so that voltage, V-minus, is tiny. What this says is this circuit, through the magic of negative feedback, manages to hold V-minus extremely close to 0. In fact, if A were truly infinite, it would be 0. If A is a million, the most V-minus will be is

15 millionths of a volt, 15 microvolts. That's tiny. Even if A was a meager 1,000, 15 millivolts would be the biggest that gets. Remember I talked about op-amps, I said if you don't connect any feedback, the op-amp is going to go flying to one limit or the other; it can't help it. But the minute you have negative feedback, you hold the difference between V-plus and V-minus—or you don't, but the negative feedback does—holds that difference to such a small level that, in fact, that huge gain multiplying that small little bit of voltage difference gives you a reasonable output. Remarkable.

That's an inverting amplifier analyzed to death, but we're never going to have to do that again. V-minus is essentially 0; very, very close to 0. That's the important takeaway message here. Because of that, we call this point a virtual ground. It's as if that point were connected to ground, which is 0 volts. It isn't, and if you did connect it to ground, the whole thing wouldn't work. But as far as this input's concerned, the other end of that resistor is at 0 volts; it's just as if it were connected to ground. The current that flows in that resistor isn't really going to ground—it's actually going through the feedback and back into the op-amp and through all those transistors and things—but as far as the external circuit's concerned, this point, once the negative feedback is working, is at essentially ground potential. It's essentially 0 volts, even though it isn't really connected to ground. That's an idea that takes a little bit of time to wrap your head around.

But there it is, a virtual ground. We draw the virtual ground with that little ground symbol in a circle to show it's not really a ground, but electrically it's as if it were a ground. This input thinks it's just connected to a resistor that goes to ground. What this input does, delivers a current V-in over R-in— because the other end of R-in is at 0—is exactly what it would do if the other end of R-in was, in fact, connected to a real ground. There's our inverting amplifier and this concept of a virtual ground.

Let's now take this concept a little bit further, because now we come up with some simple rules that allow us to analyze op-amps. Let's look these rules. Here's our inverting amplifier again that we just analyzed. Here are the two rules, and the rules are really simple. When I teach electronics, these rules, I write them down on a big piece of paper and I hang them up in the lab and they just sit there the whole semester, and the students really know them by the time they're done. First rule: No current flows into the op-amp inputs. Again, these rules are approximate, but that's a pretty good approximation because these op-amps have very high input resistances. They have huge input resistances, especially if they're field effect transistor input op-amps. Even the meager 741, they're in the megohm region, millions of ohms. Furthermore, in the presence of negative feedback, particularly in this configuration, the inverting configuration, when that minus input is at virtual ground, there's no incentive for any current. Even if there were a fairly low resistance inside the op-amp to ground, because the voltage there is so small, held there at this very low level by the negative feedback, there'd be very little incentive for current to flow. The huge input resistance is effectively enhanced in the presence of negative feedback, particularly in this configuration.

That's op-amp rule number one, and we've sort of already seen that rule. Here's op-amp rule number two: Whenever there's negative feedback and the negative feedback is working—we say it's in control—then V-plus and V-minus are essentially the same; they're almost identical. Amazing result from negative feedback: Negative feedback forces those two inputs to be at the same voltage. In this configuration, where we forced the plus input to ground, to 0 volts by grounding it, that means the minus input is also at 0 volts essentially. That's our virtual ground. With negative feedback, the two inputs are going to be forced to the same voltage.

How does that happen? Think about what would happen if V-minus were greater than V-plus. Op-amp says take V-plus minus V-minus; if V-minus is greater, that's negative. Multiply it by a big value, you're going to get a big negative output. That big negative output, through the feedback resistor, is going to be felt at the minus input, and that's going to make the minus input voltage go down, and it'll no longer be bigger than V-plus. If it goes down so far that V-minus is less than V-plus, that makes the output positive and that tries to bring V-minus back up.

That's, in fact, exactly what happened with our servo voltmeter, except there the feedback was by the mechanical motion of that motor, and it took a little bit of time. What happened was the needle continued to kind of bobble back and forth as the amplifier first went to one limit and drove the motor in one direction, and went to the other polarity and drove the motor in the other direction as the op-amp swung between its two extremes of possible outputs because the feedback was slow. The only difference between that and the circuit we're now talking about, the inverting amplifier, is here the feedback's fast; it's electronic. Electronic signals and resistors and wires travel at a good fraction of the speed of light. That's faster than the amplifier itself can react. Here, the feedback is slower. Even with the intelligent lightbulb I showed you, the feedback was slower because the lightbulb had to cool down a little bit and warm up and the photoresistor responded slowly. But here, with the feedback provided electronically, it's very fast. Instead of the op-amp rapidly oscillating back and forth between its two limits, it simply settles down to an appropriate level in between, appropriate to the conditions set up by that circuit and under the conditions that the plus and minus inputs are essentially equal.

Those are the op-amp rules, and those rules make it very easy to analyze circuits with op-amps. We'll go back in the next lecture, for example, and redo this inverting amplifier with a gain of whatever it is, R-F over R-in, and we'll do it very quickly with very little math. We certainly won't need a math alert, and we'll just apply these rules. You can apply these rules to any circuit involving an op-amp, and the rules will work as long as you provide conditions that keep the op-amp in control; that is, allow the negative feedback to do its magic.

Let's take a look, with a little more mathematical detail, about exactly why it is that V-plus is equal to V-minus. Let's do some numbers. Let's consider a case where we have an op-amp whose gain, A, is 1,000. That's pretty low for an op-amp, but that'll do for us. V-out is always A V-plus minus V-minus. In this inverting configuration with V-plus equal to 0, that's the positive input, the non-inverting input, to the op-amp grounded. We have V-out is simply minus A V-minus, so we'll consider that. Let's suppose that V-minus were 1 volt. I've argued that it can't be; I've argued that V-minus, in this case, has to be a tiny, tiny, tiny fraction of a volt and I've tried to convince you algebraically. But suppose that wasn't true; suppose V-minus was equal to 1 volt: 1 volt times 1,000 is 1,000 volts. In some sense, if I can anthropomorphize the op-amp, V-out wants to be minus 1,000 volts, because the V at the minus input is 1 volt; that's the inverting input. We're going to try to get minus 1,000 volts. Of course we can't do that unless this was an op-amp whose power supplies went to plus/minus 1,000 volts, and that would be some op-amp. Typically they go to plus/minus 15. What happens is the output voltage tries to head toward minus 1,000 volts, but it can't go that far; it heads toward the negative supply voltage of minus 15 volts. It heads down; V-out heads down. Remember, V-minus was plus 1 volt. As V-out heads down, again through the magic of that feedback resistor, V-minus also heads down. If V-minus is 1 volt, it's going to quickly be reduced, and because of the feedback resistor is why it gets reduced. V-minus also heads downward. As soon as V-minus goes negative, V-out turns around and heads back toward plus 15 volts, because now if V-minus is negative, V at the plus input, which is 0, is greater than V-minus and the amplifier output wants to go to big positive number, and it had stored plus 15 volts. That drives V-minus positive again, and that cycle keeps repeating.

That's exactly what's happening in the bobbling of the servo voltmeter, but the reason it doesn't happen very much here is we don't actually get those outputs swings is because the feedback is effectively instantaneous, almost instantaneous. What happens instead is V-minus assumes a small average value for its voltage. That's how this circuit or any other circuit with negative feedback in control of an op-amp manages to keep the plus and minus inputs the same. Op-amp rule number two is when you have an op-amp with negative feedback, you'll get an output voltage that's not at one of the limits, but within those limits. How does it manage that? Because it holds the difference between V-plus and V-minus to a very, very tiny value. That tiny value is appropriate to the inputs, and as we saw in that complicated mathematical analysis, that's what makes this particular circuit work.

OK, enough theory. Let's move on and take a look at an actual op-amp circuit that does this. Over here I have on my board, the part we're interested in is this little black chip. It happens to be a 741 op-amp. I've got some standard resistor values. I picked a 10 kiloohm resistor and a 22 kiloohm resistor; those are standard values. R-F is the 22 kiloohm resistor. Here it is. The output of the op-amp is over here on the right-hand side of the feedback resistor, the minus input is on the left-hand side. Meeting at that point, the minus input, are the feedback resistor, 22 kiloohms, the input resistor that's 10 kiloohms; so according to my derivation of the gain of this thing, we ought to have a gain of about 22 over 10 or 2.2, just a little over 2. This is effectively a gain of 2 or a little bit more. Inverting amplifier; so it's a gain of actually minus 2. I have the input to the amplifier coming in through this red wire from this coaxial cable that's connected to my function generator, and my function generator is producing a voltage that, well, let's see what it does. The voltage is displayed on the yellow curve here. The yellow is at 2 volts per division. We've got just about one division there, and consequently, we're looking at a display of about plus or minus 1 volt. It goes to minus 1 volt, it goes to positive 1 volt. That's the input. By the way, you're seeing some little glitches on there. Those are probably due to switching of the light dimmers in the studio lighting. They're electronic noise. We could get rid of them with a suitable filter, but I haven't bothered to. That's the input, the yellow curve.

The green curve, which is coming along this probe, this probe is connected to the output of the op-amp. That's the other side of the feedback resistor—that all-important feedback resistor—so we're looking at the output. Let's look at what we see at the output. Both of these channels are on 2 volts per division, so these curves are displayed in their actual relationship. I think you can see that the second curve is actually just a hair bigger than twice the amplitude of the first curve. I can see that because here's going about two divisions, and we're going just a little bit above two divisions, a little bit below two divisions. We've got a gain of a little bit more than 2 and that was, in fact, our resistor ratio: 22 kiloohms to 10 kiloohms, a gain of about 2.2. It's just right. Furthermore, you can see that this is an inverting amplifier because where the input peaks, the output troughs; where the output peaks, the input troughs. It's a gain of minus 2; it's in inverting configuration in this amplifier.

That's what you're interested in if you're building an amplifier. It would be trivial for me to change the value of that feedback resistor and make any gain I wanted within reason. I could even make a gain of less than 1. If I wanted to reduce a voltage by 1/2, I could use 10 kiloohms at the input and 5 kiloohms at the feedback. We can do whatever we want there.

However, what we probably don't usually look at if we're using an op-amp, but that's interesting for our purposes in learning about how the op-amp works is we want to know what's happening at that minus input. Let's go take a look at the minus input on the oscilloscope. I have the minus input of the op-amp right here, the junction of that input resistor and the feedback resistor. I have that connected through this probe, and that's going to the red channel of the oscilloscope, which I'll now turn on. There you see it. Let's talk about what we're seeing there. Here you see a very noisy signal that looks like it's pretty small; and it sure is small, because where those other two channels are on 2 volts per division, this one's on 50 millivolts per division. One division on here means 1/40 times as much voltage as one division on the yellow or green input or output curves. This is a very tiny variation. It's on the order of going up or down maybe 20 millivolts. That's the V-minus input; that's the input that's supposed to be 0 volts, and it's almost 0 volts. The fact that it's not quite 0 volts tells us that the gain isn't quite infinite.

In fact, if we wanted to work out what the gain is, we've got maybe 20 millivolts in and we've got something on the order of 2 volts out. We've got a gain—let's see, 20 millivolts times 10 would be 200 millivolts, times 100 would be—we've got a gain of about 100 or something. I thought it was supposed to be thousands or millions. The gains of these amplifiers are quite frequency-dependent, and as you go up in frequency the gain of the amplifier actually goes down. I chose a 5 kilohertz signal here, because if I'd taken a 1 kilohertz or the 440 Hertz A above middle C, we wouldn't have even been able to see the variation at the minus input, it would've been so close to 0. I went up to 5 kilohertz where the gain of the amplifier has fallen off. A 741, actually by the time you're at a kilohertz, it's fallen off in gain to about 1,000. This is actually a bit less than 1,000; somewhere between 100 and 1,000. We're looking there at the minus input and we're seeing op-amp rule number two realized. That's effectively at 0 volts.

Before I wrap up, let me just do one dramatic thing to this op-amp. Let me pull out the feedback resistor. The feedback resistor is that all-important resistor, and I pull out the feedback resistor. You can see several things happening. I had lost my probe there on the output, but let's not worry about that. What you see happening is—the output went to 0 because I disconnected the probe—but you see the minus input is now swinging dramatically back and forth. In fact, if I lower the gain on the input channel here, you can see that the minus input is doing basically what the input to the whole circuit is doing because the op-amp is completely out of control. Let me see if I can reconnect to the output here; that pin. You can see that the output is swinging between plus and minus limit as the input sine wave goes up and down. This op-amp is no longer in control. It's not in control because it doesn't have any negative feedback because I'd pulled the feedback resistor out. If you're actually building circuits with op-amps and you find your op-amp has gone to limit, chances are your feedback loop is broken or you forgot to put a component in or something like that. Once the feedback is working, it guarantees that the V-plus and the V-minus are essentially the same voltage.

Let me wrap-up. What we've done here is use negative feedback to tame that enormous gain that's intrinsic to the op-amp and allow us to make amplifier circuits with any gain we choose by choosing the right resistors, and the gain we get doesn't depend in the least on that intrinsic great big A of the operational amplifier. We'll build many more interesting circuits with op-amps over the next few lectures. If you want to get a feel for them, do the project for this lecture that involves your designing a simple inverting amplifier and picking the resistors, and then checking it out in your circuit simulator.

Amplifier Circuits Using Op-Amps
Lecture 13

Al the math we did in the preceding lecture enables us to analyze a number of interesting op-amp circuits because we now understand the two simple rules that govern the behavior of op-amps: (1) There is no current flowing into the op-amp inputs, and (2) with negative feedback, the two inputs—inverting and noninverting—must be at the same voltage. With this lecture, then, we'll begin looking at increasingly complex circuits involving op-amps and circuits that perform other functions than simply amplify. Important topics include the following:

- The inverting amplifier via op-amp rules
- Summing amplifiers
- Current-to-voltage converters
- Application: a light meter.

The Inverting Amplifier via Op-Amp Rules

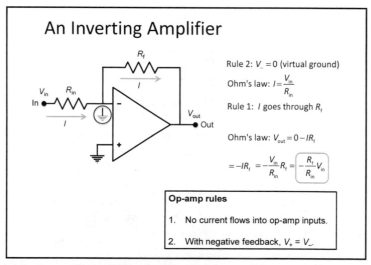

An Inverting Amplifier

Rule 2: $V_- = 0$ (virtual ground)

Ohm's law: $I = \dfrac{V_{in}}{R_{in}}$

Rule 1: I goes through R_f

Ohm's law: $V_{out} = 0 - IR_f$

$$= -IR_f = -\frac{V_{in}}{R_{in}}R_f = \left(-\frac{R_f}{R_{in}}\right)V_{in}$$

Op-amp rules

1. No current flows into op-amp inputs.

2. With negative feedback, $V_+ = V_-$.

We begin by analyzing the same inverting amplifier we looked at in the preceding lecture, now using the simple op-amp rules. Recall that those rules

243

came from the complicated math we did in the previous lecture, followed by the assumption that the intrinsic gain (A) of the op-amp was much greater than 1. That gave us the second rule: $V_+ = V_-$. Note that it now takes only a few steps of simple math to get to the result that the gain is $-R_f/R_{in}$.

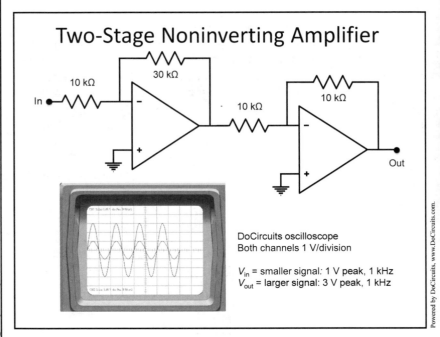

Two-Stage Noninverting Amplifier

10 kΩ

30 kΩ

In

10 kΩ

10 kΩ

Out

DoCircuits oscilloscope
Both channels 1 V/division

V_{in} = smaller signal: 1 V peak, 1 kHz
V_{out} = larger signal: 3 V peak, 1 kHz

If we don't want our amplifier to invert, we can add a second inverting amplifier. Here, we see one op-amp stage with a 10-kΩ input resistor and a 30-kΩ feedback resistor; it has a gain of –3. The output of this stage goes into a 10-kΩ input resistor of another op-amp in the same inverting configuration but with a feedback resistor equal to its input resistor; it has a gain of –1. (This second stage is called a *unity-gain inverter*.) Thus, the second amplifier simply inverts the output of the first amplifier. We had –3 for the gain of the first amplifier and –1 for the gain of the second amplifier; the whole two-stage amplifier has a gain of +3.

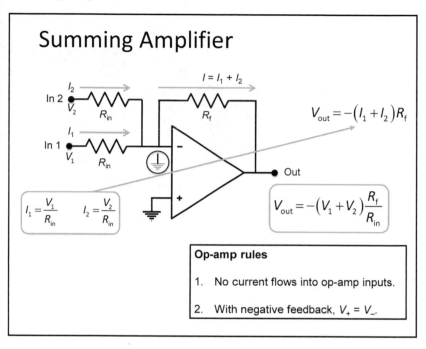

Summing Amplifier

$I = I_1 + I_2$

In 2
V_2
R_{in}
R_f

$V_{out} = -(I_1 + I_2)R_f$

I_1

In 1
V_1
R_{in}

Out

$I_1 = \dfrac{V_1}{R_{in}}$ $I_2 = \dfrac{V_2}{R_{in}}$

$V_{out} = -(V_1 + V_2)\dfrac{R_f}{R_{in}}$

Op-amp rules

1. No current flows into op-amp inputs.

2. With negative feedback, $V_+ = V_-$.

A summing amplifier begins as the same inverting amplifier we've just seen. It has an input voltage, V_1, and an input current, I_1, flowing through the lower resistor marked R_{in} to virtual ground at the op-amp's inverting input. We then connect a second, identical input resistor with a second input voltage, V_2, driving current I_2 to the same virtual ground. The currents that come in through those two resistors act as if they're going to ground, so each is completely independent of the current flowing in other input resistors.

According to op-amp rule 1, no current can flow into the op-amp. Thus, the currents I_1 and I_2 actually flow through the feedback resistor, and that feedback-resistor current is the sum of the two input currents. Applying Ohm's law then shows that the output is basically the sum of the input voltages negated, inverted, and multiplied by R_f/R_{in}.

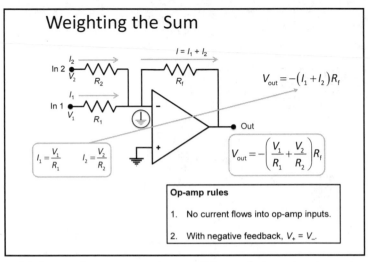

We could also configure a summing amplifier so that one input counts more than the other, giving an amplifier that produces a weighted sum.

Current-to-Voltage Converters

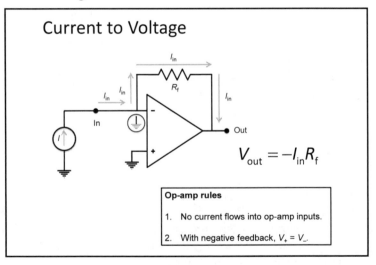

A current-to-voltage converter is like an inverting amplifier, except that it doesn't have an input resistor because it gets a current directly at its input.

Application: A Light Meter

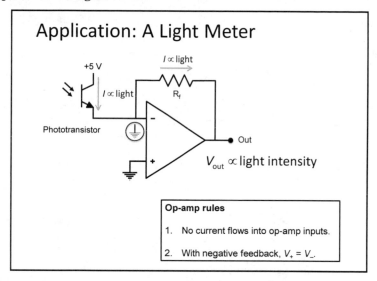

Application: A Light Meter

+5 V

$I \propto$ light

R_f

$I \propto$ light

Phototransistor

Out

$V_{out} \propto$ light intensity

Op-amp rules

1. No current flows into op-amp inputs.

2. With negative feedback, $V_+ = V_-$.

A light meter is built in the same configuration as the current-to-voltage converter, except that the current source is a light-sensitive *phototransistor*, which produces a current proportional to the light intensity. Thus, at the output, we get a voltage proportional to the light intensity. Light meters are used in cameras and many other applications.

Suggested Reading

Introductory
Brindley, *Starting Electronics*, 4th ed., chapter 9, pp. 147 to end.

Lowe, *Electronics All-in-One for Dummies*, book III, chapter 3, pp. 385 to end.

Advanced
Horowitz and Hill, *The Art of Electronics*, 3rd ed., chapter 4, sections 4.2.2, 4.2.3, 4.3.1.3, 4.3.1.4.

Scherz and Monk, *Practical Electronics for Inventors*, 3rd ed., chapter 8, section 8.4, pp. 641–642 (top).

Temperature Sensor

A temperature sensor produces an output voltage whose value in mV is equal to the temperature in degrees Celsius. Design an op-amp circuit with this sensor at the input that produces an output whose value in mV is 100 times the temperature in degrees Fahrenheit (i.e., when it's 70°, the circuit's output should be 7000 mV, or 7.0 V). Your op-amp(s) use ±15-V power supplies, and these are the only power sources available to you. As an option, simulate the circuit and verify its operation; try the DC sweep simulation.

1. In the current-to-voltage converter, a source of current is connected directly to the inverting input of the op-amp. Where does the current actually go?

2. Why is it important that the resistors in a summing amplifier be precisely matched?

3. What's a phototransistor, and why doesn't it have a connection to its base?

Amplifier Circuits Using Op-Amps
Lecture 13—Transcript

Welcome back.

All that math we did last time was worth it, because now we're going to be able to analyze all kinds of interesting op-amp circuits without having to do a lot of math because we understand those two simple rules that govern the behavior of op-amps: Namely that there's no current flowing into their inputs, and with negative feedback the two inputs—the plus input, the non-inverting input; and the minus input, the inverting input—have to be at the same voltage. The negative feedback guarantees that, and it'll keep things that way.

Let's plunge into the first of several lectures in which we look at increasingly complex circuits involving op-amps and circuits that do other things than simply amplify. I'm going to begin by looking at the circuit we had last time, the simple inverting amplifier, and I'm going to analyze it again using the op-amp rules. But no math alert this time because it's very straightforward, and I'm doing this because I want to show you how simple the op-amp rules make it to analyze circuits involving op-amps, and then we'll use the rules in the same way to develop other circuits.

Here again is our inverting amplifier—an op-amp, an input resistor, R-in, a feedback resistor, RF—but now we know the op-amp rules. Remember where they came from: They came from doing all that complicated math, but we never have to do it again. The op-amp rules came from doing the complicated math and then making the assumption that the gain A—the intrinsic gain of that triangle, the intrinsic gain of the op-amp—was much, much bigger than 1. That gave us rule two, that V-plus is equal to V-minus; it also helped us with rule one, although rule one is also a feature of the op-amps and the way they're designed. Let's see if we can understand how to apply these rules to our inverting amplifier, and in a very short time get the same result we got last time with much more complicated steps.

Rule two: Rule two tells us that the inverting input and the non-inverting input are at the same voltage. We forced the non-inverting input to be at ground, 0 volts, and that means the inverting input is also at effectively 0 volts. We saw that very tiny variation last time, essentially 0 volts; down in the millivolt range or lower. Effectively, rule two tells us that the negative input is as if it were at ground. Once again, it's not really connected to ground, and that's important for the workings of the circuit. But as far as that input resistor and whatever's connected to the other end of it, on the other side of the input resistor is ground. Not really, but that's what the circuit acts like. Rule two tells us that V-minus is 0. We've got a virtual ground there. The input voltage thinks it's just applied across a resistor, the other end of which is grounded. Ohm's law tells us that the current is simply V-in over R-in. Because we've got V-in on one side and 0 in the other side, the voltage across the resistor is simply V-in. Ohm's law: I is V over R-in.

Now we go to rule one. Rule one tells us—that's not quite what it says; that's what it infers—no current flows into the op-amp inputs. That means this current we've drawn that's flowing through R-in can't go into the op-amp through its input. The only place it can go, then, is through RF. Our inference is from rule one that I goes through RF. There it goes; same current that came in is on its way through the feedback resistor. What happens to it after that, if you're wondering? Some of it might go to the output if there's other stuff connected there. The op-amps, transistors will sink it in and will eventually deliver it back to the ground and the power supplies and so on; so the op-amp is doing its magic to make this happen.

Ohm's law tells us if I have 0 volts here, and I've got a current I flowing through RF—I've got a voltage drop across RF of IxRF—and this side is more positive than this side. But the side's at 0, because that's the way the current's flowing. This side's at 0; so this side is negative, and it's negative by 0 minus the current times the feedback resistor. V-out is 0 minus I RF; and that's simply minus I RF, because 0 minus I RF is minus I RF. But we used Ohm's law to figure out I; so V-out is minus V-in over R-in—that's the current—times RF. If we rearrange that, it's simply the ratio of the feedback resistor to the input resistor times V-in. It's a negative gain, a gain RF over R-in, but negative RF over R-in because we're coming in to the inverting amplifier.

There's a way of developing the inverting amplifier very, very, very quickly using the two op-amp rules. We'll analyze all subsequent op-amp circuits, even much more complicated ones, basically by applying the op-amp rules.

We can also make complicated circuits by building up different renditions of this particular circuit. Before we do that, I want to say a few things about this inverting amplifier circuit in this particular configuration. First of all, the feedback enhances the 0 volts of that virtual ground, as I said last time, because with that voltage at almost 0, there's no incentive for current to go into the op-amp even if it didn't have a big input resistance. That's one nice aspect of this. On the other hand, remember, an amplifier would like to have a big input resistance; well, the op-amp does. But this whole circuit is the amplifier we've now built, and in this configuration there's typically significant current flowing at the input. We had a 10 kiloohm resistor there, for example, in the circuit we built in the previous lecture.

One disadvantage of this circuit is that it doesn't really provide very, very high input resistances. If you had to make this input resistance, if you needed to make it 100 kiloohms and you wanted a gain of 10, that would need to be a million ohms, and that's beginning to get a little too big for this op-amp, comparable to the intrinsic resistances inside it. You can't really use the inverting amplifier configuration to make an amplifier that has a very large input resistance. In many cases that's not a problem, but it could be.

The other problem with the inverting amplifier—might be a problem in some applications—is it's an inverting amplifier and you might not want it to invert. How do you solve that problem? You solve that problem by adding another stage of inversion. Two negatives make a positive, and so let's look at a very simple circuit that does that. Here I have a two-stage non-inverting amplifier; two stages because I've got two op-amp stages. The first one, if you look at it, has a 10 kiloohm input resistor and a 30 kiloohm feedback resistor; so that's a gain of minus 3. The output of that one goes into a 10 kiloohm input resistor of another op-amp in exactly the same inverting configuration. This one's got a 10 kiloohm feedback resistor, too. I don't want to say it has no gain, because it doesn't have 0 gain, it doesn't give 0 at the output; it has unity gain, we say. It has a gain of 1. The output is equal to the input, but it's a gain of minus 1 because it's the inverting configuration, and so that second amplifier simply inverts the output of the first amplifier. We had minus 3 for the gain of the first amplifier, minus 1 for the gain of the second amplifier, and so this whole two-stage non-inverting amplifier that I've constructed here has a gain of positive 3.

There's the output from DoCircuits; you can actually put a simulated oscilloscope, connect it to your circuit, and do circuits. There it is; and you see in the red curve the input, and you see the output is in phase with it—that is, they peak together; it's not inverting—and you see that the output is three times the input. So we have a two-stage non-inverting amplifier with a gain of 3, a gain of positive 3, which is the same as saying it's not inverting.

Let's take a look at the real thing instead of just looking at things we can do in simulation or by drawing circuit diagrams. What you see are a couple of op-amps. There's the first one; that's the very same op-amp we used in the previous lecture, I've just changed the resistors a little bit. I've got a 10 kiloohm resistor at the input, and the closest I could get in the standard resistor values was a 33 kiloohm, so this is more like a gain of 3.3. By the way, later on we'll be working with some very precision resistors you can buy that are good to 1%, and then I could've bought a 10k and a 20k, or a 10k and a 30k, but I'm using just standard resistors here.

Here's the op-amp. The output of the op-amp is over here. The output goes back through that 33 kiloohm feedback resistor to the input to that first op-amp. The input resistor feeds in and here's the input signal coming in from our function generator; this time it's a 1 kilohertz signal. The input resistor and the feedback resistor join at the inverting input of the first op-amp. The output not only goes back to the input through the feedback resistor—there's the negative feedback—but it also goes into the inverting input of the second op-amp through that other 10 k resistor, and then there's a 10 k feedback resistor. The output of the second op-amp is over here.

Right now we're looking on the big screen at the input to the circuit and the output of the first op-amp, which we see is an inverting amplifier. Where that's a trough that's a peak; where that's a trough that's a peak; it's an inverting amplifier with a gain of about 3. These both are on two volts per division, so they're in the same scale, and that's about three times the amplitude of that one. We've got a gain of 3, inverting amplifier in our first stage. Let me just move the oscilloscope probe to the output of the second stage and lo and behold, we've now … oops, that's a problem, OK? This is a good educational point; I didn't mean that to happen, but let's go on with it. What on earth is going on? I can see what's going on; I can see that the output of the second amplifier isn't doing what it's supposed to be doing. In fact, if you look at it, it's swinging up to some very large value, which is, in fact, at two volts per division, about 12 volts. That looks like it's close to the op-amp's limit and it's going down to about minus that; so somehow this op-amp isn't in control. The feedback isn't working, and chances are what's happened is the feedback resistor has somehow become disconnected. Let's see if I can fix that and get that feedback resistor back in there.

There it is. I had a bad connection at the feedback resistor, and when I connected the probe it pulled it out the little connector in that board. So we fixed it, and now we have a gain of 3—3 times that—and it's non-inverting. Peak here, peak here, trough here, trough here. You want a gain of 3 non-inverting amplifier? There it is. Make it out of two inverting amplifiers, one with a gain of minus 3, the other with a gain of minus 1. That's what we did; works great. We could've, by the way, had a gain of the square root of 3 in each amplifier if we'd wanted and we still would've come out with 3; but I chose to make a gain of minus 3 and a gain of minus 1.

There's a simple embellishment on our idea of inverting amplifiers. Let's move on and look at some other amplifiers we can develop here with these circuits. Let's take a look now at a slightly more complicated situation: a situation in which we're going to talk about an amplifier that's going to have two inputs. Let's see what we get. Here's an amplifier called a summing amplifier. But it's not a summing amplifier yet, it's simply the same inverting amplifier we've just been talking about. I've drawn it with an input voltage. I'm calling N1 this time because we're going to have two inputs. There's an input current I'm calling I1. The current is flowing to virtual ground; it isn't really flowing to ground, it's going through the feedback resistor. But as far as that input resistor is concerned, its right end is basically connected to 0 volts to ground. Here's the beauty of this situation: I can connect a second input resistor with a second input voltage, input number two, to that same point; that virtual ground point. For that reason, that point is sometimes also called the summing point, because what happens at that point is currents that are coming in through those two resistors all think that they're going right to ground. They don't know anything about the other currents that are flowing in the other resistors, and they become completely independent of each other. That wouldn't be true if that point weren't virtual ground but were connected to ground through yet another resistor, because the voltage drop across that resistor would be determined by both currents and it would affect the currents through each resistor. But because of that virtual ground, each of these resistors just thinks it's connected to ground on the right-hand side. The current that flows in I1 and the current flows in I2, according to op-amp rule number one—no current can flow into the op-amp—both those currents together flow through the feedback resistor, and the current in the feedback resistor is the sum of the two currents coming in through those two resistors.

We have what's called a summing amplifier, and let's work out in a little bit more detail: V-out; what's V-out? Just as in our simple analysis before, we've got that virtual ground at the inverting input, 0 volts. We've got a drop now of I1 plus I2 times the feedback resistor, across the feedback resistor. The left-hand side is 0, so the right-hand side, which is V-out, is minus I1 plus I2 times the feedback resistor. What are I1 and I2? They are V1 over R-in and V2 over R-in. For now I've given those two resistors, input resistors, the same value. If I take those quantities V1 over R-in (the lower resistor R-in), V2 over R-in (the upper resistor is also R-in), and I plug them in for the Is there and I work things out, I get that V-out is minus the sum of V1 and V2 multiplied by this factor R feedback over R-in.

This is a summing amplifier. It's an inverting summing amplifier. Its output is basically the sum of the input voltages negated, inverted, and then multiplied by this gain factor RF over R-in. We can build a summing amplifier. I mentioned earlier that op-amps are called op-amps partly because they do mathematical operations. Here's an op-amp doing a mathematical operation; it's adding two voltages, and the output is basically the sum. If you don't like it to be the negative sum, put a unity gain inverter on the back of it just like we did with the gain of 3 amplifier before. If you don't want it to change the voltage but actually simply add them, make R feedback equal to R-in and you'll get an output voltage whose magnitude is the sum of the input voltages. This is a summing amplifier; a very useful circuit. It works because we have this virtual ground at the minus input of the op-amp, which means that the two currents can flow independently without either one knowing what the other one's doing.

Before we move on and actually look at one of these amplifiers in the flesh, as it were, let's look at two other possibilities: Let's first of all look at what we could do if we didn't want an actual sum but we wanted to weight the sum; we wanted one input to count more than the other. Then we could take the input resistors and make them different. Then I1 would be V1 over whatever R1 was, and I2 would be V2 over whatever R2 was, and when we put those in we get a weighted sum, weighted by those two different resistors; weighted by two different gains. We don't have to simply add the inputs; we can add them and give them different weights if we want.

To see if you've got this whole concept, and that's a lot coming at you pretty fast, let me give you a quick quiz. Here's a circuit in which I have an inverting summing amplifier at the beginning. It's got a 10 kiloohm input resistor for V1; it's got a 15 kiloohm resistor for V2; it's got a 30 kiloohm feedback resistor; and then it flips into a unity gain inverter. What's V-out in terms of V1 and V2? Press your pause button and figure this out, and we'll come back and look at it.

Here we have what input 1 sees. It sees 10 kiloohms and 30 kiloohms, so it basically has a gain of 3. Let's look at the other two resistors. The V2 resistor is 15 kiloohms. It goes to the same 30 kiloohm feedback resistor. That's a gain of two, or rather minus 2. Then we have the unity gain inverter. Before we get into the unity gain inverter, we've got minus 3 V1 plus 2 V2. We invert that with the unity gain inverter that doesn't change values but just changes basically the sign, so there's the unity gain inverter and V-out is 3 V1 plus 2 V2.

Let's take a look at what this circuit does. First, if you'd like to have a little bit of lab time, you could actually simulate, say, a gain of 2 summing amplifier that weights both inputs equally. Do it in your circuit simulator and verify that it works. We'll use inputs of 1 volt and 2 volts and then replace them with sine waves at slightly different frequencies and see what you get. I've suggested some parameters if you're using CircuitLab; DoCircuits usually defaults to parameters that'll work for this. I'd like you to describe how the output would sound. If you don't want to do this lab time, don't bother. But if you do want to do it, pause, do the lab, and let's come back and look at what we get for results from this.

Here are the results. On the left I'm doing DC results. I've got 1 volt at V1 and 2 volts at V2. The output is supposed to be a summing amplifier, an inverting summing amplifier, which weights the two equally, and it has a gain of 2 here. Indeed, we have 1 volt plus 2 volts is 3 volts; double that, you get 6 volts; invert it you get minus 6 volts; and that's indeed what the voltmeter reads.

In the case of the AC, things get very interesting. I've got an input voltage sine wave at one frequency, and a sine wave at a slightly different frequency. Whenever that happens, you get situations where sometimes the crests of the sine waves line up, and sometimes the troughs of the sine waves line up. When the troughs line up they cancel each other, and when the crests line up they reinforce. The closer the two sine waves are in frequency, the longer you have to wait until they come into phase or go out of phase. You see the output of CircuitLab's graphing showing what that pattern looks like.

I'm asking you to imagine what that would sound like. I'm going to play you that sound, not for the frequency shown here, but for a case where they're even closer: a 440 Hertz sound and a 442 Hertz sound. Here's what they sound like. What you're hearing is a phenomenon called beats, in which these two very close frequencies come into synchronism, and then go out again, and in, and out. You see a very slowly varying intensity of sound, varying at roughly the difference frequency of those two waveforms, which is just 2 Hertz. One interesting example of what you can do with a summing amplifier was summing two waveforms that are slightly different.

By the way, you may remember that I've used my function generator on several occasions to add a couple of signals; to add, for instance, noise to a sine wave when I wanted to demonstrate a filter. What the function generator has in it is a summing amplifier that can be switched in at its output, and it has two different channels. It can generate two different waveforms, and you can use those two different waveforms to add them. You can use the summing amplifier to add them, and then you get that output that's the sum, and in this case, this is an example.

That's several examples of circuits we can make with op-amps. I want to go and show you one other circuit, which is somewhat similar to but not exactly the same as the circuits we've built already. Then we'll actually use it to end by building a really useful circuit. Here's another use of the inverting configuration: Sometimes you have not a sensor or an input signal that's a voltage, but something that actually produces a current. One example is a transistor, because the collector circuit of a transistor produces a current that depends on the base circuit, and there are many other sensors that produce a current. But you'd like to be able to read that current with some kind of easy to read meter like a voltmeter, so you need a current to voltage converter. A current to voltage converter is even simpler than the kinds of circuits we've been dealing with already.

What this circuit shows you on the left is a circle with an arrow in it that represents a current source. It shows you that that source is going to provide current flowing in the direction shown. We're going to apply the op-amp rules; so the op-amp rules say with negative feedback V-minus is equal to V-plus. We've got a feedback resistor there, so we've got negative feedback. That means the negative input, the inverting input, of the op-amp is at virtual ground. The current source is very happy; it's happy to pump its current, it thinks, to ground. But rule one says it doesn't really go to ground, it goes through the feedback resistor. Same current going through the feedback resistor and around, and then to the output or back into the output of the op-amp or out to some external circuit; the op-amp takes care of all that.

But there's what happens. Just as with the analyses we'd been doing before for the voltage amplifiers, we can say the output is minus I times the feedback resistor, because the left-hand side of the feedback resistor is at 0 volts. There's a current I flowing through it, and that current I is multiplied by the feedback resistance to get the voltage across the feedback resistor; so V-out is minus I-in times RF. A current to voltage converter is like one of these inverting amplifiers we've been building, except it doesn't even have an input resistor because it's getting a current at its input directly, rather than a voltage, which, if you will, produces a current through the input resistor. It's even simpler.

Let me now reconfigure my set up a little bit and show you a very practical application for a current to voltage converter. Here's a practical application for that current to voltage converter that I just discussed. It looks pretty similar to that circuit, except instead of that arbitrary current source, I've got some particular practical device. I need to talk a little bit about that device so you can understand what it does. On the left is a phototransistor. What's a phototransistor? If you look at this device, it looks sort of like a transistor. It's got a collector, it's got an emitter; does it have a base? Yes, it has a base. But it's got no connection to that base, and instead it's got a couple of arrows coming in. They look like the arrows we used to talk about a light emitting diode; they were the light going out. This is light coming in to the phototransistor.

What a phototransistor does is it uses light to produce electron hole pairs at the base. When you connect a voltage to its collector—this is an NPN phototransistor, similar to the NPN transistors we've been using—you'll get a current that depends not on a base current, because there's no connector to the base, but depends rather on the intensity of the light falling on that base. These transistors are built in transparent plastic cases; that base is exposed to light. When light falls on it, the energy of the light creates electron hole pairs. It's very similar in some ways to the operation of a photovoltaic cell, although a PV cell is just a diode and this is a little more complicated because it's a transistor. That transistor produces a current that's dependent on the intensity of the incoming light, and now we take that current and we run it into a current to voltage converter. Here comes the current, which is proportional to the amount of light, the light intensity. That current, by the op-amp rules, can't flow into the op-amp, so it flows through the feedback resistor where it causes the voltage to drop across the feedback resistor; so the right-hand end of the feedback resistor is at a lower voltage than the left-hand end. The left-hand end is at 0 volts by op-amp rule number two because that says with negative feedback, which we have because of the feedback resistor, the two inputs, the inverting and non-inverting input of the op-amp, will be at the same voltage. This configuration, where the plus input is grounded, gives us a virtual ground at the inverting input. Same configuration as the converting amplifiers we've been building; same configuration as the current to voltage converter. In fact, this is the current to voltage converter, except our current source is now a light sensitive, a light intensity, current source that produces a current proportional to the light intensity. We get at the output an output voltage that's proportional to the light intensity.

With the circuit hooked up as I've shown it here, we'd get a negative voltage for current flowing down in the direction I've shown here. I'm going to go to a circuit that's very similar, but I want to tell you a few differences in this circuit. The main difference is I'd like to do something that I think is really silly in this digital age, but to me it makes sense. There are times when I really like to see an analog needle move to show me what's happening in a circuit, especially when changes are occurring. It's a little harder to grasp the changes when you're looking at a digital display, especially because a digital display is usually bobbling around in the last few digits as things vary slightly and you miss the point that the big variations are what you want to look at. Rather than hook up my fancy digital voltmeters, I've taken a very old fashioned analog voltmeter here, and I've connected it to the output of this circuit. Now I have a way of reading the light intensity; that is, the output voltage of the op-amp.

The other difference is because this voltmeter reads only positive—I could've switched the leads, but that would've looked confusing—I've actually turned the phototransistor around so it's going the other way. I've connected the emitter at the top, and I've connected that to minus 15 volts, in fact. I've connected a resistor to limit the current so the phototransistor doesn't get too much current through it. But it's basically the same circuit I've just shown you. Let's take a look: Here's the phototransistor. It's a little transparent looking package with two leads. It's only got two leads because it's got no base connection; a collector and an emitter. The base connection is the light coming in making the electron hole pairs; I think you saw the needle wiggle a little bit as I was playing around with it. The transistor goes into the inverting input of the op-amp, and here's the all-important feedback resistor that comes back and gives us our negative feedback and lets this whole circuit work. If I put my hand over the transistor, you could see the voltage go down. Right now the transistor is responding to the ambient light in the room and it's producing a significant voltage, about 3 volts at the output.

If I take a flashlight and turn on the flashlight, you can see that the voltage goes up. We don't have as much variation as we might like because we got such intense studio lights; I could perhaps increase that by changing some of the resistor values. But there we have it. If I were to work underneath shade here—you probably can't see that too well; there we go—I can go all the way down towards 0 and then sneak that light in and it goes up quite a bit.

We have a light meter; what's this good for? Most cameras have a light meter in them. You don't see the light meter in a modern camera, but it's in there, and it does things like calculate how long the exposure ought to be. That's a light meter, and it uses a current to voltage converter with a phototransistor as the current source, making a current that's proportional to the intensity of the light falling on it.

We've seen quite a few applications in this lecture of op-amps. They're very versatile; they're wonderfully easy to analyze once you understand the op-amp rules. We're going to go on in the next couple of lectures and look at some more applications of op-amps.

If you're interested in working with such an application for yourself, I have a rather challenging project for you with this one. It's to make basically a circuit that becomes an electronic thermometer. You've got some challenges because I want you to convert from a sensor that produces a voltage output that corresponds to Celsius temperatures, and you need to convert it to Fahrenheit. That's a challenge because you've got to do two things: You've got to convert the size of the degrees because they're different, and you've also got to account for the fact that there's that 32 degree offset; 0 Celsius is the freezing point, but 32 Fahrenheit is the freezing point. That's a challenging project, probably the most challenging one you'll have encountered, but I urge you to try it. There's no one right answer, but if you do that and then we talk about the project, you'll see what the solution is.

More Fun with Op-Amps

Lecture 14

S
o far, we've used op-amps to build inverting amplifiers, configurations in which the input of the circuit is brought in, ultimately, to the inverting input to the amplifier. Although these amplifiers are versatile, there are a few disadvantages. For example, their input resistance is determined by the input resistor and tends to be fairly low; thus, they draw significant current from their input source. That can be a problem with weak input sources. Further, another inversion is required if we don't want an inverting amplifier. In this lecture, we'll look at a wide range of circuits using op-amps, beginning with noninverting configurations. Important topics we'll cover include the following:

- The unity-gain voltage follower
- Followers with gain
- Difference amplifiers
- Peak detectors
- Schmitt triggers
- Application: Freezer alarm.

The Unity-Gain Voltage Follower

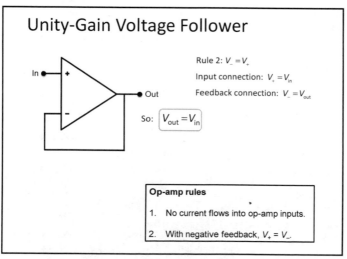

Unity-Gain Voltage Follower

Rule 2: $V_- = V_+$

Input connection: $V_+ = V_{in}$

Feedback connection: $V_- = V_{out}$

So: $V_{out} = V_{in}$

Op-amp rules

1. No current flows into op-amp inputs.

2. With negative feedback, $V_+ = V_-$.

The unity-gain voltage follower consists of just one op-amp. A quick analysis of this circuit shows that $V_{out} = V_{in}$; the output is equal to the input. This device might be useful in situations in which we have a very weak signal source, but that source can't, for example, power a loudspeaker or drive a large meter. Because the input goes directly to the op-amp's input, op-amp rule 1 shows that this circuit draws essentially zero current from its input source. A unity-gain voltage follower doesn't change the voltage, but it does allow us to deliver more current to the output. We should note that this circuit is subject to *common mode response* issues because neither input is at ground.

Followers with Gain

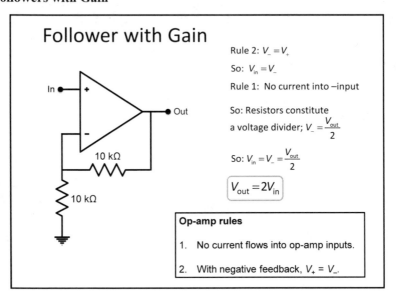

This circuit is called a *follower with gain*. It's similar to the unity-gain follower, except it multiplies the voltage. Notice that it takes two equal resistors to make a follower with gain of 2. In the inverting configuration, two equal resistors would give us unity gain.

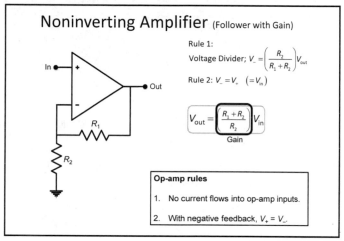

Noninverting Amplifier (Follower with Gain)

In ● →[+ op-amp −] → ● Out

R_1

R_2

Rule 1:

Voltage Divider; $V_- = \left(\dfrac{R_2}{R_1 + R_2} \right) V_{out}$

Rule 2: $V_- = V_+ \quad (= V_{in})$

$$V_{out} = \left(\dfrac{R_1 + R_2}{R_2} \right) V_{in}$$

Gain

Op-amp rules

1. No current flows into op-amp inputs.

2. With negative feedback, $V_+ = V_-$.

We can analyze this circuit with arbitrary resistances R_1 and R_2. Using the voltage divider equation, we find that the output and input voltages are related by the ratio of the sum of the resistors divided by the first resistor. That gives us a gain, which we can set to anything we want simply by choosing appropriate resistors.

Difference Amplifiers

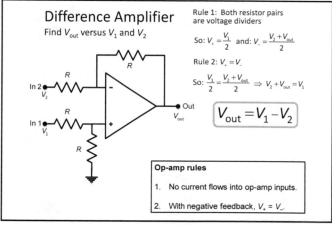

Difference Amplifier

Find V_{out} versus V_1 and V_2

In 2 ● —[R]— ● V_2

In 1 ● —[R]— ● V_1

R

R

Out V_{out}

Rule 1: Both resistor pairs are voltage dividers

So: $V_+ = \dfrac{V_1}{2}$ and: $V_- = \dfrac{V_2 + V_{out}}{2}$

Rule 2: $V_+ = V_-$

So: $\dfrac{V_1}{2} = \dfrac{V_2 + V_{out}}{2} \Rightarrow V_2 + V_{out} = V_1$

$$V_{out} = V_1 - V_2$$

Op-amp rules

1. No current flows into op-amp inputs.

2. With negative feedback, $V_+ = V_-$.

In a difference amplifier, the output voltage is the difference between the two input voltages.

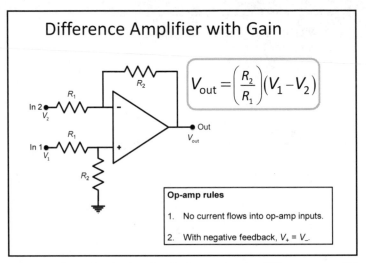

Difference Amplifier with Gain

$$V_{out} = \left(\frac{R_2}{R_1}\right)(V_1 - V_2)$$

Op-amp rules

1. No current flows into op-amp inputs.
2. With negative feedback, $V_+ = V_-$.

We can expand on the basic circuit for a simple difference amplifier by making the two input resistors the same but the feedback resistor and the resistor to ground different; that provides gain given by R_2/R_1.

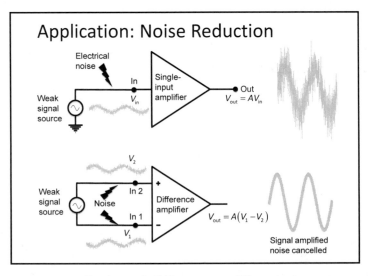

Application: Noise Reduction

Single-input amplifier $V_{out} = AV_{in}$

Difference amplifier $V_{out} = A(V_1 - V_2)$

Signal amplified noise cancelled

One common application of difference amplifiers is in very sensitive circuitry where it's necessary to eliminate any electronic noise that might creep in to the input signals. If the noise is on both input lines, a difference amplifier will cancel it while amplifying the desired signal.

Peak Detectors

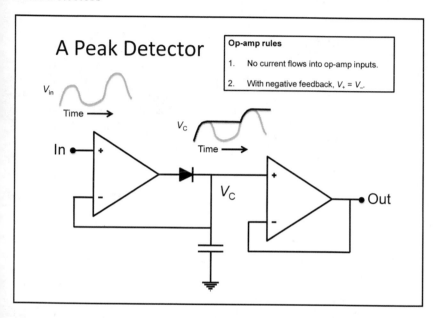

A peak detector is useful when we have a signal that's varying and we would like to know its maximum value. It's an op-amp connected basically in the unity-gain follower configuration, except it has a diode. That means that the amplifier feedback works only if the input voltage goes positive relative to the voltage at the negative input. Otherwise, the diode is reversed biased; we don't have feedback; and the output of the op-amp will go to its limit. A capacitor is added to serve as a memory device of sorts, and the second op-amp, in a voltage follower configuration, presents a very high input resistance that allows the capacitor to hold its charge for significant times.

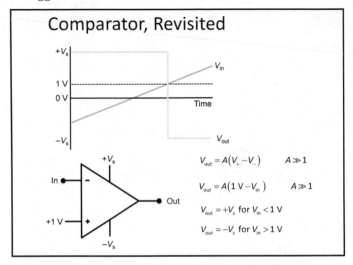

Earlier, we saw an op-amp as a comparator; it looked at its two inputs and went to one limit or the other, depending on which input was greater.

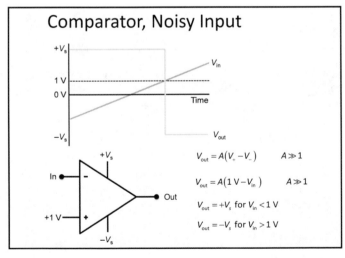

Imagine we have a voltage that is slowly rising with time, and we'd like to know when it reaches 1 V. Ideally, as soon as V_{in} reaches 1 V, the output should switch abruptly down and become the negative limit. But if V_{in} has some noise on it, instead of giving us a clean break from one level to another, it wiggles back and forth.

We can deal with this problem by using a circuit called a *Schmitt trigger*, which incorporates some positive feedback.

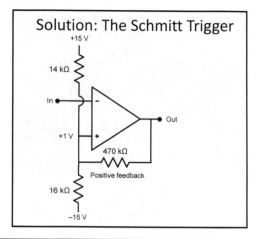

Solution: The Schmitt Trigger

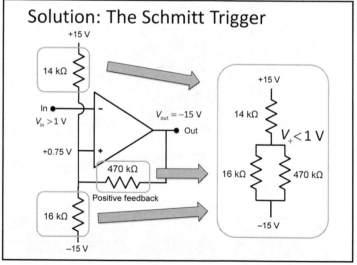

An extra resistor lowers the comparison voltage to about 0.75 V when the output is negative. That means that the voltage must fall below 0.75 V before another transition takes place. Once it falls below 0.75 V, the output swings to +15 V, which raises the comparison voltage at the + input to 1.22 V.

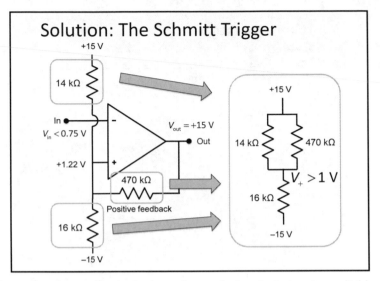

Solution: The Schmitt Trigger

What we've done, effectively, is to change the levels that we're switching to two slightly different levels. That phenomenon is called *hysteresis*.

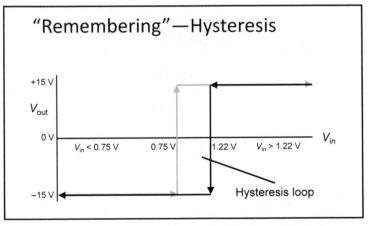

"Remembering"—Hysteresis

If we have a circuit that's supposed to switch back and forth between two levels, such as a thermostat in a home, hysteresis tells us not to always switch right at the same level; we wait until one level gets a little too high and the other gets a little too low. This allows us to avoid the problem of multiple switching.

Application: Freezer Alarm

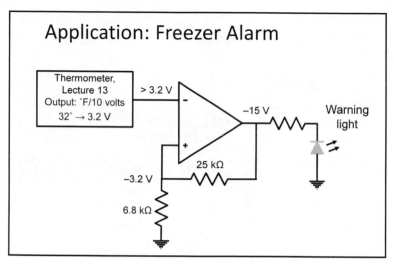

One practical example of hysteresis is an alarm that would let you know if the temperature in your freezer goes above 32° if you're away from your house for an extended time. Here, the warning light comes on when the temperature goes above freezing, and the positive feedback provided by the 25-kΩ resistor prevents it from turning off again even when the freezer temperature returns to normal.

Suggested Reading

Introductory
None of the introductory books covers the material in this lecture.

Advanced
Horowitz and Hill, *The Art of Electronics*, 3rd ed., chapter 4, sections 4.2.4, 4.3.2.2, and 4.5.1.

Scherz and Monk, *Practical Electronics for Inventors*, 3rd ed., chapter 8, section 8.4, p. 642 (bottom).

Schmitt Trigger
Verify the 0.75-V and 1.22-V thresholds in the Schmitt trigger circuit shown in this lecture.

Difference Amplifier
Design a difference amplifier with a gain of 10 and very high input resistances (such as 10^{12} Ω, typical of an op-amp). You'll need to combine several circuits from this lecture. Simulate it and verify that it works.

Questions to Consider

*1. In the circuit below, what is V_{out} in terms of V_1 and V_2?

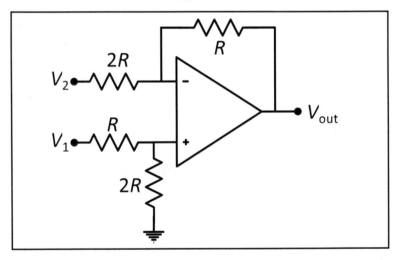

2. Explain how a difference amplifier can be used to reduce electrical noise.

3. What is meant by the term *hysteresis*?

More Fun with Op-Amps
Lecture 14—Transcript

Welcome to yet another lecture on the wonders of operational amplifiers. So far what we've been doing with op-amps is to build inverting amplifiers; configurations where we brought the input of our circuit in ultimately to the inverting input to the amplifier. Although these amplifiers are wonderful, versatile, and easy to build, there were a few disadvantages. One is because they have an input resistor that tends to be typically fairly low—tens of kiloohms perhaps—they don't present a very, very high input resistance to the input source that we're connecting, and sometimes that's important. The other problem with them is we need another inversion if we don't want an inverting amplifier. Today we're going to go through a whole zoo of circuits using operational amplifiers, and the first ones we're going to start with are non-inverting configurations of the operational amplifier.

Let's go over to our big screen and look at a very simple configuration. Here we are, and here's what's called a unity gain voltage follower. The voltage follower consists of just an operation amplifier. We're going to take our input right into the plus input of the amplifier. Remember, an operational amplifier has very, very high input resistance; 10^6 ohms for bipolar junction transistor op-amps, 10^{12}, maybe a teraohm for FET input op-amps. This is a simple, simple circuit. The output is simply connected back to the minus input, and that's the feedback. That's all there is to this circuit.

Let's analyze it, which is going to be very quick, by using our op-amp rules. Rule number two tells us that V-plus and V-minus are the same. The negative feedback works to keep 0 volts, essentially, between these two. We don't have virtual ground in this case. The reason we don't is because the plus input isn't connected to ground, neither is the minus input; what happens is the two inputs become equal to each other. Rule two tells us that V-minus is V-plus. The input connection tells us that V-plus is the input signal we want to amplify. Finally, we have the feedback connection. The feedback connection tells us that V-minus is the same as V-out because they're just connected by a piece of wire. The end result is V-out equals V-in.

Huh? What good is this? This is a unity gain voltage follower. It's a voltage follower because the output follows the input. It's exactly equal to the input, in fact. We say, "Why on earth would you want that?" There are many situations, and we'll see one in the next lecture, where you have a very weak signal source that you'd like to do something significant with, and the weak signal source can't, for example, power a loudspeaker or even drive a big meter or something. You put a unity gain voltage follower that doesn't change the voltage all, it doesn't change anything, but it allows you to deliver more current to the output. You can think of this thing as a bit of a power amplifier or a current amplifier, but it doesn't amplify voltage. It gives you exactly the same output as you had at the input. That's a unity gain voltage follower.

Before I praise it up too much, I should point out that it has an issue with it that you should be aware of in the real world, an issue I've been ignoring. Remember, any difference amplifier is supposed to look at only the difference between the two input voltages, but difference amplifiers actually respond a little bit to the actual value of the voltage as well. If you put 0 volts and 0 volts in, you should get 0 volts out; if you put 2 volts and 2 volts in, you should get 0 volts out; but you get something slightly different because of what's called the common mode response. It's not a big deal, especially if you're dealing with reasonable voltage levels here, but it's something to be aware of. In this circuit, you're subject to common mode issues because neither input is at ground. In the inverting configuration, both inputs are at 0 volts effectively, and so that's not an issue. It's a slight issue for these followers, but not a big issue, and the follower is a quick and simple way to make an amplifier that doesn't invert.

Let's move on and look at some other circuits in this zoo of circuits we're going to deal with. Let's begin with this one, and let me ask you to pause a moment and see if you can figure out what this circuit does.

Let's come back and take a look at this circuit. Let's look at what rule number two says: Rule number two says, again, that the two inputs to the op-amp have to be at the same voltage as long as we have negative feedback. We do, indeed, have negative feedback coming from the output through that 10 kiloohm resistor to the minus input. What that means is V-plus equals

V-minus in this case; and V-plus is V-in, so V-minus is equal to V-in. Rule number one says that no current flows into the inputs of an op-amp; so even though it looks like there's a third path for current to go after it goes through the upper one of those 10 kiloohm resistors, there really isn't because current can't flow into the op-amp. The 10 kiloohm resistors constitute a nice true voltage divider, and since they have the same resistance they divide the voltage in half. The voltage at the minus input is half of the out. The resistors constitute a voltage divider; V-minus is, therefore, half of V-out. But V-minus is equal to V-in, as we saw by rule number two, and that means V-in is equal to V-out over 2; or turn that around and V-out is 2 V-in.

This circuit is called a follower with gain. It's like our unity gain follower, except it multiplies the voltage, in this case by two. We have now a follower with gain. Notice that it took two equal resistors to make the follower with gain. In the inverting configuration, two equal resistors would give you unity gain. Here, the two equal resistors give us a gain of two.

We can analyze this circuit with arbitrary resistances R1 and R2. We still have a voltage divider. The analysis still works just the same way. The inputs are still equal, and consequently the minus input is equal to the plus input, which is equal to V-in. If we work out the voltage divider equation from very early in the course, we find that the output voltage and the input voltage are related by this fraction, this ratio, of the sum of the resistors divided by the second resistor. That gives us a gain, and we can set that gain to anything we want simply by choosing those resistors.

Don't take my word for it; let's go and build these circuits. Here we are. We'll turn on our oscilloscope so we can see what's going on. What I have here is a 1 kilohertz signal coming out of my function generator going into this connector and onto my board, and it's going into the plus input of this op-amp. You'll notice there's just nothing else there, except this all-important red wire, which is the feedback connection. Right now I've got this thing set as a unity gain non-inverting amplifier; a unity gain follower. I turn it on and you can see that the output looks exactly like the input; the output is green. It's following the input perfectly. If I change the input waveform or change the input frequency, the output will do exactly the same thing as the input.

Now let me just take a few minutes and reconfigure the circuit. It isn't going to take much work to do this to make it into a follower with gain. I've got a couple of 10 kiloohm resistors, and I'll just bring them in here and make a voltage divider. I haven't had to do much to alter this circuit. What I've done is to make a voltage divider of two 10 kiloohm resistors going from the output of the op-amp to the beginning of another 10 kiloohm resistor, which goes to ground. The only other change I've made is to take the feedback loop and run it not from the output right back to the input, but from the junction of those two resistors back to the input. Let's look at what we see on the screen, and lo and behold we have a gain of two amplifier. The output is exactly twice the input; otherwise it's following the input perfectly.

There's a voltage follower with gain—a gain of two, voltage follower with gain—one of many, many circuits we can make with op-amps. Let's move on and look at some other circuits now.

Here's a circuit; again, if you'd like to pause and think about this one, you can. It's a little bit more complicated to analyze this one, but not too much. But notice, this circuit has two inputs. I want to know what V-out is in terms of those two inputs labeled one and two. In this case, you'll notice there are four resistors and they're all exactly the same. Pause, look at it, and let's do an analysis; again, using our op-amp rules.

Let's take a look at how this circuit works and what it does. Rule one tells us no current flows into the op-amps, and that implies, as it did before in the follower with gain, that both those strings of connected resistors are basically voltage dividers. If we look at the lower string, that tells us that V-plus is simply V1 divided by 2, because we've got a string of two equal resistors going from V1 to ground; so the midpoint, which is V-plus, is V1 over 2. V-minus is a little bit more complicated, but you can see that it's the average of the V2 input voltage and the output voltage because the feedback resistor there is connected from the minus input to the output. Rule two tells us that V-plus is equal to V-minus, so those two voltages have to be equal. I'll set them equal, and I'll do a little bit of algebra, and I'll work on that algebra a little bit more, and I find that V-out is simply the difference between V1 and V2.

What have I got here? I've got a difference amplifier. Its output voltage is the difference between the two input voltages. I'm now exploiting the fact that the basic op-amp is itself a difference amplifier to produce this amplifier. If you want to build a difference amplifier, you've got to be a little bit careful. You've got to use resistors that are real really carefully matched; otherwise you won't get that nice V1 minus V2 situation.

We can take this basic circuit for a simple difference amplifier, and we can expand on it by making the two input resistors the same, but making the feedback resistor and that resistor to ground different, and then we'd get some gain. V-out would be this factor of the ratio of those resistors multiplied by the difference between the two voltages. If you wanted to weight the two voltages in the difference separately, you'd change the input resistors as well.

That's a difference amplifier with gain or without gain. Before we go on and talk about what you might use these things for, let's take a look at what one of these things does. I have a difference amplifier already wired on my circuit, and I just have to make a few changes in order to get it to work. I've made a few changes here. I've disconnected the input that I had connected to my simple non-inverting follower amplifier, and I've taken it down here into my difference amplifier, which is this op-amp right here. You can see these brown resistors; those are 1% precision resistors. They're more expensive than the usual kind, but I've used those so they're pretty well matched so I get a true difference amplifier. One of the inputs is through that resistor; the other input is through this resistor. The first input is coming in through the same red wire that I used to bring in the input to my voltage follower. The second input is coming in on this yellow wire from this blue cable coming out of my function generator.

On the big screen, you see the outputs of the function generator; it has two channels. The yellow is that same sine wave we were just looking at. The function generator can produce different waveforms and now it's producing a square wave. Down, up; constant voltage down, constant voltage up; constant voltage, and so on. I'm going to ask you to think a minute about what the difference between those two should look like; that's what the difference amplifier should be producing. Think about that a minute. If you want to pause and scratch your head a little bit go ahead; otherwise I'm going to go now and connect the probe from the oscilloscope to the output of this difference amplifier.

Is that crazy? Not so crazy, actually; if you think about it, you can convince yourself that what we're seeing is, in fact, one of these waveforms subtracted from the other. There's the difference waveform. You can see the jumps associated with the square wave; you can see the down going part of the sine wave superimposing on the constant part of the square wave; and so on. This is a difference amplifier. It amplifies the difference between two signals. Maybe it just gives it unity gain; maybe it gives it some additional gain. In this particular case, it's a unity gain difference amplifier. It's just taking the difference between those two voltages, and they came in with equal magnitudes. There it is; so that's a difference amplifier.

What might you use a different amplifier for? Let me give you one example. One thing you can do is reduce noise. If you have an ordinary so-called single ended amplifier—one input, one output; both of them understood to be relative to ground—you send in an input signal. You have some quantity that you're trying to measure, and you get an output. But often, along the way, electrical noise comes in. We've seen that in this studio where we had noise problems in a previous lecture on the big screen due to lights flashing on and off in here or some other kind of circuit switching. If that electrical noise gets into the very small input signal to your amplifier, it can either overwhelm it or at least dramatically alter it, so you get a noisy signal and the noisy signal gets amplified, and you get a noisy signal at the output.

That's no good. We saw one solution to this earlier, which was to use a filter. A filter won't work if the noise is at almost the same frequency as the signal you're looking at, though. But here's how a difference amplifier can help: Here's a difference amplifier whose output is proportional to the difference between the two inputs. What you do is connect your source so that it's connected to the two inputs of the difference amplifier. This isn't just an op-amp; it's that whole circuit that I just showed you.

Now what happens? Suppose electrical noise gets on both those wires? The electrical noise is the same on both wires. The difference amplifier takes the difference between those two inputs. They're at different voltages according to that source that's connected between them, but the noise is the same on both; so in the subtraction, the noise get subtracted out. Both input waveforms have noise, but the noise is subtracted out and we get a clean output signal. One common application of difference amplifiers is in very sensitive circuitry where you need to make sure you eliminate any noise that might creep in to your input signals. Noise reduction is one example of what we can do with a difference amplifier.

I've argued that this is going to be a bit of a circus, looking at other applications of op-amps. Let's move on to yet another one. Here's something called a peak detector. There are times when I have a signal that's varying and I don't really care about what it's doing as a function of time; I'd like to know what the maximum value it obtains is. I'm going to show you a simple circuit that will "remember" the maximum value of something. I should caution you ahead of time: Don't go and make a maximum/minimum thermometer, for example, with this circuit and expect it to hold the temperature for a week. You'd use digital circuitry and digital memory that we'll cover in later lectures. But as I'm going to show you in an application of this in just a moment, maybe we have a situation where something, over a few seconds, some quantity's changing and we want to find out what the peak was.

What I have here is an op-amp connected in basically the unity gain follower configuration that I introduced at the beginning of this lecture, except there's that diode. That means the amplifier feedback only works if the input voltage goes positive relative to the voltage at the minus input. Otherwise, the diode is reversed biased and we don't have feedback, and the output of the op-amp will go to one of its limits. I'm going to now add a capacitor. That's going to be my sort of memory device that's going to hold this information. Since I don't want the capacitor to discharge easily, I'm going to throw a unity gain inverter on the other end of it, using the virtuous fact that the unity gain inverter has this very, very high input resistance. That's going to be a peak detector.

What I'd like to do now is imagine putting in an input signal that varies with time. What I get at the output of the diode and into the capacitor is when the input is going up, the output of the diode goes up. But when the input goes down, it can't go down because of the diode, so it stays there. Then another higher peak comes along and it locks in at that height peak, and that's how it works as a peak detector. Let's actually build a peak detector now. The peak detector we're going to build is, in fact, a second use of the phototransistor that I showed you with the current to voltage converter in a previous lecture. It's only the second of three examples we're going to use with that same light meter. We're now going to build a peak detecting light meter.

So here I have a different circuit on the board. In fact, this part of the circuit is exactly what we worked with yesterday. There's the little photo transistor; there's the op-amp; there's the resistor in series with the phototransistor; there's the feedback resistor. That's the same circuit we had in that previous lecture; I talked about it as a current to voltage converter. It became a light meter when the current source was a phototransistor. We have the phototransistor; the only difference now is I've built exactly the peak detecting circuit that I just showed you. The peak detecting circuit, remember, had two op-amps; the first one in a basically voltage follower configuration, except with a diode, and the diode's right there. It had a capacitor also that was going to store the charge we put on it and that was going to hold the peak level for us. Then it had a second unity gain inverter, so we'd have a very high input resistance and the capacitor wouldn't discharge very quickly. The size of the capacitor and the input resistance of that second op-amp determine how quickly the capacitor discharges, or equivalently how long it can store the information.

You might say you don't see two op-amps there, but you do. You see a dual 741 op-amp here, which is two op-amps on one package. In fact, actually, that's not the 741; yes, that is the 741 op-amp. A dual 741 op-amp; two 741s in one package. Now I'm going to demonstrate this circuit. I've got to be a little bit careful because the studio lights are quite bright, and remember, what this circuit is supposed to do is record the peak of the brightness. What I'm going to do is take a flashlight, and I'm going to start with the flashlight shielding the phototransistor from the studio lights. I'm going to turn on the circuit and watch the old analog voltmeter here. The old analog voltmeter jumps up a little bit. I now remove the shielding and it jumps up a little bit more as it sees the studio lights. I turn on my flashlight and it jumps up still more. Now, when I take a flashlight away, or even completely shield the phototransistor, nothing happens. The voltage there is an indication of what the peak voltage was, or equivalently the peak light level that that detector saw. We have a peak detector. A varying voltage was coming out of the first op-amp, and that varying voltage was representing the varying amount of light falling on the phototransistor. That varying voltage went into the peak detector, which used the diode, the first op-amp of the capacitor, to store the maximum voltage as the charge on the capacitor. Then that was read out by our voltmeter, which has a rather low resistance. That's one case where we would use a unity gain inverter to avoid the low resistance of the voltmeter from draining the capacitor.

There's a peak detector, and it's still reading that peak level. That's another addition to our zoo of possible op-amp applications. Let's move on to one more before we wrap up this lecture, and it's a little bit more complicated; take a little bit more time to explain it. I've got to go over and use the big screen briefly to talk about what we're doing here.

We're going to actually go back to an earlier application of op-amps before we talked about negative feedback, and that's the op-amp as a comparator where it simply looked at its two inputs and it went to one limit or the other depending on which input was bigger. Here I've got an op-amp hooked up as a comparator, and I've got plus 1 volt at the plus input, so what I'm doing is comparing the minus input with 1 volt. If the minus input is less than 1 volt, I'm going to get one value for the output; if it's greater than 1 volt, I'm going to get the other value. That's all it's going to be able to tell me: Is the input there greater or less than 1 volt? Is the battery good or bad? Did I turn on the green LED or the red LED? Is the output plus 15 volts or minus 15 volts? Let's take a little look at how this comparator works. V-out for any op-amp is A V-plus minus V-minus, and A is a great big number. V-out in this case is A times 1 volt minus V-in, and A is a big number. V-out goes to plus the supply voltage, 15 volts, typically. If V-in is less than 1 volt, because then V-plus is bigger than V-minus, and it goes to minus the supply voltage, typically minus 15 volts, if V-in is greater than 1 volt.

Now let's imagine we have a voltage that's slowly rising with time, and we'd like to know when that voltage reaches 1 volt. We're going to use this circuit, this comparator, to catch that point when the voltage reaches 1 volt.

279

Here's a graph of our voltage versus time at the input, and also the voltage that the comparator produces. Remember, if the input voltage is less than 1 volt, we're going to see an output of plus VS. If the input voltage is more than 1 volt, we're going to see an output of minus VS. Here's what happens: As soon as V-in reaches 1 volt, the output switches abruptly down and then becomes the minus limit. That's how this circuit should work.

That's ideal. There's always noise. We just discussed the use, for example, of a difference amplifier to help eliminate noise. What if that V-in has a little noise on it? Then things are going to look a little bit different. Let's look at the same situation now, but V-in is going to have some noise, and I'm going to actually exaggerate the noise to show you what happens. Here's V-in, but it's got some noise on it. The first time V-in crosses 1 volt, the output of the comparator, the op-amp without any feedback, falls to its negative limit. But then the input with its noise wiggles back down. It crosses below 1 volt again, and so it shoots back up. It's little noise wiggles another time. We shoot down again, and then we cross again, and up we go. Finally, V-in has risen to such a level that it's not going to go below the 1 volt, and then we're good. But what's happened is instead of getting a nice clean break from one level to another—from green LED to red LED, or whatever—we get this back and forth, back and forth business. You can imagine if the noise has higher frequency, that's going to be more significant. That's a problem for us, and we need to deal with that problem. How do we deal with it?

We deal with it with a circuit called the Schmitt trigger. The Schmitt trigger looks a little bit complicated, but bear with me. The Schmitt trigger actually has positive feedback. It has a feedback from the output back to the plus input. What's going on here? I've taken a string of resistors from plus 15 volts to minus 15 volts, a 14 kiloohm and a 16 kiloohm, and if you work out that voltage divider, you can convince yourself that the voltage at the plus input would be 1 volt; that's the 1 volt I want to compare with. It would be 1 volt, but for the presence of this extra resistor. That resistor's sensing what the output voltage is, and let's see what the effect of that positive feedback is. I have all these resistors. Remember, if it's 14 k and 16 k that comes out right on the 1 volt that I want to compare with. But I don't just have 16 k, because what I now have is a combination of resistors that looks like 15 volts, 470 kiloohms, and it's connected to minus 15 volts because I'm assuming at this point the output is at minus 15 volts. The input voltage here is less than 1 volt.

The effect of this extra resistor here is actually to lower that comparison voltage a little bit below 1 volt. In fact—I'm going to let you work this out if you want to do the project—it's down to about 0.75 volts. What that means is as the voltage falls, it now needs to fall below 0.75 volts before we get that transition. On the other hand, once it does fall below 0.75 volts, the output swings to plus 15 volts, and now that 470 kiloohm resistor is connected up to the plus 15 volts, and now I have this configuration. That pulls this point up above the 1 volt. In fact, as you can work out, it happens in this case to pull it up to 1.22 volts. What we've done, effectively, is to change the levels at which we're switching to two slightly different levels, and that phenomenon is called hysteresis. It's a phenomenon that says if we have a circuit, something that's supposed to switch back and forth between two levels. Don't switch always right at the same level. Wait until you get a little too high, and then wait till you get a little too low.

Something much simpler that has the same effect is the thermostat that I showed you when I first talked about feedback. If you set the thermostat to 68 degrees, you don't want the furnace to turn on as soon as it drops to 67.9999 degrees, because immediately it'll get above 68 degrees and it'll be 68.0001. It'll turn off and the furnace will be cycling on and off. Very bad for the furnace motor; a racket in your house. Instead, the thermostat has a built-in hysteresis. It maybe switches the furnace on when it gets to 67 degrees, and then it lets it heat up to 69 degrees before it shuts off. That's called hysteresis, and we represent hysteresis with a graph that kind of shows what happens, in this case of the circuit I just described, switching on the way up at 0.75 volts but then coming back down, switching at 1.22 volts and that avoids that switching. If we make the hysteresis big enough, bigger than that noise we were worried about, then we'll avoid that problem of multiple switching. That's the rather sophisticated phenomenon of hysteresis.

I want to end with one practical example of hysteresis. Here's an example: Here's a freezer alarm you might build. Suppose you have a freezer in your house, and you go away for a while and you're worried your food might spoil if the temperature goes above 32 degrees. Here I have the output of that thermometer that was part of the project in a previous lecture. Remember, it put out a voltage that was 1/10 of the Fahrenheit temperature. It's connected to the minus input. I have a simpler hysteresis set up, a simpler situation involving negative feedback. Here's what happens in this circuit: 32 degrees is 3.2 volts in this case, and so if we got below 32 degrees we're at plus 15 volts on the output. We got 3.2 volts; I've divided that voltage divider to make that be true at the plus input. But the problem is this: What if the temperature goes high, above 32 degrees? Then the thing switches. We turn on that LED, that red warning light, which you'll notice is connected with its forward bias direction upward so when the op-amp output goes to minus 15 volts, we get the light on. But at that point, the feedback, the positive feedback, swings the comparison to minus 3.2 volts, and that corresponds to minus 32 degrees, and that light won't turn off again until your freezer gets back down to below minus 32 degrees. No freezer gets down that low; freezers get down to about 0 degrees Fahrenheit. Once your freezer goes above 32, that red light turns on, and the positive feedback locks it on and it tells you that even if the freezer goes back to normal, the food may have spoiled because it was above freezing. It doesn't tell you how long; that could be another thing you could add to this.

Let me wrap up now with what we've got here. We've done many, many, many circuits here. We started with a voltage follower; we added a follower with gain; we did a difference amplifier; we did a peak detector and connected it to our light meter; we looked at the Schmitt trigger and this whole phenomenon of hysteresis; and we developed out of that this freeze alarm.

Here's a project if you want to do it; if you don't want to do it, fine. First part of the project is sort of mathematical and circuit analysis: Verify those 0.75 and 1.22 volt thresholds in the Schmitt trigger that I showed you in this lecture. The second part of the project is to design a difference amplifier that has a gain of 10 and very high input resistances like the teraohm, 10^{12} ohms, typical of an op-amp. A hint about that one is you'll need to combine several circuits from this lecture, and I'd like you to simulate this circuit and verify that it works.

Using Op-Amps with Capacitors
Lecture 15

In this lecture, we'll look at capacitors used with op-amps. In the previous lecture, we used a capacitor to detect the peak of a voltage, but the capacitor wasn't really active with the op-amp. The op-amps were present to help the capacitor get and hold the charge. Now, we want to use capacitors actively with op-amps, which means putting them in the feedback loop of the op-amp. Among the key topics we'll cover in this lecture are the following:

- Capacitors: charge, voltage, current
- A capacitor as a feedback component
- An op-amp integrator

- Application: an integrating light meter
- A triangle/square wave-function generator.

Capacitors: Charge, Voltage, Current

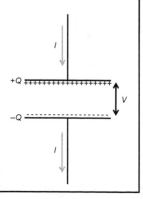

Capacitor Review

- Capacitor: a pair of conductors separated by an insulator
- Charge on capacitor plates is proportional to voltage between them: $Q = CV$
- Charge delivered by current I flowing in wires leading to capacitor
- Rate of charge buildup on capacitor proportional to I
- Voltage proportional to charge, so rate of voltage increase proportional to I

As you recall, a capacitor is a pair of conductors separated by an insulator. Remember that a capacitor is not like a resistor. It doesn't obey Ohm's law, because capacitor voltage and current aren't proportional. Rather, current is proportional to the *rate of change of voltage*.

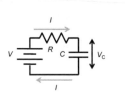

Charging a Capacitor with a Battery

- Capacitor initially uncharged; $V_C = 0$
- Current flows, delivering charge to capacitor
  - Current depends on voltage across R
  - As V_C increases, voltage across R decreases
- So rate of charge buildup on capacitor decreases
- So rate of increase of capacitor voltage decreases
- Eventually $V_c \rightarrow V$
  - Timing depends on "time constant" RC

A simple way to put charge onto a capacitor is with a battery. We then get a voltage that rises rapidly at first, but as the capacitor charges and the voltage across the resistor decreases, the rate of voltage rise decreases. There are many situations in which we would like to have a steady voltage rise, as in a cathode ray tube oscilloscope.

A Capacitor as a Feedback Component

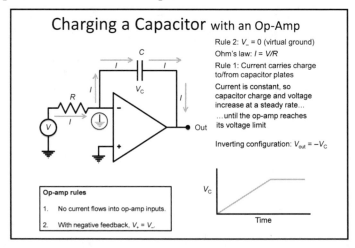

Charging a Capacitor with an Op-Amp

Rule 2: V_- = 0 (virtual ground)

Ohm's law: $I = V/R$

Rule 1: Current carries charge to/from capacitor plates

Current is constant, so capacitor charge and voltage increase at a steady rate...

...until the op-amp reaches its voltage limit

Inverting configuration: $V_{out} = -V_C$

Op-amp rules

1. No current flows into op-amp inputs.
2. With negative feedback, $V_+ = V_-$.

A simple inverting configuration shows what happens when we put a capacitor in the feedback loop of an op-amp and supply a constant input voltage. The circuit gives an output voltage that increases steadily with time.

An Op-Amp Integrator

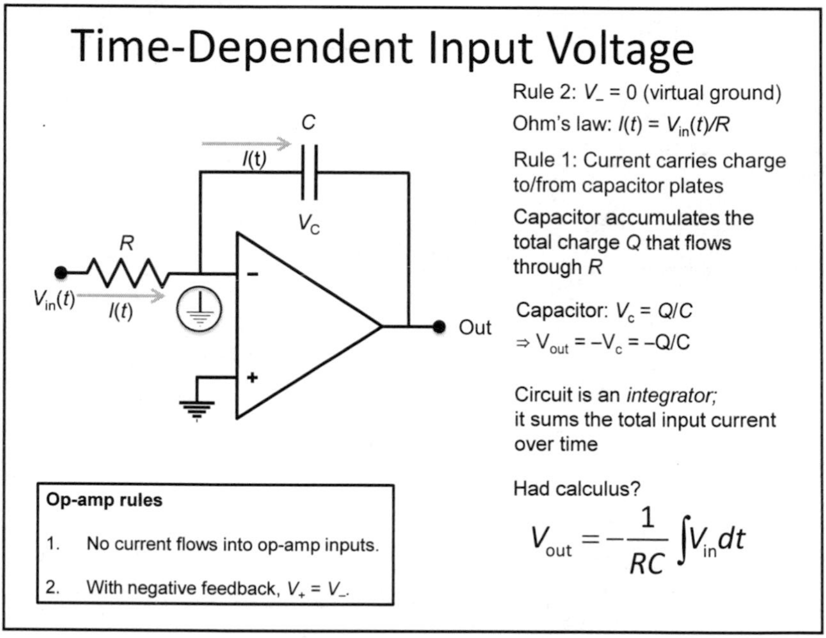

Time-Dependent Input Voltage

Rule 2: $V_- = 0$ (virtual ground)

Ohm's law: $I(t) = V_{in}(t)/R$

Rule 1: Current carries charge to/from capacitor plates

Capacitor accumulates the total charge Q that flows through R

Capacitor: $V_c = Q/C$

$\Rightarrow V_{out} = -V_c = -Q/C$

Circuit is an *integrator*; it sums the total input current over time

Had calculus?

$$V_{out} = -\frac{1}{RC} \int V_{in} dt$$

Op-amp rules

1. No current flows into op-amp inputs.

2. With negative feedback, $V_+ = V_-$.

Instead of a constant voltage at the input, which results in a linear rise or fall (depending on the sign of the input voltage at the output), we can use time-dependent input voltage. Here, as long as current flows in the direction shown, the capacitor charge keeps building up; the capacitor accumulates the algebraic sum of all the current. Consequently, we get a situation in which the output voltage of the circuit is proportional to the total charge that has accumulated on the capacitor. This circuit is an *integrator*; it integrates, or adds up, the algebraic sum of the charges delivered by all the currents over time.

Application: An Integrating Light Meter

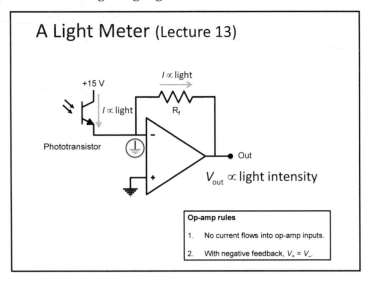

We've seen a light meter twice so far: first, as a device that gave an output proportional to the amount of light falling on it, then, as a peak detector that recorded the maximum light intensity.

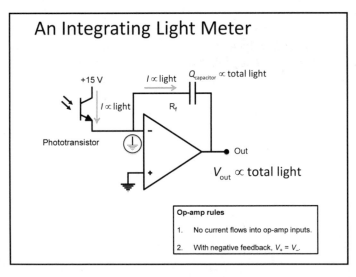

An integrating light meter is a third type; this light meter records the total amount of light that falls over time. Such a device might be used to determine the amount of solar energy accumulated over a certain period or to close a camera shutter after a specific amount of light had come through the camera's lens.

A Triangle/Square Wave-Function Generator

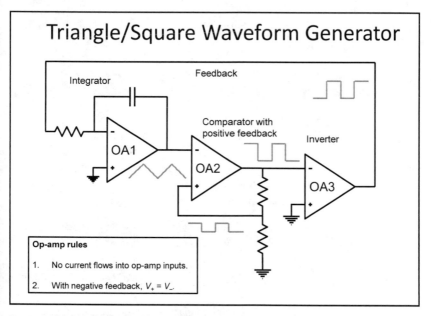

In general, to make a circuit that undergoes some kind of periodic back-and-forth oscillation, we typically need a combination of both positive and negative feedback, as we see in this simple triangle/square wave-function generator. This waveform-generator circuit consists of three op-amps: an integrator, a simple version of a Schmitt trigger, and a comparator.

Introductory
None of the introductory books covers the material in this lecture.

Advanced
Horowitz and Hill, *The Art of Electronics*, 3rd ed., chapter 4, sections 4.2.6 and 4.3.3.

Scherz and Monk, *Practical Electronics for Inventors*, 3rd ed., chapter 8, section 8.4, pp. 643–644.

Project

Triangle/Square Generator
Simulate the triangle/square generator described in the lecture. Use 10 kΩ for all resistors, 0.1 μF (100 nF) for C. Use 741 op-amps in CircuitLab, with explicit ±15-V power supplies. Simulate for 5 ms with 0.01 ms step; choose "Yes" to Skip Initial. Look at the outputs of the integrator and either other op-amp. As a challenge: Can you add a control for frequency adjustment?

Questions to Consider

1. Why, when charging a capacitor from a battery and a resistor, does the capacitor voltage not rise at a steady rate?

2. How does using a capacitor in the feedback loop of an op-amp allow the capacitor to charge at a constant rate?

3. What is meant by an *integrating* light meter? (Although the term *integrate* comes from calculus, you don't need to know calculus to answer this.)

Using Op-Amps with Capacitors
Lecture 15—Transcript

Welcome to the final lecture on operational amplifiers. In fact, this is the final lecture on analog electronics exclusively, although we'll come back to some analog electronics when we mix it with digital electronics toward the end of the course. Today's lecture, I want to look at what happens when we use capacitors along with operational amplifiers. You might say, "We did that in the previous lecture when we built that peak detector; we used the capacitor to detect the peak of a voltage," but the capacitor wasn't really active with the op-amp. The op-amps were sort of there to help the capacitor get the charge and hold the charge. Now we want to use capacitors actively with op-amps, and what that means is putting them in the feedback loop of the op-amp.

Before we do that, let me begin by reviewing what a capacitor is. Remember that a capacitor is a pair of conductors separated by an insulator, so we represent it by these two plates with the wires connecting in. We've put positive charge plus Q on one plate, negative charge minus Q equal in magnitude, opposite sign, on the other plate. We get a voltage across the two plates, and that voltage is related to the charge by the charge being the capacitance—the measure of what this capacitor can do—and farads times the voltage.

How does the charge get there? It gets delivered by whatever current is flowing in the wires leading to the capacitor. The rate of charge build up, as we found way back in the lecture when we worked with filters, is proportional to the current. The faster the current is coming in, the more current that's flowing, the faster charge is being delivered onto the plates of the capacitor. The voltage is proportional to the charge, and so the rate of voltage increase is proportional to the current. In other words, a capacitor isn't like a resistor. It doesn't obey Ohm's law. It doesn't have a proportionality between voltage and current; rather, it has a proportionality between rate of change of voltage and current.

Let's look a little bit further about what happens when we try to put charge onto a capacitor. One of the simplest ways to do that is in a circuit with a battery. I've got a battery, a resistor, and a capacitor; they're all connected in series. I start with the capacitor uncharged, and I want to ask what happens. The capacitor's initially uncharged, V sub C; the capacitor voltage is 0. But current can flow because that capacitor is acting at that instant almost like it's a short circuit because it has no voltage across it. All the battery voltage is dropped across the resistor, and so a current, V over R, flows and that current delivers charge to the capacitor. That current depends on the voltage across the resistor, and initially that voltage is big; it's the full battery voltage. As the voltage across the capacitor increases, however—which it does because as charge goes on the capacitor the voltage is proportional to the charge—that means the voltage across the resistor decreases, because the capacitor voltage and the resistor voltage have to add up to the battery voltage. The rate of charge buildup on the capacitor decreases because that rate is determined by the current, and the current is determined by the voltage across the resistor, and that's going down. The rate of increase of the capacitor voltage also decreases, and eventually what happens is the capacitor voltage becomes equal to the battery voltage, but that takes a very long time typically. That timing, how long it takes, depends on the so-called time constant of the circuit, the product of the resistance and the capacitance. If you work that product out, you'll find it actually has the units of seconds. If we were to plot the voltage across the capacitor versus time as a function of time measured in this time constant R times C, you'd find that there's a sort of exponential charging curve. The voltage rises rapidly at first and then it tapers off, and only asymptotically does it approach the battery voltage.

That's what happens when you try to charge a capacitor with a battery. Don't take my word for it; let's do it. Over here, I have a very simple set up consisting of a 6-volt battery, a 6.8 kiloohm resistor, a 2,200 microfarad capacitor (2.2 millifarads), and our good oldfashioned analog voltmeter connected directly across the capacitor. Right now I have this black wire, which is short circuiting the capacitor, so the capacitor is guaranteed to have 0 volts across it. What I'm going to do is remove that wire, and the capacitor will charge, and let's watch it charge on the voltmeter. The voltage goes up gradually; it goes up fairly rapidly at first. Now it's approaching 2 volts; going up a little slower. It's going up slower and slower; it's approaching 3 volts. It's going up slower and slower; it's approaching 4 volts. It's going to go up very, very slowly until it eventually gets fairly close to 6 volts, although because this is a simple old-fashioned analog voltmeter, it has a pretty low resistance itself, and it's acting a bit like a voltage divider so it'll never quite get all the way to 6 volts. But the point is this: When you try to charge a capacitor with a battery or another constant DC source through a resistor, you get a voltage that rises rapidly at first and then the rise decreases; the rate of rise decreases.

There are many situations in which you would like a steady voltage rise, a voltage that rises steadily with time. I can give you one good example and that's an old-fashioned cathode ray tube oscilloscope where the electron beam would sweep across the face of that tube, and it did so with a voltage that rose steadily on some electrodes in that tube, and you needed a circuit to produce a voltage that rose steadily with time. That's not what happens when you simply tried to charge a capacitor with a battery and a resistor.

We'll end this little experiment by shorting the capacitor back to 0 again, and talk about how we might achieve a steady rise in voltage. How we might achieve that is with feedback. Feedback is going to come to our rescue and we're going to put capacitors in the feedback loop of an op-amp. Let's see how that works. We'll go over to our onset first and do a pretty careful analysis of what happens when we put a capacitor in the feedback loop of an operational amplifier. Here I have a configuration, and you'll recognize it as the inverting configuration because the plus input of the op-amp is grounded, the minus input of the op-amp is connected through a resistor to some input—we'll talk about that in a minute—but instead of a feedback resistor, we have a feedback capacitor. As long as the system is in control— let's put a voltage across that input; let's assume for now it's a fixed voltage like we'd have from a battery—rule two says the plus and minus inputs are equal. In this case, the plus input is at ground, so the minus input is also at 0 volts. It's not really grounded. It's at that virtual ground, again being held there actively by the op-amp's action. I have to caution you: In circuits with capacitors, it's very easy for the op-amp to get out of control and then that's no longer true. But while the op-amp is doing its thing correctly and we're in control—we have the negative feedback working—that will be at virtual ground. That's what rule two says. Ohm's law says $I=V/R$, then, because as far as the resistor is concerned, the right hand end of it's at ground, the lefthand end of it is at V.

Rule number one says the only place the current can go is through the feedback loop. What does the current do? In this case, it delivers charge to the plates of the capacitor. If it's a positive voltage, it puts positive charge on that plate, and positive charge leaves the other plate and negative charge goes on the other plate; the capacitor charges. The current is therefore carrying the charge to and from the capacitor plates because it can't go in to the input of the op-amp.

Here's the beauty of this circuit: The current is constant. Why is it constant? Because the op-amp is working to keep this point at virtual ground, and so the current is constant and the capacitor charge and voltage therefore increase at a steady rate, and they do so—and it just goes up and up and up—but it can't go up forever because the op-amp's output voltage can't exceed the supply voltage of the op-amp. When we get to that point, we've saturated the op-amp. The feedback is no longer in control, and the output is at one or the other of the limits. In fact, as I've drawn it here, this graph isn't quite correct because I've got current flowing in this way and that means this end of the capacitor is positive and the output would actually be negative because this is an inverting configuration. But you get the idea; the voltage increases in magnitude. It would actually go down and then it would reach the minus limit; if I reverse that voltage it would go up. The beauty of a capacitor in the feedback loop of an op-amp is it allows us to make a circuit in which we can make steadily increasing voltages; voltages that increase steadily with time. There are many, many occasions when we want to do that. We have this inverting configuration; V-out is, in fact, minus DC.

But that's not all we can do with capacitors in the feedback loop of an op-amp. Here's what happens if we put in a time-dependent voltage at the input. Instead of just a steady voltage at the input, a constant voltage, which results in a rising or falling linearly, depending on the sign of the input voltage at the output, let's put in a voltage that varies with time. I've marked it explicitly V-in(T). V-in is a function of T, time, and I've marked the current I also to be a function of time. Let's analyze what happens with this circuit. Rule two says, as always—in this case with the plus input grounded—that the minus input is at virtual ground. Ohm's law says, as it did before, that the time-changing current is equal to the time-changing voltage over that resistance. Rule one says no current flows into the inputs of the op-amp, and therefore all the current is flowing to carry charge onto or off the plates of the capacitor.

What does the capacitor do? It therefore accumulates the charge from the current that's flowing in the wires leading to the capacitor. If current is always flowing in the direction shown in this diagram, the capacitor charge just keeps building up. If current ever flows the other way, charge is removed from the capacitor, but it accumulates, if you will, the algebraic sum of all the currents. It, in other words, builds up current over time. VC, the capacitor voltage, is proportional to the charge on the capacitor, and consequently we get a situation in which the output voltage of this circuit is proportional to the total charge that's accumulated on that capacitor, not like in our peak detector because now we're counting charge that both flows off and flows on.

The circuit is what we call an integrator. It integrates or adds up the algebraic sum of the charges delivered by all the currents over time. Have you had calculus? If you have, you'll recognize that what this circuit is doing basically is integrating. The output voltage is the integral of the input voltage, and the constant in front of the integral is RC, that time constant. Once again, if you haven't had calculus don't worry about this; but if you've ever seen calculus, a capacitor integrates. An op-amp circuit with a capacitor in the feedback and a resistor at the front is an integrator and performs that difficult mathematical operation of integration, which is why this kind of circuit was once used in analog computers to solve complicated mathematical equations. We no longer do that, but it's been done. By the way, just another hint for those of you who've had calculus: If I switch the C and the R, put the capacitor at the input and the resistor in the feedback loop, I'd have a differentiator, and the circuit would take the derivative of whatever signal I was sending at the input.

A circuit with a capacitor in the feedback loop is an integrator. It sums up the charge delivered by all the currents that are coming in to the input of that circuit. Let's take a look, in fact, at how that works. Let's take a look at what one of these circuits actually does. Over here at this point on the board I have an op-amp again, a 741 op-amp; I have a 1 kiloohm input resistor; I have a small capacitor between the output and the input; and that's it. The integrator circuit coming in is a sine wave coming out of my function generator, and also showing on channel one of the oscilloscope, the yellow channel. Let's take a look at the same time at the output. The output is coming through this probe. It's connected to the output of the op-amp, which is also one side of the capacitor.

Let's take a good look. First of all, the output is also a sine wave. Its amplitude isn't the same; those two channels are on two different scales, so the amplitude of this is actually a little bit less than half of that. That's determined by a combination of the frequency of the sine wave and also the values of that R and C. I've chosen them so that they work fairly well for this particular circuit. Notice something else, though. Here's the peak of the sine wave coming in, but the peak of the sine wave going out is in a different location. It isn't completely out of phase; if it were, the peak of the green would be below the trough of the yellow. But it isn't; it's sort of halfway along. In fact, the green is peaking right about where the yellow is crossing 0 on the way up. We've sort of seen that before because we talked about how capacitors change phase, but now we can think about it a little bit more mathematically. In fact, if you know about calculus, again, you know that the integral of sine is cosine. What's a cosine? A cosine looks just like a sine except it's shifted. Cosine peaks at 0 when its argument is 0, the function cosine, and sine peaks when its argument is 90 degrees, and is 0 when its argument is 0. We have exactly that same thing right here: This little triangle marks the time the oscilloscope thinks is time; T=0, it calls it. That's when the sine wave that was our input is rising on its way up, and that's when the cosine wave, which is the integral of the sine, is peaking. This, starting here at T=0, this is a cosine, one cycle of a cosine; this is one cycle of a sine. This device has actually performed this mathematical operation we call integration.

Even if you don't know about calculus, what the circuit has done is accumulated the currents that are being pumped in as a result of this input voltage applied across that input resistor, and those currents flow on and off the plates of the capacitor, and they result in a voltage across the capacitor and consequently an output voltage of the op-amp that's out of phase by 90 degrees, by a quarter of a cycle, relative to the sine wave.

That's all well and good, and we could think about the mathematics of that some more, but let's look at some more interesting waveforms. I haven't done much with this, but again this function generator is capable of many different waveforms. I'm going to go over here and press the waveform button. Right now it's set on sine, and I'm going to change it to a square wave. Whoa, so let's think about what's happening here. Now the input is a square wave; again, that means it starts out at some low value. In this case, we're at 500 millivolts per division, and so we're coming in at a low voltage of about minus 250 millivolts, going up to a high level, down, up, down. We're staying at these two levels.

What happens? What happens is as the level is low—remember this is an inverting configuration, this integrator—the voltage rises, the voltage across the capacitor rises, at a steady rate. When the input voltage is negative, we get a rise because of the inverting configuration. When the input voltage swings positive, we take charge off the capacitor plates; we get a steady fall. When the square wave goes negative again, we rise, and we've generated a beautiful triangle wave. Well, not quite beautiful; there's this little spike at the top. It has a little trouble handling this rapid change that's going on. But that's turned the square wave into a triangle wave; the integration of a square wave is a triangle wave. We don't need to know calculus to do that; we can just see that by seeing if the voltage is steady, we know we get a steady rise or fall in the capacitor voltage. That's what we're seeing here, and that alternates as the square wave at the input alternates sine.

We can do more than that. We can go back and look at other waveforms. Here's a waveform called a ramp. That ramp waveform has inside it a circuit very much like the one we've been talking about that produces a steadily rising voltage, then it switches right back to where it started and rises again and just keeps repeating that cycle. Lo and behold, you can convince yourself that what happens if the voltage is steadily rising, the charge on the capacitor then rises at first and then begins to level off and then goes down, and it repeats that process.

By the way, I didn't show it here, but if I'd made this a symmetric ramp that went up and down and up and down, we would've seen that this looked almost like a sine wave. It wouldn't be a sine wave; it would be a series of parabolas because the integration of the function x, a rising function, is $x^2/2$, if you know calculus. We would get something that looks like a sine wave. But that's sort of a cheap way to make something that is approximately a sine wave, and very cheap function generators might attempt to do that. It's not a good way to make a sine wave, but it would almost look indistinguishable from a sine wave.

There's this circuit performing the process we call integration; what do we want to do with this circuit besides do calculus? Let me show you one useful example. Let's go back to our light meter again.

We've used the light meter twice already. We used it first to make just a simple light meter that gave an output that was proportional to the amount of light falling on it; then we made a peak detector that detected the maximum amount of light. Now we're going to make an integrating light meter, a light meter that records the total amount of light that's fallen on this thing. You might imagine making such a device to determine something about the amount of solar energy you get over a period of 15 minutes or something as the sun fluctuates going in and out of clouds, or something like that. We're going to build an integrating light meter. It's going to determine the total amount of light that falls over time. You could imagine making a camera that looks at the total amount of light that comes in and says "I'm going to close the aperture when that reaches a certain level that gives me the right exposure." An integrating light meter is a useful thing.

We're going to start with a light meter we had before. It was based on a phototransistor. The phototransistor sent a current into an op-amp connected in the inverting configuration at first with a feedback resistor, and we got an output voltage that was proportional to the light intensity. Now we're going to do something different: We're going to replace that resistor with a capacitor. That capacitor is going to now charge, and instead of an output voltage that's proportional to the current that's flowing, we're going to get an output voltage that's proportional to the total amount of charge that that flowing current has delivered. That's why this will be an integrating light meter that will add up exactly the total amount of light that's fallen on that phototransistor. Let's go ahead and build that thing with our current proportional to the light, the current going onto the capacitor plates proportional to the light, and therefore the charge is proportional to the total amount of light that's fallen on the phototransistor; and so the output voltage also becomes proportional to the total light. Let's look at a circuit that does that.

Here we are. I have the same circuit I've just been using. By the way, you may wonder why I keep stopping to reconfigure. It's because I want to show you that wiring op-amps is really simple. Once I've got an op-amp sitting on my board, it's very simple to reconfigure it into a totally different circuit. All I've done with my integrator circuit I showed you before is to replace that relatively small capacitor with a very large capacitor, a 2200 microfarad capacitor. That's because I'm going to be concerned here with relatively slow variations. Before I had a 1000 Hertz variation, and this capacitor works better to accumulate the large amounts of charge that are going to come flowing on here. By the way, you've noticed some LEDs. Don't worry about those; we'll talk about those shortly.

Right now I've changed the circuit so that instead of connecting the input from my function generator, I've got this orange wire here looping over to a phototransistor, which you can just barely see; a little clear plastic phototransistor, the same one we used before. The output of the circuit is looking at my old-fashioned analog voltmeter. Because this is an inverting configuration and this time I'm not turning the transistor around, you'll notice I have the red into the black and vice versa so the voltmeter will still read the negative voltage as a positive voltage. But let's not worry about that. The other thing you'll notice is I have this big yellow looping wire, and that wire is at this moment short-circuiting the capacitor. I'm going to pull that wire out, and the phototransistor is going to begin to pass current depending on how much light is falling on it. That current is going to accumulate charge on the capacitor, and the capacitor is going to begin to build up charge.

I'm first going to pull this out and we'll let the ambient light add some charge to the capacitor, and then we'll go with our flashlight. Here we go. Capacitor begins to charge from the ambient light, accumulating ambient light. We'll put the brighter light on, it goes up faster. We'll take the light away; it still continues to go up. This isn't a peak detector; it's adding up the total amount of light that's fallen on this over time. Put some more light in, up it goes further. Put some more light in, up it goes further.

That's in integrating light meter. That's a circuit you might use to determine over some time interval of your choice—it could be a thousandth of a second; it could be, as in this case, a few seconds—the total amount of light that's accumulated in this system. We have an integrating light meter, a very practical device using this idea of an integrator: a capacitor in the feedback loop of an op-amp that accumulates a total amount of charge depending on how much current has flown into that system.

Here we are for our last analog circuit, and probably our most sophisticated. You'll notice that I've intentionally removed that function generator. I've used that a lot throughout this course, and you may have wondered how it works. The reason I've removed it is to emphasize that our final circuit is a function generator that we're going to build ourselves. We're going to build a generator, in fact, mixed with triangle waves and square waves, and we're going to see how it works. Then if you want to do the project, you can build one yourself.

Let's talk a little bit about how you make a waveform generator. In general, to make a circuit that undergoes some kind of periodic back and forth oscillations, you typically need a combination of both positive and negative feedback. That's what we're going to have in this simple triangle square generator that we're going to build. How does a function generator work? It has a more sophisticated circuit than what I'm showing you, but here's a lot of the basic idea. Here's the circuit I'm going to build and demonstrate and that you can also build and play with in the project if you choose to do it. It consists of three operational amplifiers. Two of them are in the inverting configuration, the one on the left is an integrator; we recognize that from this lecture. The one in the middle is that circuit with positive feedback, a kind of Schmitt trigger-like thing, although a little more simple. It simply has a voltage divider string at its output, and its plus input is tapping off a portion of that. This one's got positive feedback, and this one is functioning as a comparator and it's comparing the output of the first operational amplifier, the one that's the integrator, with whatever's at the plus input of the second operational amplifier. Whatever's at the plus input of that second one is half of what's at its output if those two resistors are the same; they don't have to be, but if they are.

What's at the output of the second op-amp? I don't know, but I do know that it's open loop; that is, it has no feedback, at least right within itself. There's kind of overall feedback going on in the whole circuit. But OA two doesn't have any intrinsic feedback, and so its output is at either plus or minus limit, plus or minus approximately 15 volts for typical op-amps. That's the situation, and so the junction of those two resistors, which is equal to the voltage at the plus input of OA number two, is either plus 7.5 volts or minus 7.5 volts. When the output of the integrator, the output of op-amp number one, reaches either plus or minus 7.5 volts, things are going to happen. That's going to switch the output of op-amp two to the opposite value.

What's op-amp three? It doesn't have any feedback either. It's got its plus input at ground, so it's also a comparator and it's just sensing whether the input to the minus input of that op-amp is either above or below 0. It's basically functioning as an inverter, because if the minus input is below 0 it's less than the plus input and the output goes positive, the positive limit. If the minus input to OA three is greater than 0 volts, greater than ground, it's going to switch the other way. OA three is simply functioning to invert whatever the output of OA two is. OA two is at plus or minus 15, and OA three is in at minus or plus 15. That's the inverter.

There's the feedback that makes this all work. I claim that's a triangle square wave generator. I claim we're going to get a square wave there. I claim we're going to get a triangle wave there. We're also going to get a square wave at the output of OA two, but it's going to be out of phase with the square wave at OA three because OA three is an inverter. We'll also get a square wave at the junction of those two resistors, and it'll have half the amplitude of the square wave at the output of either OA two or OA three. There it is: a lower level output square wave. This is a triangle square wave generator. Does it work? Let's try it out and see.

Here I have that circuit wired. Three op-amps in a row, one, two, three: the integrator, you can see the capacitor, which is its feedback loop; the second one with that voltage divider string going across, and down here, that's the voltage divider; and the third one, which inverts. At the output of the third one, I have a resistor and two LEDs, and those LEDs look like they're just lit all the time. But they aren't lit all the time; let's take a look at what we actually see at the output of those circuits.

I have the oscilloscope on. You'll notice now I'm using only one probe on the oscilloscope because we've got a circuit that makes its own waveform. This circuit isn't something that needs an input and then makes an output; this is a circuit that makes its own output. I'm going to connect this first to the output of the last op-amp, the one that's functioning as an inverter, and there's our square wave. That square wave is swinging between about we've got 5 volts per division. Here's 0; one division, two division; about 12 1/2 volts, which is about the most you can expect out of a 741, swinging down to about minus 12 1/2 volts, a nice square wave. Now we're going to connect to the output of op-amp number one, which is at one side of the capacitor; let's just clip on to the capacitor. There we go, and we have a beautiful triangle. If I put both of these on at once, you'd see the triangle switching at the peak and minus just as that square wave that's coming out of those other op-amps switches. There's a beautiful triangle square wave generator. Very simple circuit: Three $0.40 op-amps, a few resistors, and a capacitor, and we've got a function generator.

What determines the frequency? Obviously the value of the capacitor for one thing, because the bigger it is the longer it takes to charge. Instead of putting that capacitor, let's switch to a much bigger capacitor and see what happens. Now I've replaced a capacitor that was a fraction of a microfarad with a capacitor that's, in fact, 100 microfarads; much bigger. It's going to take much more time to charge. Let me turn the circuit on, and you can see what's happening. Now you can see those LEDs that are connected at the output of the square wave part of the function generator are flipping back and forth between red and yellow as the output goes alternately toward plus limit, and then toward minus limit.

That's a pretty sophisticated circuit we've developed, and that's where we're going to wrap up with analog electronics. I do recommend that if you'd like to do the project, you simulate the triangle square wave generator we just described; I give you some parameters. I caution you that doing this circuit is really pushing the limits of some of these simulated circuit software that you've been using. For some values of components, some kinds of op-amps, they may simply not work. If you do the project, I'll give you some more hints about how to make it work.

Digital versus Analog
Lecture 16

To this point in the course, we've been working with analog electronics, but with this lecture, we switch to digital electronics. Digital circuits use voltages to represent two distinct states: 1 and 0. Strings of 1s and 0s represent binary numbers or encode other information, such as letters, and these digital quantities can be manipulated and combined with simple logic operations. This lecture will lay the foundation for understanding digital electronics before we move on to developing digital circuits. Important topics include the following:

- Digital versus analog
- Binary numbers
- Logic operations: AND, OR, NOT
- Truth tables and Boolean notation

- Logic gates and symbols
- The NAND and NOR operations
- Everything from NAND or NOR!

Digital versus Analog

Analog	Digital
Continuous range of voltages and currents	Two states
Electrical quantities are analogs of physical quantities	Values of physical quantities are encoded as binary numbers
Irreversible degradation in transmission and copying	Error-free transmission and copying

Here are some important differences between analog and digital. Analog is the old way of doing things in electronics, and, increasingly, digital is the new way.

Analog quantities take on a continuous range of values. These include most physical quantities, such as temperature, voltage, speed, weight, time, intensity of sound, and so on. In fact, most of the quantities in the physical world are analog quantities. We can't get away from the fact that we live in a largely analog world, but today's electronics are largely digital.

© Stockbyte/Thinkstock.

Some real-world quantities are intrinsically digital, taking only discrete values. One example from everyday life is money.

Continuously varying analog quantities can be converted to digital. For example, an electronic thermometer "discretizes" the temperature. It might read 98.6° or 98.7°, but it won't read 98.7392° because it doesn't have that many digits, and it can't cover the continuously varying range of temperature. We represent these analog quantities as digital quantities with limited numbers of digits. And, ultimately, those digits are reducible to 0s and 1s.

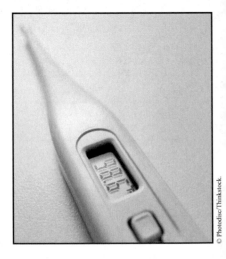

© Photodisc/Thinkstock.

Binary Numbers

Binary Numbers

- Two binary digits (bits)
  - 0   "zero," "no," "false," "low"
  - 1   "one", "yes," "true," "high"

Places: 8 4 2 1

$$1001_2$$
$$= 1 \times 8 + 0 \times 4 + 0 \times 2 + 1 \times 1$$
$$= 9_{10}$$

Every quantity in the world of digital electronics is represented with two binary digits: 0 and 1.

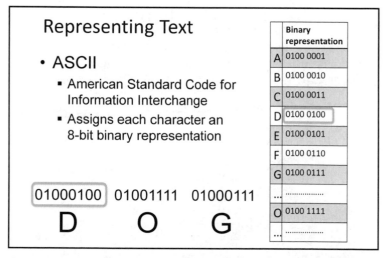

Representing Text

- ASCII
  - American Standard Code for Information Interchange
  - Assigns each character an 8-bit binary representation

01000100 01001111 01000111
D   O   G

	Binary representation
A	0100 0001
B	0100 0010
C	0100 0011
D	0100 0100
E	0100 0101
F	0100 0110
G	0100 0111
...	....
O	0100 1111
...	....

Text is represented with the American Standard Code for Information Interchange (ASCII).

Logic Operations: AND, OR, NOT

Logic Operations

- **AND**
  - Output is 1 only if both inputs are 1
  - Boolean algebra: A·B

  A
  B ─── A·B

  AND logic symbol

- **OR**
  - Output is 1 if either input is 1
  - Boolean algebra: A+B

  A
  B ─── A+B

  OR logic symbol

- **NOT**
  - Output is opposite of input
  - Boolean algebra: $\overline{A}$

  A ─▷○─ $\overline{A}$

  NOT logic symbol

At the heart of processing digital information is the science of digital logic—a combination of mathematics and philosophy that governs everything in digital electronics. Just as we add, subtract, multiply, and divide in mathematics, we can do simple operations, known as AND, OR, and NOT, in digital logic. These operations are represented with the logic symbols shown above. The output of an AND operation is 1 only if both inputs are 1. The output of OR is 1 if either input is 1. The output of NOT is the opposite of the input.

Truth Tables and Boolean Notation

Logic operations can be represented with *truth tables*.

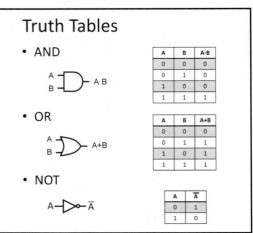

Truth Tables

- **AND**

  A
  B ─── A·B

A	B	A·B
0	0	0
0	1	0
1	0	0
1	1	1

- **OR**

  A
  B ─── A+B

A	B	A+B
0	0	0
0	1	1
1	0	1
1	1	1

- **NOT**

  A ─▷○─ $\overline{A}$

A	$\overline{A}$
0	1
1	0

Logic Gates and Symbols

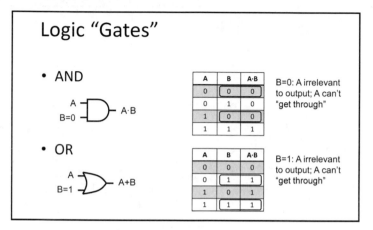

Circuits that implement logic operations are called *logic gates*. Consider an AND operation in which the B input is 0. If that's the case, then the output must be 0 because for an AND operation, the output is 0 if either of the inputs is 0; both inputs must be 1 for the output to be 1. Thus, the A input is irrelevant to the output in this case; whatever A is, it doesn't have any effect on the input. The operation is like a gate that's closed; A can't "get through."

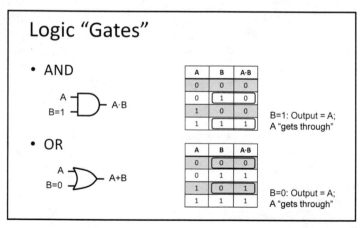

On the other hand, if B = 1, then the output is the value of A. Now A "gets through" and the gate is open. A similar gating happens with the OR operation.

More Logic Gates

- ## NAND (NOT AND)

  - Output is 0 only if both inputs are 1

  - Boolean algebra: $\overline{A \cdot B}$

NAND logic symbol

A	B	A·B	$\overline{A·B}$
0	0	0	1
0	1	0	1
1	0	0	1
1	1	1	0

- ## NOR (NOT OR)

  - Output is 0 if either input is 1

  - Boolean algebra: $\overline{A+B}$

NOR logic symbol

A	B	A+B	$\overline{A+B}$
0	0	0	1
0	1	1	0
1	0	1	0
1	1	1	0

- ## NOT from NAND, NOR

NAND and NOR gates are somewhat more obscure to think about but much more useful than AND and OR gates. NAND stands for NOT AND, and its output is 0 only if both inputs are 1; it's exactly the opposite of the AND gate. NOR stands for NOT OR, and its output is 0 if either input is 1; the only possible state in which the NOR gate has a non-0 output is when both inputs are 0. Once we have NAND or NOR gates, we can also make NOT gates by connecting together the two inputs of a NAND or a NOR gate.

Connecting the two inputs in the AND gate truth table rules out the middle two rows, which have different values for A and B. We have either 00, in which case we get 1 at the output, or 11, in which case we get 0; thus, the AND has become a NOT gate. NANDs and NORs are useful because we can use them to build other gate operations.

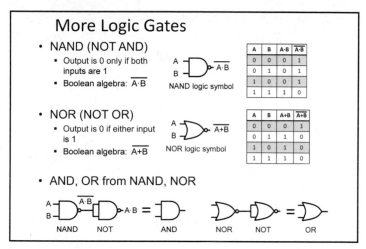

More Logic Gates

- NAND (NOT AND)
  - Output is 0 only if both inputs are 1
  - Boolean algebra: $\overline{A \cdot B}$

NAND logic symbol

A	B	A·B	A̅·̅B̅
0	0	0	1
0	1	0	1
1	0	0	1
1	1	1	0

- NOR (NOT OR)
  - Output is 0 if either input is 1
  - Boolean algebra: $\overline{A + B}$

NOR logic symbol

A	B	A+B	A̅+̅B̅
0	0	0	1
0	1	1	0
1	0	1	0
1	1	1	0

- AND, OR from NAND, NOR

NAND and NOR can make NOT; if we want to make an AND gate, we start with a NAND and add a NOT. If we invert the output of a NAND gate, we get back to an AND gate. We could build that structure with two NANDs, with the second one connected as a NOT gate. Equivalently, we can make an OR gate from NOR and NOT.

Everything from NAND or NOR!

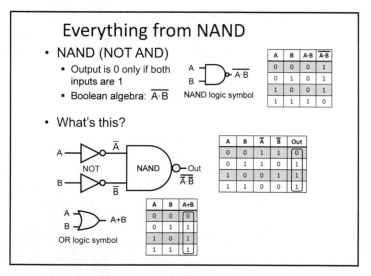

Everything from NAND

- NAND (NOT AND)
  - Output is 0 only if both inputs are 1
  - Boolean algebra: $\overline{A \cdot B}$

NAND logic symbol

A	B	A·B	A̅·̅B̅
0	0	0	1
0	1	0	1
1	0	0	1
1	1	1	0

- What's this?

A	B	A̅	B̅	Out
0	0	1	1	0
0	1	1	0	1
1	0	0	1	1
1	1	0	0	1

OR logic symbol

A	B	A+B
0	0	0
0	1	1
1	0	1
1	1	1

We can build everything from the NAND operation. For example, we can make NAND gates into inverters, put the inverters on the front of the input to another NAND gate, and get an OR gate. We can also get NOT from a NAND gate, and we can get NOR from NAND by inverting the OR gate we just built.

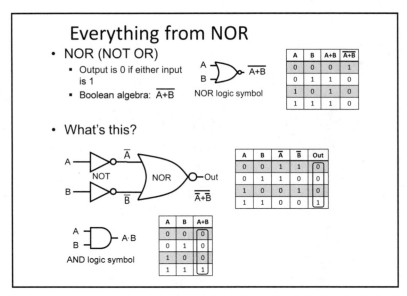

Similarly, we can get everything from NOR if we put inputs into the NOR and invert them. These versatile NAND and NOR gates are the basic building blocks of digital circuits.

Introductory

Brindley, *Starting Electronics*, 4th ed., chapter 10.

Lowe, *Electronics All-in-One for Dummies*, book IV, chapters 1–2.

Shamieh and McComb, *Electronics for Dummies*, 2nd ed., chapter 7 through p. 151.

Advanced

Horowitz and Hill, *The Art of Electronics*, 3rd ed., chapter 10, section 10.1.

Scherz and Monk, *Practical Electronics for Inventors*, 3rd ed., chapter 12, sections 12.1–12.2.2.

Projects

Digital Logic

How many two-input logic operations are possible? How many were introduced in this lecture?

Using any combination of AND/OR/NAND/NOR/NOT, design a three-input AND gate.

Using only NAND gates, design an Exclusive OR gate—whose output is 1 if one of its two inputs is 1, but not both. Challenge: What's the minimum number of gates you can use?

Questions to Consider

1. Why are NAND and NOR more useful logic operations than the more familiar AND and OR?

*2. Construct the truth table for the following logic circuit and write a symbolic expression for the output in terms of the two inputs, A and B.

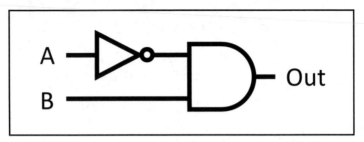

Digital versus Analog
Lecture 16—Transcript

Welcome to the first of our lectures that deals with digital electronics.

Just before this lecture I went out to reheat my cup of tea in the microwave oven and I discovered, to my surprise, that I actually had to turn a knob to set the time on the microwave oven. I thought how rare that is these days, how old-fashioned that is, because today almost everything we deal with electronically seems to be digital. Digital is all the thing; all the rage.

Can you remember what happened on June 12, 2009? Think a minute about that. That was a famous day in the history of electronics because that was the day that analog television went completely off the air, and television from then on is digital. What does it mean to be digital versus analog? We've been doing analog electronics until now; we're switching now to digital electronics. What's the difference? Let's take a look at the contrast between analog and digital.

In analog, we deal with a continuous range of voltages and currents and other electrical quantities. In digital, in contrast, there are really only two states. That's a slight understatement; there could be many, many states. But there are a discrete number of states, and they're ultimately based on just two possible voltages or currents. The electrical quantities we deal with in analog electronics with the amplifiers we've been building and so on, these are analogs of physical quantities. The voltage might correspond to the speed, or it might correspond to the intensity of sound, or it might correspond to the intensity of light, and we actually built meters in which we saw a voltage that corresponded to the intensity of light, for example. In digital electronics, the values of the quantities we want to measure or talk about or characterize are encoded as binary numbers. They're not done analogously; the electrical quantities aren't analogous to the physical quantities. Instead, we encode the physical quantities as numbers, binary numbers, based on just those two possible states.

Another aspect of analog electronics is there's an irreversible degradation when we transmit or copy information in the analog way. You've seen this if you're old enough to remember VHS tapes, which most people are. VHS tapes, if you ever copied one, it got worse and worse and worse. You can see the same thing if you go to a photocopier and try to keep copying something over and over again. Errors creep in, and they multiply, and they get worse.

In the digital world we have error-free transmission and copying; which, by the way, is why publishers, movie makers, and so on are very, very, very worried about piracy in the digital age, because you get perfect copies. Analog is the old way of doing things; digital is the new way of doing things. I want, in this lecture, to explore in detail the difference between analog and digital. This will be kind of a meta electronics lecture. We're not going to do any actual electronics. We're not going to hook up any oscilloscopes. We're not going to make any circuits. I'm not even going to talk about any transistors; that will come later. I want to lay the foundation for understanding digital electronics.

Let's look at some examples of what it means to be digital versus analog. In analog quantities, just to give you some examples of analog quantities, they take on a continuous range of values. They include most physical quantities; examples: temperature, voltage, speed, weight, time, intensity of sound, and so on and so forth. In fact, many of the quantities in the physical world are analog quantities; we can't get away from the fact that we live in a largely analog world. But today's electronics is largely digital. A couple of examples: A thermometer with a column of alcohol or some other liquid—it used to be mercury, now it's something else—that rises up. The height of that column is a measure of temperature, and that's a continuously variable quantity, as is the temperature. A speedometer on your car, a dial-type speedometer, is an example of a measurement of a continuously varying quantity: the speed of the car. Finally, an ordinary old-fashioned analog clock with hands is an analog measurement of a quantity that's continuously varying with time.

On the other hand, many of the quantities we want to measure either are intrinsically digital or get converted to digital quantities. They take only discrete values. A couple of examples from everyday life are: cash money. If you have a pocketful of change, there are a discrete number of cents in there. You don't have half a cent; you don't have a tenth of a cent; you don't have pi cents. You have $0.25 and then a penny, so you've got $0.26 or whatever.

You buy eggs at the grocery store, you could buy a dozen eggs; maybe you could buy half a dozen eggs; maybe you can even convince them to sell you an egg; but you can't buy half an egg, or 3.7 eggs, or anything like that. Eggs and cash money come in discrete quantities.

On the other hand, many other quantities that we convert to digital quantities are then approximations of those analog quantities that are continuously varying. Here are just a few examples of those: An electronic fever thermometer discretizes, if you will, the temperature. It might read 98.6, it might read 98.7, but it won't read 98.7392 because it doesn't have that many digits and it can't cover the continuously varying range of quantities. A digital clock reads the time in numerical values, and a digital speedometer reads the speed as a digital number. Those are examples where we've converted analog to digital. We represent these analog quantities as digital quantities with limited numbers of digits. Ultimately, those digits are reducible to 0s and 1s, and the rest of this lecture is going to be about how we do that reduction to 0s and 1s, and how we process those quantities. Again I emphasize, because everything is reducible to 0s and 1s, and you can tell the difference easily between a 0 and a 1, but you can't easily tell the difference between 1.4279 and 1.4278. You can tell the difference between a 0 and a 1, and that's why you can make perfect copies with digital electronics.

Let's look a little bit more detail about this distinction between analog and digital electronics. Here I've got kind of a tale of two speedometers. At the top I've got an analog speedometer, as many cars still have, but many don't. If I were to blow up the dial of that speedometer again, and again, and again, I'd see that that needle could move to any position at all. As I show the speedometer blowing up, I show more and more fine gradations of the values of the speed. That's an analog situation in which the speed could be any quantity; it takes on a continuous range. On the other hand, below we have a digital speedometer. It's reading 63 miles per hour. Let's just take a look at that 3; that 3 is the digit that this particular speedometer can read. That 3 is representing the binary number 011, and we don't get any finer than that. This particular speedometer can't tell us whether we're going to 63.1 or 63.2 miles per hour because it's got only these two digits.

We could look at the case where that digit was, in fact, a 2. If it were a 2, it would be the numerical representation 2, and it would be 010. We go back to 3, we're at 3, 011. If we go to 4, then we get the number 4, and that's represented by the binary 100. We'll talk more about that in just a moment.

There's a tale of two speedometers, an analog and a digital one. Let's take another look at these binary numbers that we use to represent every quantity in the world of digital electronics. There were two binary digits; there were only two. There's 0; write it with the symbol "0," the number "0." We call it "zero." We could also call it "no," we could call it "false," sometimes we'll call it "low," representing a low voltage level. Then there's 1, which we can call "one." Sometimes we'll call it "yes," sometimes we'll call it "true," sometimes we'll call it "high." All those things are ways of talking about the distinction between the 1 and the 0: 0 and 1, no and yes, false and true, low and high. Those are binary numbers.

Think about a digital number like, for example, a decimal number like 127. You know what that means. It means there one 100 in that number, there's two 10s (one, two), and then there are seven 1s; and so the number is 127. Binary numbers work in the same way. You probably studied binary numbers sometime back in your math education, but let me just give you a quick review here.

Here's a binary number: 1001. Just like the 1 in 127 is the hundreds place, 10^2; and the 2 in 127 is the 10s place, 10^1; and the 7 in 127 is the ones place, 10^0 (anything to the 0 is 1). In binary, the individual places represent different powers of two. Here I have the binary number 1001, a four-bit binary number. The first 1 is in the eighth place, 2^3. This tells us we've got one 8. The second 0 is in the four's place, so this tells us we don't have any fours; 4 is 2^2 squared. The third 0 is in the twos place, 2^1; this tells us we don't have any 2s. Then there's a 1 in the ones place, so this number is 1x8+0x4+0x2+1x1, or the decimal number 9. Sometimes we'll actually put a subscript, 9_{10}, to tell us we're representing this number in the base 10 notation. We could put a similar subscript on the 1001, a 2 to represent this number as base 2.

Numbers aren't the only things we need to represent. We also need to represent text. You write a term paper and it's represented digitally in your computer. With text we have something called the ASCII code, the American Standard Code for Information Interchange. It assigns each character an 8-bit binary representation, and here's a table of just a few of these representations for the capital letters A through G, and then I've jumped down two O down there. You can see A is 0100 and then 0001, and then B is 00010, which is the second number in the binary sequence, and so on. Here's the binary representation of a word, an actual word. If we look up what this word is, let's look at the first character in the word. That character, if we look on our table, corresponds to the letter "D." That's the letter "D," and the reason I included the "O" is because the second character is the letter "O," and the third character is the letter "G," so there's the word *dog* spelled out, represented by this sequence of 1s and 0s that's the basis of all binary information.

What we need to do now is think about how to process this binary information; how to do things with it. How to decide if two binary numbers are equal; how to add two binary numbers; how to compare things. How to ask, "Is this bit the same as that bit?" How to ask, "What's happening with these bits?" We need to process this information. At the heart of processing all this digital information is the science of digital logic. It's sort of a branch of mathematics; it's sort of a branch of philosophy. It's not exactly a branch of electronics, but it's going to govern everything that goes on in electronics, digital electronics. I'm going to spend the rest of this lecture looking at the way we think logically about these yes and no bits.

We do it with some simple operations. Just like in mathematics and arithmetic, you know how to add, you know how to multiply, you know how to subtract, you know how to divide. With these logic symbols we need to know some simple operations on them. The simple operations we're going to deal with at the beginning are AND and OR, and then we'll add another one called NOT. We represent these operations with symbols. The first one we're going to deal is AND, and AND means, as it says, "and." What it means is this: The output of an AND operation, the result of an AND operation, is 1 only if both inputs are 1. We'll call the inputs A and B. If A and B are both 1, the output, which we'll write A.B, surprisingly—you might think

A+B would be for AND, but no it's A.B—and the reason for that is it makes normal algebra work out if you treat that as sort of multiplication. But it's not normal algebra; this is called Boolean algebra. It's the algebra of logical levels: trues and falses, 1s and 0s. The AND operation says the result, the output, of an electronic circuit that does this operation is a 1, a yes, a true, a high, only if both inputs are 1. Here's our symbol for the AND operation, it's this kind of stretched out D-like thing. On the left are the two inputs; on the right is the output, A.B.

The other logic operation we want to deal with is OR, and this has two inputs also. By the way, both of these could have more than two inputs, and the extension of what I'm talking about here would apply. The output of OR is 1 if either input is 1. Is this true or is that true? That whole statement; here's a statement: The dog is blue and the cat is orange. The dog isn't blue, but the cat is orange; the combined statement about the dog and the cat, if you say the dog is blue or the cat is orange, that's true if the cat is orange even if the dog isn't blue. That's what OR is doing. By the way, people often ask what if they're both true, for OR as we're going to define it here? That's still one or the other is true, and so the output is still 1. We'll talk later about something called exclusive OR that's a little bit different. In Boolean algebra we write OR as A+B; that's the OR, the plus sign. We give it a symbol that looks like this, not quite the same D as the "and" symbol. It's pointed on the right end and it's kind curved on the left end. But it has two inputs, A and B, and the output is A+B, and the output is 1 if either input is 1.

Finally, we have one more operation that's very straightforward: that's the NOT operation. That says if you're a 1, give me a 0 as a result. If it's true, give me a false as a result. If it's 0, give me a 1, and so on. NOT gives an output that's the opposite of the input. We write that with a bar over the symbol; so when I write A bar, A with a bar over it, I'd read that as NOT A. The logic symbol is a little triangle with a circle at the end. The triangle should remind you a little bit of amplifiers, and it's sort of like that in a way. The little circle is the all important thing here; that little circle is actually what makes it a NOT. The NOT gate has only one input, and its output isn't A. We'll be using these ideas a lot from now on.

Those are the AND, OR, NOT symbols. We can represent what these things do by what are called truth tables. Let's build up the truth table for AND. Here I'm going to list A and B. There are actually four possible combinations of the two inputs A and B. They could both be 0; A could be 0, B could be 1; A could be 1, B could be 0; or they could both be 1. What I do is make a table with those four combinations of A and B, and in the right-hand column I list A and B. If A and B are both 0, the output is 0 because both of them have to be true or 1 for the output to be true or 1. If A is 0 and B is 1, the output is 0. If A is 1 but B is 0, the output is still 0. The only time we get a 1 at the output of the AND gate is when both inputs are 1. Here we have the truth table for AND.

We can write a truth table for OR. The truth table for OR, we do it the same way: We write the same four combinations. But remember, an OR gate has a true output if either one or both of its inputs are true, and so we have a different situation. The only time the OR gate doesn't have a 1 at its output is when both inputs are 0, and otherwise it's got an output of 1. In the right-hand column here we have 0111.

Finally, for our NOT, we could make a truth table. It's a little bit silly, but there it is. There's only two possible input states because the NOT gate has only one input. A can be 0 or 1; and if it's 0 or 1 the output A bar or NOT A is the opposite.

These are the truth tables for our simple operations. These operations, when we implement them electronically, are called logic gates. I want to explain why they're called logic gates. Here are our AND and OR, and here are the truth tables we just developed for them. Let me consider a case where on the AND gate I set B equal to 0; the second input is 0. If that input is 0, the output has to be 0, because for an AND gate, the output is 0 if either of the inputs is 0. Both inputs have to be 1 for the output to be 1. A is irrelevant to the output in this case. In other words, whatever the input is, whatever A is, it doesn't have any effect on the output, which just remains 0. This is like a gate that's closed; A can't get through.

You can do the same thing with the OR gate if you set B equal to 1 and you work out the truth table, and you'll see that in that case the output is

1 regardless of what A is. Again, A can't get through. This is like a closed gate, and so this is why these things are called "gates." How would I open the gate? I'd open the gate now by setting B equal to 1 for the AND gate. B is 1; if A is 0, the output's still 0. On the other hand, if A is 1, the output is 1. The output is whatever A is. A gets through in this case; the gate has been opened and A gets through. The output is A if B is 1, and if B is 0 the output is blocked. That's why these are called logic gates, and you can work out the same thing in the case of the OR gate. If B is 0, then the output A gets through. Whatever A is, that's what the output is. That's why these things are called logic gates.

You might think that takes care of pretty much everything we need to do logically, AND, OR, and NOT. But actually, there are some gates that are much more useful, although a little bit more obscure to think about, and I need to introduce those. These gates are called NAND and NOR. NAND stands for "NOT AND," and its output is 0 only if both inputs are 1. It's exactly the opposite of the AND gate; its output is the opposite. We write its symbol as A and B with a bar over it; NOT A and B. Even though this sounds like a slightly odd thing to do, the NAND gate proves to be much more useful than the AND gate, and I'll show you why.

Here's the truth table for an AND gate and then for a NAND gate. A, B, I've got the four combinations. I've got the combination that's the output of an AND gate, A and B. I've added a fourth column that simply inverts, takes the opposite, of the A and B to make NOT A and B, and that's the 1110 of the NAND gate. There's the truth table for a NAND gate. A NOR gate is analogously not OR; it's an OR gate in which we then negate the output. Remember, the output of an OR gate was 1 if any of the inputs were 1. The output of the NOR gate is 0 if either input is 1. There's only one possible state in which the NOR gate has a non-zero output, and that's when both inputs are 0. The Boolean algebra symbol is A plus B with a bar over it, NOT A or B; and that's the NOR gate.

We can make NOT gates from NAND and NOR. By the way, we can't make NOT gates; no way. I can give you a million AND gates, you can't make a NOT gate, but you can easily make it from NAND or NOR. You just connect the two inputs together. If you connect the two inputs together, in the AND gate's truth table, that rules out the middle two rows, which have different values for A and B. You have either 00, in which case you get 1 at the output, or 11, in which case you get 0, and so that has become a NOT gate. You can make NOT from NAND; A at the input, NOT A at the output, which is equal to a NOT gate. I'll let you convince yourself that you can do exactly the same thing with a NOR gate. Once you have NAND or NOR gates, you could also make the NOT operation. That makes them useful.

NAND and NOR can make NOT. NOT is useful, because if we wanted to make, for example, an AND gate, well, we could start with a NAND gate and add a NOT to it. If we invert the output of a NAND gate, a NOT AND gate, we'll get back to an AND gate. We could build that structure with a NAND and another NAND connected as a NOT gate. Equivalently, we could make an OR gate from NOR and NOT, and the NOT could've been made from another NOR gate or it could've been made from a NAND gate; it doesn't matter. But the point is these NANDs and NORs are useful because we can build them; we can use them to build other gate operations. We can't do that with the AND and the OR. That's why we use NAND and NOR a lot of the time.

You can go out when we're doing electronics and buy an AND gate and an OR gate, but you'll find it's more useful many times to work with the NAND gates and the NOR gates because you have this flexibility of being able to do different things with them. Let me convince you a little bit more of that flexibility by saying if I have a NAND operation, I can actually build everything from the NAND operation.

Here's a NAND gate again. Its inputs are A and B; its output is NOT A and B. There's its truth table. I've shown both columns: the column for A and B, and the column for NOT A and B. Its Boolean algebra symbol is A dot B with an overscore. Let me show you that we can do basically everything from it. Let me ask you what this thing is: This thing is a NAND gate with a couple of the inverters or NOT gates—we also call NOT gates inverters—in front of it, on the way in. What do we have? Let me convince you what we have.

The first inverter, the upper inverter, is a NOT gate. It takes A at the input and turns it into NOT A, A with a bar over it. The second inverter, the lower inverter, takes the other input B and it turns it into B with a bar over it, NOT B. Then we're feeding NOT B and NOT A into a NAND gate, so let's see what we get.

If we've had an AND gate, we would've had NOT A and NOT B. But we have a NAND gate, so there's another bar over that whole NOT A and NOT B. What does that work out to? Let's work out the truth table. Here I have again the four possible combinations: 00, 01, 10, 11. I have A bar and B bar because that's what I'm actually bringing into the gate. Then if you look at what the output will be and you convince yourself the output will be 1 if the two inputs NOT A and NOT B are 0. In other words, if the inputs at the far left, A and B are 1s, it will be 1 in either of those other two middle situations where the outputs are 0 and 1 and 1 and 0. But if the two inputs A and B are 0, then the NOT A and NOT B will be 1, and the output of the NAND gate will be 0. This is beginning to sound a little bit complicated, but let's move on.

Let's compare that truth table with another truth table we know. That truth table's output for the four combinations of A and B is 0 or 00 at the inputs, and 1 for all the rest. That's exactly what we get with an OR gate. What this funny combination with the NOTing at the two inputs does, and then the NAND gate, is produce an OR gate. If I give you NAND gates, you can make them into inverters, put the inverters on the front of the inputs to another NAND gate, and you can get an OR gate.

You can get OR from NAND; you can also get NOT from AND; and you can also get NOR from NAND because you could invert this OR gate that we've just built. You can get everything from NAND. Similarly, if I give you NOR gates, you can get everything from NOR. If you put inputs to the NOR and invert them, and you work through the same argument we just did, A goes to NOT A at that inversion of the input; B goes to NOT B. Work out what the output is. We've got a truth table. Again, the four combinations: 00, 01, 10, 11 at the inputs. We invert them by those two inverters, so we have 111, 001, 00. We look at what a NOR gate does to that. It produces the output column

0001. The only time a NOR gate's output is 1 is when both its inputs are 0. That's that last row.

If you do a comparison, you'll find that's exactly what the AND operation does. The AND operation produces an output of 1 only when its two inputs are 1. The A and B inputs at the far left of this funny circuit with the inversion and then the NOR gate; when they're both 1 the output is 1, and that's the only time. That's what an AND gate does. Here we've made AND from NOR.

The bottom line here is that these NAND gates and NOR gates are truly versatile. I could give you a bin of NAND gates and you could build every logic circuit. In principle, you could build a whole computer out of just the NAND gates. I could give you a bin of NOR gates and you could build a whole computer. You couldn't do that even if I gave you infinitely many AND gates; you couldn't get that negation that it would take to make the other gates, but everything can be built from NOR and NAND. These are going to be the basic building blocks of our digital circuits.

Again, we haven't done any electronics in this lecture; we've just looked at the logic behind digital electronics. What we're going to do in the next lecture is take these operations and understand how to implement them, first electrically and then electronically. By the end of the next lecture, we'll actually be designing circuits using these things.

I can't give you a project that's all about circuitry, but I can give you a project that thinks in a little more detail about these operations. Let's move to a project, and let me ask you the following questions in this project, and I urge you to go and do the project if you can and try to figure this stuff out on your own. But even if you don't want to work it all out, take a look at the project to see where we've gone with it.

The first question is how many two-input logic operations are possible? I introduced some operations in this lecture that had two inputs and produced an output. But how many possible two-input logic operations are there, and how many were introduced in this lecture? I'll give you one hint. There are far more possible than I introduced.

Project number two question is: Using any combination of AND, OR, NAND, NOR and NOT, design a three-input AND gate. That is, a gate that has three inputs—A, B, C—and whose output is 1 only when all 3 of those are 1. Finally, I mentioned this earlier in the lecture: Using only NAND gates, design a so-called exclusive OR gate; a gate whose output is 1, if one of its two inputs is 1, but they aren't both 1. There are many ways to do this. You can use lots of gates; you can use a few gates. The challenge is to use the minimum number of gates possible, and you can figure out what that number might be.

Electronics Goes Digital
Lecture 17

In the preceding lecture, we were introduced to logic operations and Boolean algebra. In this lecture, we'll transfer that knowledge to electronics. For electronic circuits to be digital, they must be able to implement the abrupt transition between the two possible states—0 and 1—with nothing in between. The fundamental electronic device that is able to do that is a switch, which is either on or off, open or closed, high or low, yes or no, true or false, or 1 or 0. We'll begin this lecture with a demonstration of how the logic operations we saw in the last lecture can be implemented using circuits with manual switches. Then, we'll move on to see how it's done with electronic switches, such as diodes and transistors. Key topics we'll cover include the following:

- AND, OR, NOT: simple electrical implementations
- Diode logic
- Transistor-transistor logic
- CMOS logic: today's mainstream
- Logic families

AND, OR, NOT: Simple Electrical Implementations

In a demonstration, we see a light bulb representing the output of three logic gates. If the light bulb is on, we're in the 1 state. If the light bulb is off, we're in a 0 state. We have a source of power and two switches connected in series. The switches represent the A and B input. An open or off switch represents the 0 state; a closed switch represents the 1 state. In the AND operation, the two switches are in series, so both A and B must be closed for the bulb to light—that is, for a 1 at the output. In the OR operation, the switches are in parallel and either A or B or both must be closed for the output to be a 1. In the NOT operation, an open

© Spike Mafford/Photodisc/Thinkstock.

switch—a 0 state—produces a 1 state at the output; a closed switch—a 1 state—produces a 0 state at the output.

Diode Logic

Diode Resistor AND Gate

Logic definitions: 0 V = Logic 0; 5 V = Logic 1

Ideal diodes: on ($R = 0$) or off ($R = \infty$); no voltage drop

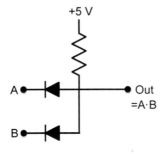

A	A	B	B	Out	Out
0 V	0	0 V	0	0 V	0
0 V	0	5 V	1	0 V	0
5 V	1	0 V	0	0 V	0
5 V	1	5 V	1	5 V	1

AND

Analogous to the gates using light bulbs, batteries, and switches is a diode-resistor implementation AND gate. The logic levels in this circuit are: 0 = 0 V and 1 = 5 V. Looking at the truth table for this circuit, we see that if both A and B are 0, both diodes are on because the 5 V can send a current through either diode into ground (0 V), which is connected to A and B. In fact, if A is 0, B is 0, or both are 0, we get 0 at the output because one or both of the diodes will be on. The only way we won't get 0 at the output is if both A and B are 1 (5 V); then, the diodes are shut off.

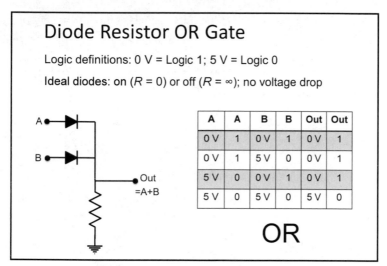

Diode Resistor OR Gate

Logic definitions: 0 V = Logic 1; 5 V = Logic 0

Ideal diodes: on ($R = 0$) or off ($R = \infty$); no voltage drop

A	A	B	B	Out	Out
0 V	1	0 V	1	0 V	1
0 V	1	5 V	0	0 V	1
5 V	0	0 V	1	0 V	1
5 V	0	5 V	0	5 V	0

OR

A diode resistor implementation OR gate works similarly. If both A and B are 0, the output is 0. If either A or B is 1 or both are 1, the output is 1.

Transistor-Transistor Logic (TTL)

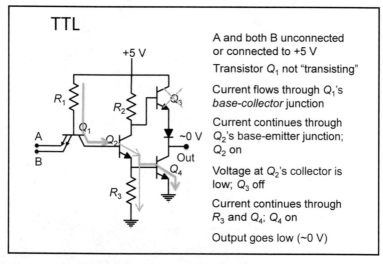

TTL

+5 V

R_1

R_2

Q_3

Q_1

A

Q_2

~0 V

B

Out

Q_4

R_3

A and both B unconnected or connected to +5 V

Transistor Q_1 not "transisting"

Current flows through Q_1's *base-collector* junction

Current continues through Q_2's base-emitter junction; Q_2 on

Voltage at Q_2's collector is low; Q_3 off

Current continues through R_3 and Q_4; Q_4 on

Output goes low (~0 V)

TTL is a form of logic that was developed in the 1970s. It's still in use on occasion, although it is being superseded by circuits based on MOSFETs.

Here, we have a transistor with two emitters, each of them behaving like the emitter of a BJT. In this circuit, the output goes high if either of the inputs is low.

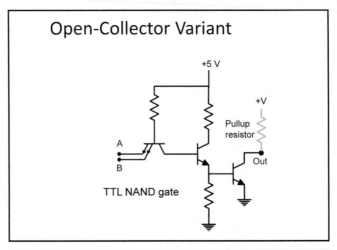

Open-Collector Variant

+5 V

+V

Pullup resistor

A

B

Out

TTL NAND gate

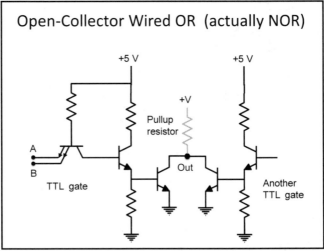

Open-Collector Wired OR (actually NOR)

+5 V

+5 V

+V

Pullup resistor

A

B

Out

TTL gate

Another TTL gate

A variant on TTL called *open collector output* is useful for connecting the outputs of two TTL gates directly. The result is a wired OR (actually NOR) situation.

CMOS Logic: Today's Mainstream

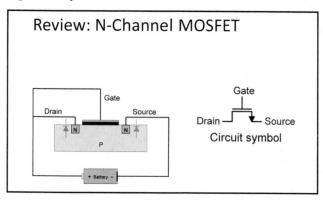

As we saw in an earlier lecture, a MOSFET consists of two electrodes (a *source* and a *drain*), which in this case, are incursions of N-type semiconductor material in a block of P-type material. There's another electrode, the *gate*, which is a thin metal layer insulated from the block of semiconductor by a layer of silicon dioxide. This transistor can be considered a switch that's turned on by putting positive charge on the gate. We can also build complementary MOSFETs that have N-type blocks and P-type incursions; they're turned on by putting 0 volts on the gate.

CMOS
Complementary Metal-Oxide Semiconductor

- Uses N-channel and P-channel MOS transistors in series combinations
- Except when switching state...
  - One transistor in series combination is off
  - Draws essentially no current
  - Consumes essentially zero power
- When switching
  - Current flows to charge gate capacitance
  - Power consumption increases with switching rate
- Ubiquitous in computers and consumer electronics

CMOS stands for *complementary metal-oxide semiconductor*. These circuits are ubiquitous in consumer electronics and computers.

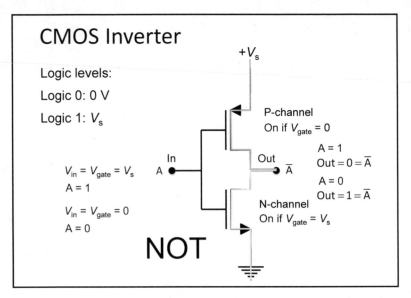

This CMOS inverter (a NOT gate) is a simple example of a CMOS circuit.

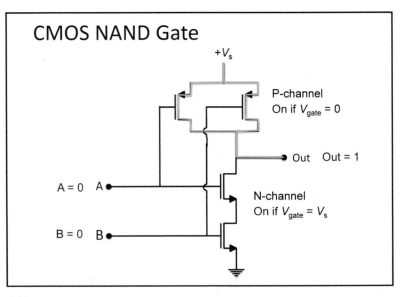

A CMOS NAND gate is slightly more complicated.

Logic Families

Logic Type	Power per Gate	Speed (MHz)	Supply Voltage	Logic Voltages	Comments
Bipolar Original TTL (7400 series)	10 mW	35	4.5–5.5 V	0–0.8 V 2–5 V	Robust, but obsolete High power consumption Relatively slow
Bipolar 74LS TTL	2 mW	45	4.5–5.5 V	0–0.8 V 2–5 V	Lower power, faster than original TTL
CMOS 74HC	3 µW quiescent 0.6 mW, 1 MHz	45	2–6 V	0–1 V 3.5–4.4 V	Logic levels listed are for 5 V supply
CMOS 74HCT	3 µW quiescent 0.6 mW, 1 MHz	45	4.5–5.5 V	0–0.8 V 2–5 V	TTL compatible variant of 74HC
CMOS 74LVC	3 µW quiescent 0.8 mW, 1 MHz	200	2–3.6 V	0–0.8 V 2–3.3 V	Increasingly popular; used with fast, large-scale integrated logic; surface mount only

The table above shows a number of logic families. The CMOS 74HCT series is compatible with the older TTL and can serve as a modern replacement in most applications.

Suggested Reading

Introductory

Brindley, *Starting Electronics*, 4th ed., chapter 11 through p. 187.

Lowe, *Electronics All-in-One for Dummies*, book IV, chapter 3 and chapter 4 to pp. 565–569.

Shamieh and McComb, *Electronics for Dummies*, 2nd ed., chapter 14, pp. 323 to end.

Advanced

Horowitz and Hill, *The Art of Electronics*, 3rd ed., chapter 10, sections 10.2–10.3.

Scherz and Monk, *Practical Electronics for Inventors*, 3rd ed., chapter 12, sections 12.4 and 12.11.

Project

Digital Circuit Design

Design a digital circuit with four inputs and one output, whose input represents a 4-bit binary number and whose output is 1 only when the input is the binary equivalent of 2, 6, or 8. Use any gates in the table of basic logic gates.

Questions to Consider

1. What does the *C* in *CMOS* stand for? How does this relate to the basic fact that there are two types of doped semiconductors?

2. Why does CMOS logic consume much less power than TTL?

3. Why is CMOS logic more sensitive to static electricity damage than TTL?

Electronics Goes Digital
Lecture 17—Transcript

Enough of that abstract logic and Boolean algebra; let's get onto electronics. Let's make electronics go digital.

For electronic circuits to be digital, they have to be able to implement that 0 and 1; that abrupt transition between two possible states with nothing in between so it's unambiguous which state you're in. We've got to be able to make those binary numbers. The fundamental electronic device that's able to do that is a switch. A switch is on or off, open or closed, high or low, yes or no, true or false, 1 or 0, and so I'm going to begin with a demonstration of how you could implement the logic operations I introduced last time using circuits with actual manual switches. Then we'll move on in the rest of this lecture to see how it's done with electronic switches like diodes and like transistors. But let's begin with this demonstration of three of the logic operations I introduced last time; the first three, in fact.

What I have here is a lightbulb. The lightbulb represents the output of each of these three logic gates, if you will. If the lightbulb is on, we're in a 1 state, a true state, a yes state, a high state. If the lightbulb is off, we're in a low state, a 0 state, a false state, and so on. I've got a battery; that's my source of power. I'm just going to hook my circuit up to the battery. In this particular circuit, I have two switches. The switches are connected, as you can see, in series from the battery. Current has to flow from the battery through this switch, and then through this switch in order to get to the light bulb. What happens if I turn this switch on? The switches represent the A and the B input. Here's A; here's B. An open switch, an off switch, represents the 0 state, the false state, the off state, the low state. A closed switch represents the high state, the 1 state. For this particular logic gate, which is what it is, I close switch A and nothing happens. I close switch B and nothing happens, having opened switch A first. Of course, with both switches open, nothing's happening. I close both switches, and the light lights. This is a situation, a circuit, in which the lightbulb is lit—that is the output is a 1, a true— only when both inputs are 1's; that is, when both switches are closed. That's clearly the operation AND. A and B, switches A and B, the two inputs, both have to be closed in order for this circuit to have an output that's positive, that's 1, that's the high state, that's true. That's my first logic gate. It clearly implements the first logic operation I introduced, namely AND.

Let's move on to the next one: You can probably guess what it's going to do. But here it is. It also has two switches, so it's got two inputs. Again, an open switch represents a 0, a false, a low state. A closed switch represents a 1, a high state. The lightbulb being on represents the high state; the lightbulb being off represents the low state. Here we go: I'm going to close switch A and the lightbulb lights. I'm going to close switch B and the lightbulb is still lit. I'm going to open switch A and the lightbulb is still lit. I'm going to open switch B also and the lightbulb goes off. This is a situation in which the lightbulb is lit—in other words, the output is a 1—if either A is a 1, or B is a 1, or both. What's that? That's clearly the OR operation.

I have an AND gate and I have an OR gate, the two most fundamental, simplest to understand, two input gates that I introduced last time. I could very easily modify these so they'd be NAND and NOR, those more useful but a little more obscure gates.

We have one other very basic operation, and that was the invert operation, the NOT operation, the operation that takes A and makes as its output NOT A. Let's take our battery, put it to the third of these gates. This one has only one switch, because the invert or NOT operation has only one input. The input is this one switch, and right now the switch is open and the light is on. Why is the light on when the switch is open? If you look at the way this circuit is connected, I've got a wire coming from the battery. I've got a resistor here, actually; it's getting pretty warm, too. This isn't a very energy efficient circuit by any means. Then current is flowing through the lightbulb also, and the lightbulb is lit. You notice if you look very carefully, it's not quite as intense as the other two because of the presence of this resistor. The switch is connected directly across, in parallel with the light bulb. When I close the switch, the current has a much lower resistance path to follow through the switch, it doesn't go through the lightbulb. The lightbulb goes off when the switch is closed. This is an operation that takes an open switch, a 0 state, and produces a 1 state at the output, and it takes a closed switch, a 1 state, and produces an off lightbulb that's a 0 state at the output. This is an inverter. Again, this is an extremely inefficient implementation of that, because right now lots of current is flowing through the resistor; it's getting quite warm. I'm wasting a lot of power, so we'll disconnect this.

But I have electrical—I won't say electronic yet, but electrical—implementations of the logic operations that I introduced, and the fundamental component that made those work is the switch. We want to move on now and do the same thing with electronics.

Remember the definition that I gave at the beginning of this course for electronics is "one circuit controlling another." Right now it was my hands controlling the switches; now we're going to move to the situation where one circuit provides the inputs to some switches and controls the output. I'm going to take you through several different ways in which we actually do implement logic circuits electronically in the commercial electronics world. Before I do that, I want to begin with a simpler example that we don't actually use, but which I'd like you to think about.

Let me ask you what this circuit is. Here's a circuit consisting of a resistor and two diodes. I've labeled the left-hand ends of the diodes A and B. Those are the inputs to this logic gate, and the output is labeled. There's a resistor there, two 5 volts. What does this circuit do? If you want to think about it a little bit, think about whether those A inputs are 0 volts or 5 volts, and what happens. If you want to pause and work that out, you can, and I'll come back and I'll work it out and we'll see what this particular fairly simple circuit does.

Let's have a look. Let's begin by defining logic levels. You always have to do that in a circuit. You have to say, "What do I mean by 'what is 0'? What is false? What is low?" It's very common in conventional to define 0 volts or something close to 0 volts as logic 0. You don't have to do that. You could define 0 volts as logic level high and 5 volts, for example, as logic level low. But I'm going to define 0 volts as low, and in this particular circuit 5 volts as high. I've chosen 5 volts because that's the voltage in one common although obsolete-growing logic family that I'll be talking about. There are my logic definitions; I have to make those definitions before I can talk about this circuit and what it does logically.

These are ideal diodes. Remember diodes? We've looked at them many times in this course already in analog electronic circuits. The ideal diode is on and has essentially 0 resistance; it could be replaced with a wire. It's like a closed switch if it's forward biased; that is, if the end of the diode coming into that little triangle is more positive than the end going out. That's how we turn diodes on or off. If it's off, if we put the bar end of the diode at a higher voltage than the triangle end, then it's reversed biased and it's an open circuit. But if it's on, R is 0; there's no voltage drop across the diode.

Let's work out a truth table for this particular circuit. I've listed here two columns for A and B, because one column I'm listing the actual voltage and in the second column I'm interpreting what that voltage means in terms of logic levels. Zero volts corresponds to 0 logic level by my definition. Here I have on the top row 0 volts at A, 0 is the logical input at A; 0 volts at B, 0 is the logical input at B. I think you can see what happens. If I have A and B both at 0, both those diodes are on because the 5 volts can send a current through either diode into ground, into 0 volts, which is connected to A and B. I'm not bothering to show you the ground or show you what's going on, I'm just saying we connect A to 0 and what happens? Current flows through the resistor and down to ground through A or through B, and that turns on one or the other of those diodes and it pulls the output down to 0. In fact, if either A is 0 or B is 0, or they're both 0, you can see that we're going to get 0 at the output because one of the other or both of those diodes are going to be on. The only way we won't get 0 at the output is if we put both A and B to 5 volts. Then the diodes are shut off. There's the same voltage as the 5 volts, we don't have higher voltage at the right end of the diodes. They're turned off. They're open circuits, they're open switches, and so the output is connected to the 5 volts through that resistor. As long as we aren't drawing any current, the output will, in fact, be 5 volts.

This circuit has a truth table whose output is 0 volts. If any one of the inputs, either of the inputs or both, are at 0 volts; translate into logic: the inputs are logic 0. Its output is 5 volts; translate: logic 1 if both of those inputs are 1. This has a truth table 0001. Its output is a 1 only if both inputs are 1, and it's clearly an AND gate. This is a diode resistor implementation AND gate. We don't actually make gates this way, but I show you this because it's quite simple and straightforward and it's quite analogous to the gates I showed you using lightbulbs, batteries, and switches. That's what that is.

Here's another logic gate made from diodes. If you want to pause and figure out what this does, that's fine. If you don't, bear with me. I'm going to come right back and we'll take a look at analyzing this circuit. Again, we're going to use the same logic definitions: 0 is a logic 0, 5 volts is a logic 1. Again, those aren't set in stone. We could do the opposite, but we're going to keep with that convention. These are ideal diodes. They're either on, they have no resistance, or they're off, they have infinite resistance.

Let's look at what happens. Again, same table: 0 volts at A, 0 volts at B. What does that do? If A and B are both at 0 volts, clearly the output has no way to be anything but 0 volts because there's nothing but 0 connected anywhere in the circuit. If A and B are both 0, the output is 0. But what if B is 1; that is, B is 5 volts? If B is 5 volts, that B diode is going to be forward biased. Current is going to flow from left to right through that diode down through the resistor. The B diode is on. It's a short circuit. It's like a piece of wire. It's connected to 5 volts. It's also connected to the output because the diode becomes basically a piece of wire, and so the output goes to 5 volts. You can see the same thing would happen if the A diode were on. The output would go to 5 volts if A were at 5 volts, or if both of them were at 5 volts. This is a circuit that gives us an output of 1 if either A is 1 or B is 1, or they're both 1. This is an implementation of OR in diode logic.

Again, we don't actually use diode logic these days, but this is a very simple implementation of the AND and OR operations. You can see how it would work.

What we do in the modern era is have a whole family of electronic circuits that, in fact, implement these logic gates. I want to spend the rest of this lecture talking about some of those logic families. I won't cover them exhaustively, I won't tell you all the details, I won't go into a lot of subtleties, but I want to give you a sense of how just a few of them work.

I'm going to begin with what's called transistor transistor logic; TTL logic, as it's called. TTL is a form of logic that was developed in the 1970s. It's still in use on occasion, although it's being superseded rapidly by circuits based on metal oxide semiconductor field effect transistors, which I'll talk about in a minute. But TTL is still with us enough that I think it's useful to see how it works. It's also useful because you know about bipolar transistors; we've spent a lot of time on them in the analog part of this course. It's useful to see how this gate works.

Here's a TTL gate with two inputs: A and B listed, an output, and a bunch of transistors and resistors. It's not a very complicated circuit. In TTL, by the way, the voltage for the power supply is always 5 volts. A logic 0 state is somewhere in the range of 0–0.8 volts. Notice it doesn't have to be perfect; that's the nice thing about TTL. The high state is somewhere on the order of 3–5 volts. As long as there's a big gap there you can distinguish, it doesn't matter whether the output is 4.5 volts, or 5 volts, or 3.8 volts, or 4.7 volts. That's a 1, and that's all you care about in a digital circuit.

How does this thing work? Let's imagine that A and B are either not connected to anything, or they're connected to 5 volts. Look at A and B: They're emitters. This is a funny transistor, the one labeled Q1. It has two emitters. Let's just get used to that. It's a transistor with two emitters, and each of them behaves like we know the emitter of a bipolar junction transistor behaves. But if A and B are either at 5 volts or open, which is the same thing effectively, transistor Q1 isn't, if I may coin a word, transisting at all; we're not using it as a transistor. On the other hand, there's a base to collector junction. We don't normally use transistors this way and we might fry them if we did, but current can flow that way through the collector through the base to collector junction of Q1. It goes on into that transistor labeled Q2, and it turns on its base emitter junction. That base emitter junction is turned on, and as a result, the voltage at Q2's collector, the upper electrode of Q2, is low. Consequently that means there can't be any base current in that upper transistor Q3; that transistor is turned off. Current does continue on through R3, and that builds a voltage across R3, and current can then also go through Q4's base emitter junction. That turns Q4 on, and as a result that pulls the output down to approximately 0 volts.

If A and B are both unconnected, which is the nice thing about TTL, I don't recommend doing this, but you can leave an input unconnected and it'll be as if it were a high state, the 5 volt state. But don't do it; it causes some problems with your circuits.

What if we ground one of these two inputs to ground? Then that first funny transistor with the two emitters does conduct, it turns on. As a result, it tries to draw current away from Q2's base emitter junction. It can't do that; it turns the base emitter junction off, Q2 is off. The collector of Q2 goes high, and because the collector of Q2 goes to a relatively high voltage we can now drive current through Q3's base emitter junction. If there's any current to be drawn at the output, that current will flow into the output. The point is that will connect the output basically to the high 5 volts situation. If we ground A connected to 0, logic 0, if we did the same thing with B, we'd have the same argument. If we did the same thing with both of them, we'd have the same argument. This is a circuit where the output goes high if either of the inputs is low.

Let's review that. A and B both unconnected or connected to 5 volts, A and B are at logic 1; that means the output is low. If A and B are both grounded, or one or the other is grounded, is connected to 0 volts, is at logic 0, the output is high. The truth table looks like 001011101. The only case where we get an output of 0 is when both inputs are high, and that's an AND gate.

That's TTL NAND. We could easily build TTL NOR, and we can make inverters, and we can make whatever we want basically. But that's how transistor transistor logic works. You don't really need to know those details to build digital circuits, but sometimes it's helpful to understand what's actually going on inside your gates because it helps you troubleshoot some of the weird anomalies that can happen.

There's one sort of variant on TTL I'll just mention briefly, because we're going to use it in a circuit we'll do a little bit later. It also makes wiring some circuits easier. Occasionally, you can get TTL gates in which there isn't that complicated output structure, there's just a single transistor sitting there with its collector hanging and you get to connect to that collector pin and connect whatever you want. That's called open collector output, and it's useful in some instances. One place where it's useful is it means you can connect the outputs of two TTL gates directly. You connect this thing called a pull up resistor, it goes up to the power supply voltage—the 5 volts it would be, or any voltage you want actually, because it doesn't have to be the 5 volts in that case—and you can connect another gate directly to it. You get what's called a wired OR situation in which the output is the output of either gate. It's actually a wired NOR because it's the inversion, but that's OK. This is a situation you'll see occasionally, and we'll have one occasion in this course to use an open collector TTL gate. We'll look at that a little bit later.

Let's move on, because TTL, as I mentioned, is getting obsolete. You can still buy them. Some of them are really hard to find—I'll tell you, because I still use them in teaching sometimes—because they're obsolete. What we use instead are the metal oxide semiconductor field effect transistors. I introduced those way back in the lecture on transistors, but let me remind you what a MOSFET is. Here's a diagram of an n-channel MOSFET. The MOSFET consists of two electrodes called a source and a drain, and they look in these simple diagrams pretty much symmetric. In this case, the source and the drain are little incursions of n-type semiconductor in a big p-type slab. There's another electrode called the gate, which is a metal gate insulated by typically silicon dioxide from the p-type semiconductor. The way this thing works—and you might want to go back and review that lecture if you're having trouble remembering this—if you put positive charge on that gate, it'll draw negative electrons, of which there are a few in that p-type material (not many). It'll draw them into that region between the two n-type incursions and it will make an n-channel, it's called (that's why this is an n-channel MOSFET) and then the thing will conduct. This is a switch that you turn on by putting positive charge on the gate. As I mentioned back in the lecture on transistors, you can build a complimentary MOSFET that has an n-type slab and p-type incursions. That's going to be really important in how we understand digital logic circuits in the modern era.

Let's move on and talk about what's called CMOS, a term you've almost certainly heard. Your digital camera almost certainly has a CMOS sensor, for example. Your computer, I guarantee you, uses CMOS circuitry as most of the circuitry for the logic operations it does. There are some very good reasons. CMOS uses n- and p-channel transistors, the complimentary transistors. That's why it's C; the "C" in CMOS is complimentary metal oxide semiconductor. We use the n-channel and the p-channel transistors in series combinations, and what that means is one of those transistors is always off. The circuits that go from the power supply to ground in any CMOS device are always open circuits because one of the two transistors is always off. They don't draw any power. Great; they don't need any power. Not quite. The only time they need power is when you're switching the state. No power when you're not switching state, but when you're switching the state of the system—when you're switching from a 0 to 1 or an 1 to a 0; when you're turning the transistor on or turning it off—current has to flow to charge that gate capacitance to dump charge onto that gate, or to pull charge off the gate. There are little bits of current flow in these circuits when you're changing state. The faster you choose to change state, the more rapidly you have to pull charge on and off those gates, and so the more power these things use. These circuits are ubiquitous in consumer electronics and computers, and the higher the speed of your computer, the more power it draws. One of the things your computer will do to same battery is to slow down. Unless you're doing real heavy number crunching, or analyzing huge images, or something else that requires all the processing power of your computer, the computer will actually throttle down the speed at which it does its everyday operations because that'll reduce the speed at which charge has to be moved on and off the gates of the transistors, and that will reduce the power consumption and make your battery last longer. That's the idea between CMOS.

Here's a very simple example of a CMOS circuit: This is a CMOS inverter. We have A at the input, and we have an output. Here's the deal: Logic 0, 0 volts; logic 1 (v supply, I'm going to call it), it could be 5 volts for CMOS; that's compatible with TTL. It could be 3 volts. Varying voltage levels are used, so I'm just going to call it the s. The p-channel transistor turns on if you make the gate 0. The n-channel transistor turns on if you make the gate equal to the supply voltage; that is, a logic 1. If A is a logic 1, the n-channel transistor turns on, the p-channel transistor turns off. The n-channel transistor is the bottom one, and therefore in that situation with A=1, the output is connected directly to ground and goes to 0; so the output goes to 0 if the input is a 1. That's why it's an inverter, a NOT gate. What happens the other way around? If vn is 0, v gate becomes 0 on both gates; A is 0 logically. In that case, the n-channel transistor is off, the p-channel transistor is on, and the output is connected directly to v supply. The output becomes a 1, or the opposite of what the input was.

A CMOS inverter is really easy to understand. You can see that point I made that the circuit consists of two transistors, complimentary transistors, in series. Unless you're changing state and flowing charge on and off those gates, there's no power being drawn in principle from the power supply. That's a NOT gate.

Let me ask you what this does. You want to work this out? Take a pause and work it out. Otherwise, I'll go right on to explain it. This one's a little more complicated. Up at the top, it has two p-channel transistors in parallel. Down below, it has two n-channel transistors and they're in series. Remember that the p-channel transistor turns on if the gate voltage is 0; so if A is 0, the leftmost p-channel transistor is going to be on. If B is 0, the rightmost p-channel transistor is going to be on. If A and B are both 0, they're both going to be on. Under those conditions, if A is 0, the n-channel transistor, the upper one, will be off. I don't care what the lower one is doing, because those are in series. If the upper one is off, no current can flow through that system; there's no connection from the output to ground.

What happens in this case? Let's say A is 0. If A is 0, I turn on the left hand transistor; it becomes a closed circuit, becomes just like closing the switch on my logic devices over there. The output is connected to the supply voltage, so I get a 1. So if A 0, I get a 1. If B is 0, similarly, I connect the right hand p-channel transistor to the output, and I get a 1 also. What if they're both 1? If they're both 1, then I turn on the lower two transistors; both the n-channel transistors come on. If only one of them is 1 and the other's 0, one transistor will come on, but it won't do anything for me because there still won't be a complete path to ground and one of the upper transistors will come on and the output will still be high. The only way to get the output low is to turn on both inputs, to put both inputs to 1. The output is 0 only if both inputs are 1; so if we work out what does this do, the only way the output is 0 is if both inputs are 1. That's the opposite of what AND would do, so this is NAND. That's a CMOS NAND gate. If you want to build a CMOS NOR gate, I'm going to have you do the project for this lecture.

Let me wrap up by talking a little bit about logic families. We have a number of families, and again I've only picked a few of them to talk about. The original TTL, these are called 700 series logic circuits. They begin with the number 74 and then some other digits: 7400, 7402, and so on. They draw about 10 milliwatts of power, which is significant in a big circuit. They can run up to 35 megahertz, 35 million cycles per second. They run on a fairly narrow range of supply voltages and the logic levels are about 0 to 0.8, as I mentioned earlier, and it's from as low as 2 to about 5 volts. They're robust, they consume high power, they're are relatively slow, and they're increasingly obsolete. But we do talk about them and we do use them. There's bipolar 74LSTTL, which is a much lower power version of TTL. What you'll find is in all these logic families, a 7400 and a 74LS00 do the same thing and you can use them in almost the same way. Then there are several CMOS families; in fact, there are many CMOS families. But the CMOS family that I'm most interested in for this course is the CMOS 74 HCT series. The "T" stands for TTL-compatible. These are CMOS circuits that are designed to run on the same supply voltage as TTL. You could, in principle, pull out a TTL circuit and plug in a CMOS circuit. There are some considerations that make that not necessarily workable all the time, but it often can work.

By the way, increasingly, you don't use these individual small logic gates except as ways of sort of patching logical connections between what are called large-scale integrated circuits that have maybe hundreds to thousands of transistors, or maybe even millions, and are designed to do specific tasks. But it's still worth knowing about these things.

Let me just give you a quick look at some of the basic logic gates. Some of the basic logic gates here are the 7400, 74 LS00, or 74 HCT 00. It's a quad; that means there are four gates in it, two input NAND gate. There's a picture of how you actually connect it in the little integrated circuit package that looks like some of the other integrated circuits we've used already. The O2 is a quad two input NOR; the O4 is six inverters, a hex inverter, six NOT operations; and a quad two input or and a triple three input NAND and a dual four input NAND; and so on and so forth. There's a whole bunch of these that do all kinds of things, and we'll meet some of them in the coming lectures. Those are some of the basic gates.

Before we go on to the project, let me just show you a quick demonstration with one of these circuits. Here I have a more complicated board than I was using with the analog part of the electronics; we'll get to the details in a minute. Ignore the mess that's on here; this is all the circuits we're going to deal with in the rest of the course. But I want to focus on this one particular device right here, a 7400 dual quad two input NAND gate. I'm only using one of the gates in it. I've got it connected to its power and ground and I have its two A and B inputs connected to two switches. What I'd like you to pay attention to are these two switches. They're supplying inputs to this gate, which doesn't look like much; not much happens that you can see with these integrated circuits. I'm taking the switch inputs and the outputs over to some lights over here. Let's look at those lights. The two lights you see down here that are green represent A and B; they represent the positions of these two switches. This light, the red one, represents the output. In this bank of lights, the output is high if it's red and low if it's green. That's how we can tell the difference between the two logic states. Notice, that's a lot simpler than a fancy oscilloscope. All we need with digital logic is to know is it a 1 or is a 0? Right now, both inputs are 0. There they are, 00. A and B are both 0, and the output is indeed 1. I turn B to 1, and it goes to 1 and the output is still a 1. I turn A to 1, and the output is still a 1. I turn them both to 1, and the output goes to 0. That's indeed a gate whose output is 0 only when both inputs are 1. That's the inverse of AND, so that is indeed a NAND gate. All the circuits on here are built with basically NAND and NOR gates, although inside them there may be more complicated arrangements of transistors that effectively implement those things.

Let's stop there. But if you want to do the project, here's what I'd like you to think about doing for this particular lecture. First of all, I've given you nothing really so far that had anything to do with using metal oxide semiconductor transistors, so your first thing would be to design a CMOS NOR gate. Look at the NAND gate and see if you can figure out how to make an NOR gate. Then the second one is a little more obscure: I'd like you to design a digital circuit with four inputs and one output. I want the input to represent any 4-bit binary number, and I want the output to be a 1 only when the input is the binary equivalent of either 2 or 6 or 8. You can use any gates you want in that table of basic logic gates, so you can use the multiple input gates if you'd like. Why on earth would you want to build such a circuit? It's actually a very useful one, and when you look at the project you'll see why.

Flip-Flop Circuits
Lecture 18

S o far in this course, the digital circuits we've considered have typically had one or two inputs. Those two inputs produced an output, which we described by means of a truth table. The output responded directly to the inputs. In this lecture, we'll move to more sophisticated circuits—circuits that "remember." These circuits constitute the core of computer memory and many other applications. We'll begin by looking at the basic bistable circuit—a circuit that has two stable states—and we'll see how we change those states. Such circuits are sometimes called *flip-flops*. Key concepts here include the following:

- The basic bistable circuit
- Gating the inputs: set and reset
- Application: debouncing

- Clocking
- Master-slave flip-flops.

The Basic Bistable Circuit

The two inverters here are in a stable, consistent state: 0 at the input to the first inverter, 1 at the output of the first inverter, and 0 at the output of the second inverter, connected back to the input of the first. We could also have 1 at the input to the first inverter, 0 at its output, and 1 at the output of the second inverter. This circuit can only be in one of those states.

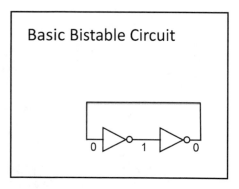

Basic Bistable Circuit

In a more symmetric drawing of the same circuit, the output of the upper inverter (called Q) is connected to the input of the lower inverter, whose output is $\overline{Q}$ (meaning *not Q*). But this circuit isn't useful because we have no way to change

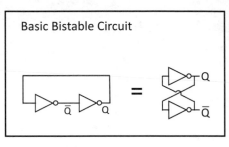

its state. To make the circuit useful, we replace the inverters with logic gates. The extra inputs to the gates will allow us to change the state of the bistable circuit.

Gating the Inputs: Set and Reset

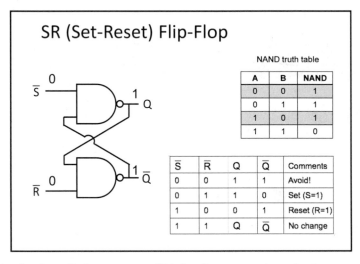

SR (Set-Reset) Flip-Flop

NAND truth table

A	B	NAND
0	0	1
0	1	1
1	0	1
1	1	0

$\overline{S}$	$\overline{R}$	Q	$\overline{Q}$	Comments
0	0	1	1	Avoid!
0	1	1	0	Set (S=1)
1	0	0	1	Reset (R=1)
1	1	Q	$\overline{Q}$	No change

In this circuit, called a *set-reset (SR) flip-flop*, we replace the inverters with NAND gates. Recall that a NAND gate has a truth table that is the inverse of AND. The only time its output is 0 is when both its inputs are 1. Note that the inputs here are labeled $\overline{S}$ (*not set*) and $\overline{R}$ (*not reset*). If we set both inputs to 0, both outputs (Q and $\overline{Q}$) have to be 1. This is not a bistable state storing a 1 in one output and a 0 in the other; thus, this is a situation that we'd like to avoid.

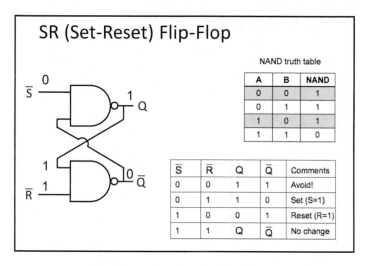

If we put a 0 at the $\overline{S}$ input and a 1 at the $\overline{R}$ input, we get a 1 at Q. That 1 is cross-connected back to the other input of the lower NAND gate. Consequently, we have 1 and 1 at the inputs to the lower NAND gate, and its output is 0. That is a legitimate state of the flip-flop.

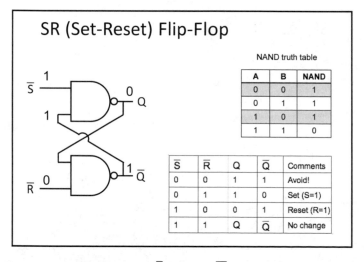

We can also do the opposite: Set $\overline{S}$ to 1 and $\overline{R}$ to 0. That gives us 0 at Q and 1 at $\overline{Q}$. We have reset the flip-flop.

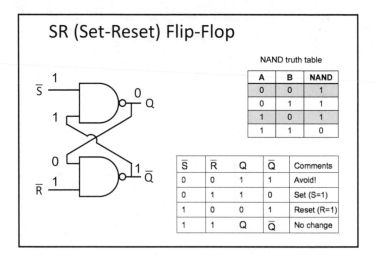

SR (Set-Reset) Flip-Flop

NAND truth table

A	B	NAND
0	0	1
0	1	1
1	0	1
1	1	0

$\overline{S}$	$\overline{R}$	Q	$\overline{Q}$	Comments
0	0	1	1	Avoid!
0	1	1	0	Set (S=1)
1	0	0	1	Reset (R=1)
1	1	Q	$\overline{Q}$	No change

Finally, we have the state in which we set both $\overline{S}$ and $\overline{R}$ to 1. This is consistent with either $Q = 1$ and $\overline{Q} = 0$ (as shown above), or $Q = 0$ and $\overline{Q} = 1$ (as shown below). So having $\overline{S} = \overline{R} = 1$ leaves the state of the flip-flop unchanged.

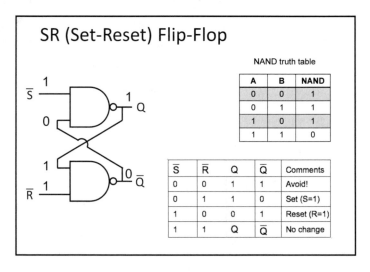

SR (Set-Reset) Flip-Flop

NAND truth table

A	B	NAND
0	0	1
0	1	1
1	0	1
1	1	0

$\overline{S}$	$\overline{R}$	Q	$\overline{Q}$	Comments
0	0	1	1	Avoid!
0	1	1	0	Set (S=1)
1	0	0	1	Reset (R=1)
1	1	Q	$\overline{Q}$	No change

Application: Debouncing

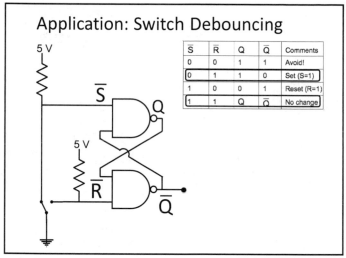

Application: Switch Debouncing

$\overline{S}$	$\overline{R}$	Q	$\overline{Q}$	Comments
0	0	1	1	Avoid!
0	1	1	0	Set (S=1)
1	0	0	1	Reset (R=1)
1	1	Q	$\overline{Q}$	No change

One of the simplest uses of an SR flip-flop is to stop the bouncing that occurs when a mechanical switch is turned on. In this case the flip-flop "latches up" in the Q = 1 state the first time the switch closes, and it can't change until the switch is again thrown all the way to its right-hand position.

Clocking

The simple SR flip-flop uses *asynchronous logic*; the output changes as soon as the input changes. But if we have large-scale circuits with numerous flip-flops, as in a computer, we need to use *synchronous*, or

Synchronous versus Asynchronous

- Asynchronous logic
  - Output changes as soon as inputs change
  - Chaos in large-scale circuits!
- Synchronous (clocked) logic
  - Central clock governs all logic switching
  - Clock is a square-wave generator
  - Today's computers: clock frequency ~several GHz

clocked, logic. Here, a central clock governs all the logic switching. The clock is a square-wave generator running at a fixed rate, with its voltage swinging between the two logic levels (0 and 5 V in the circuits we're discussing).

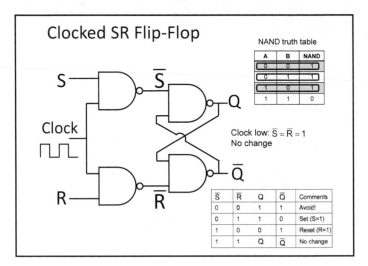

Clocked SR Flip-Flop

NAND truth table

A	B	NAND
0	0	1
0	1	1
1	0	1
1	1	0

Clock low: $\overline{S} = \overline{R} = 1$
No change

$\overline{S}$	$\overline{R}$	Q	$\overline{Q}$	Comments
0	0	1	1	Avoid!
0	1	1	0	Set (S=1)
1	0	0	1	Reset (R=1)
1	1	Q	$\overline{Q}$	No change

To make our simple SR flip-flop into a synchronous circuit, we add two NAND gates in front of it. Whenever we have a NAND gate, if both inputs are high, the output is low, and if either input is low, the output is high. Thus, if the clock is low (the clock constitutes one input to the NAND gates), the outputs of both NAND gates are high. That's the state in which the SR flip-flop doesn't change, so the flip-flop can't change state unless the clock is high. That's what makes all the flip-flops in a circuit change synchronously.

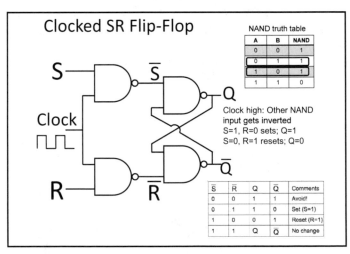

Clocked SR Flip-Flop

NAND truth table

A	B	NAND
0	0	1
0	1	1
1	0	1
1	1	0

Clock high: Other NAND
input gets inverted
S=1, R=0 sets; Q=1
S=0, R=1 resets; Q=0

$\overline{S}$	$\overline{R}$	Q	$\overline{Q}$	Comments
0	0	1	1	Avoid!
0	1	1	0	Set (S=1)
1	0	0	1	Reset (R=1)
1	1	Q	$\overline{Q}$	No change

When the clock is high, the left-hand NAND gates act as inverters, and their S and R inputs become the $\overline{S}$ and $\overline{R}$ inputs to the bistable circuit at right.

Master-Slave Flip-Flops

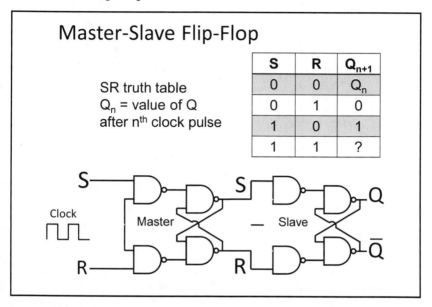

Master-Slave Flip-Flop

SR truth table
Q_n = value of Q
after n^{th} clock pulse

S	R	Q_{n+1}
0	0	Q_n
0	1	0
1	0	1
1	1	?

There are still some issues with the clocked flip-flop. The whole time the clock is high, the flip-flop is subject to change. Thus, if there's any noise in the system, it's possible that during the high phase of the clock, the flip-flop could flip back and forth several times. A more orderly situation would be if it changed only once per cycle. Further, a flip-flop that involves feedback from $\overline{Q}$ back to the set input could be subject to oscillation. One approach to solving that problem is a master-slave flip-flop.

We started with a basic bistable circuit. We then replaced the inverters with NAND gates so that we could set and reset. Then, we added another pair of NAND gates for the clocking. To make a master-slave flip-flop, we build the exact same circuit again.

Master-Slave Flip-Flop

To this configuration, we add one more gate—an inverter. It goes from the clock input to the master flip-flop into the clock input of the slave flip-flop, but there's an inversion in between.

Master-Slave Flip-Flop

Clock high: Master accepts inputs
$\overline{\text{Clock}}$ low: Slave isolated; can't change state

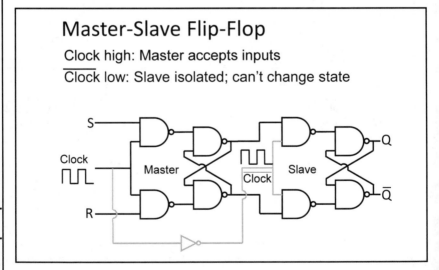

When the clock is high, the master flip-flop can change state. But at the same time, the clock input to the slave flip-flow is low, so the slave can't change state. Only when the clock goes low does the clock input to the slave go high, and at that point, the contents of the master are transferred to the slave.

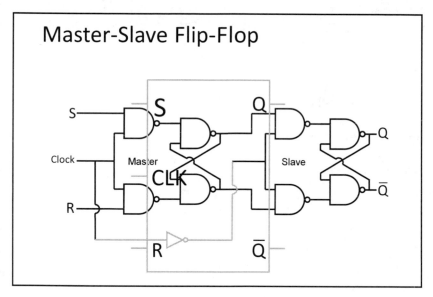

The master-slave flip-flop is a device that gives us increasing order in a complicated electronic system. It allows the clock to control the states when the flip-flops can change state. Further, the master-slave arrangement allows that change to occur only once per clock cycle. For our purposes, we'll represent the eight NAND gates and one inverter in this master-slave flip-flop as a single rectangle with the appropriate inputs and outputs. Similar rectangles will be our symbols for different kinds of flip-flops.

Suggested Reading

Introductory
Brindley, *Starting Electronics*, 4th ed., chapter 11, pp. 187–194.

Lowe, *Electronics All-in-One for Dummies*, book IV, chapter 5 through p. 595.

Advanced
Horowitz and Hill, *The Art of Electronics*, 3rd ed., chapter 10, section 10.4.1.

Scherz and Monk, *Practical Electronics for Inventors*, 3rd ed., chapter 12, section 12.6 through 12.6.2.

SR Flip-Flop Design
Design an unclocked SR flip-flop using NOR gates. Be sure to label the inputs correctly. Determine its truth table. How does it differ from the NAND-based flip-flop presented in this lecture?

SR Flip-Flop Simulation
Simulate either your design or the NAND-based SR flip-flop from this lecture and verify that it works.

Questions to Consider

1. In the unclocked $\overline{\text{flip-flop}}$ introduced at the start of this lecture, we labeled the inputs $\overline{S}$ and $\overline{R}$. Later, after adding gates for clocking, the inputs became S and R. Why the change?

2. Why do we use clocked logic in complex circuits, such as those in computers?

3. What advantage do master-slave and edge-triggered flip-flops offer over the simpler clocked flip-flop without the master-slave configuration or edge triggering?

Flip-Flop Circuits
Lecture 18—Transcript

Welcome to a lecture in which we're going to find ourselves doing a lot of flipping and flopping; maybe not a good thing for a politician, but an interesting thing for a digital electronic circuit. If you think about the digital circuits we've considered so far, they've had typically one, or more likely two, inputs, and those two inputs produced an output, which we described by means of a truth table. The output responded directly to the inputs; change the inputs, and the output changed.

We're going to move to more sophisticated circuits now; circuits that basically remember. Circuits that have two possible states they can be in, and once you put them in one of those states they may be able to stay there until you do something to change that state. Those circuits are going to constitute the core of computer memory and many, many, many other applications that we're going to be involved with for the remaining lectures of the course, in fact. We're going to spend this lecture on the basic, bistable circuit, the circuit has two stable states, and how we change those states. We're going to build up to some fairly sophisticated circuits just by the end of this lecture, and then we'll go on and build even more sophisticated circuits out of these later.

These circuits are sometimes called flip-flops because they can flip and flop from one state to another. Strictly speaking, the term flip-flop is usually reserved for the more sophisticated of these circuits that I'll get to later, but I'm going to use the term generically. We want to build a circuit that has two stable states. How do we do that? We start with the simplest possible logic operation: the NOT operation, the inversion. Here's a picture of a single converter; remember, it's a triangle that looks sort of like an amplifier, with that circle at the end that says invert. I'm going to connect a second inverter to it, and then I'm going to do a kind of feedback thing, taking the output of the second inverter, the right hand one, and bringing it back to the input of the left hand one.

Let's consider now the possible states of this system; what could it possibly do? Suppose there happens to be zero—low, false, whatever we want to call it—at the input to the first NOT gate. Because these are NOT gates, they invert, they change the state; the output of that first one is a 1. If the output of that one is a 1, that's the input to the second inverter, so the output of the second inverter is a 0. It's that 0 that's connected around back to the input of the first inverter. This is a stable, consistent state: 0 at the input to the first one, 0 at the output of the second one, and a 1 at the output of the first one; stable. It's a basic circuit; it can be in that state. On the other hand, what if there happened to be a 1 at the input to the first gate? Then there would be a 0 at its output, because it's a NOT gate, or an inverter. Then there would be a 1 at the output of the second gate, and that 1 is coming back around to be the 1 at the input of the first gate. That's also a stable state. That's why I call this thing a basic, bistable circuit. It can be in one or two of those states.

If I built this circuit and turned it on random differences in the transistors inside those gates and so on, it would cause it to go into one of these states and I wouldn't be able to determine which state it was. Sometimes it would just happen randomly depending on how the power was applied; maybe in other cases it would be prejudiced toward one or the other of those states. Not very useful, because I can't change the state. I want to now move on and develop circuits in which I can change that state. But before I do that, I want to redraw this circuit. It's going to be exactly the same circuit, but I'm going to draw it in a way that looks more symmetric, so we don't have one gate at the beginning and one gate at the end.

Here's a more symmetric version of this circuit. I'm going to call the output of one of the gates Q, the output of the other gate Not Q, and I'm calling them that because we've just seen this is a bistable circuit. One of the outputs has to be a 1 and the other one has to be a 0 in order for this circuit to be truly bistable, which I just showed you it is. I'm going to redraw the circuit now in this much more symmetric way in which we see the two inverters, and we see them cross-coupled with this kind of x-like connection. You'll notice the little jump in the wire that shows the two wires aren't connected where they cross. We have arbitrarily called the upper one's output Q and the lower one's output Not Q. As we've just seen, one of those Q or Not Q will be a 1 and the other one will be a 0.

This, again, still isn't useful. We can't change the state of the circuit. Let's see what we can do now to add some gating—remember the idea of logic gates—to change the state of the circuit. Now I'm going to go to the big screen and I'm going to analyze a circuit in which I'm going to replace those inverters with something a little more sophisticated, namely NAND gates. Let's go and look at what's called a set-reset flip-flop, an SR flip-flop. So what I have here is instead of the first inverter, the upper inverter, I now have a NAND gate. Instead of the lower inverter, I have another NAND gate. I still have that cross-coupling back from the output of the upper one to the upper input of the lower one, and from the output of the lower one to the lower input of the upper one, still very nice and symmetric.

Let me remind you how a NAND gate works. A NAND gate has a truth table that's the inverse of AND. In other words, the only time its output is 0 is when both its inputs are 1. We're going to leave that on there and use that fact to understand how this SR flip-flop works. By the way, notice that I've labeled these inputs Not S (that stands for "not set") and Not R (that stands for "not reset"). Those names will become obvious as we move on, and eventually we'll change them to set and reset in a more sophisticated circuit.

We want to build a truth table for this SR flip-flop. Here we have S and R listed, four possible states, 0, 0, 0, 1, and so on; the states for Q. I'm leaving a column for comments about that also. Let's begin by considering the case where both of the inputs are 0. Look on the NAND truth table: If any input to a NAND gate is 0, the output has to be 1. We know immediately S and R are both 0; both outputs have to be 1, Q and Q bar. This isn't a bistable state storing a 1 in one output and a 0 in the other; this isn't doing quite what we want this circuit to do. That's a situation we'd like to avoid, and later we'll build more sophisticated circuits that can avoid that more easily. We don't want that state. What if, on the other hand, we put a 0 at the S bar input and a 1 at the R bar input? Again, any NAND gate that has a 0 at any one of its inputs has a 1 at its output. As a result, there's a 1 at Q; there has to be. That 1 is cross-connected back to the other input of the lower NAND gate, and consequently, we have 1 and 1 on the lower NAND gate. The output of the lower NAND gate is 0, and that's all entirely consistent. That's a legitimate state of that flip-flop. If we do that, if we set S bar to 0—that's why we're calling it Not Set, because we'd set Set to 1 to try to set it—we're setting Not

Set to 0, and we get Q going to 1. That's called setting the state of a flip-flop, setting Q to 1; and so that's called Set. S=1; we don't have an S input; we have a Not S, which is a 0.

We do the opposite very obviously. Put S bar to 1 and R bar to 0; there's the opposite state there. You can convince yourself very easily that the one with a 0 has to have a 1 at the output. The two 1s make Q0, and so that makes Q0 and Q bar 1, and we've reset the flip-flop. We've set Q back to 0, and Q bar in the same time goes to 1.

Finally, we have the state in which we set both S and R to 1, or S bar and R bar to 1; so there's that state. What happens then? Let's suppose that Q happens to be 0. If Q is 0, we have a 0 to the input of the lower gate. Since a NAND gate with any input 0 gives us a one of the output, Q bar in that case is 1. Q is, in fact, 0 because we have 1 and 1 at the inputs of the upper gate. We have the state Q is 0, and Q bar is 0, which is just the state we had before. In other words, there's been no change.

What if we have the opposite situation? What if we have a 1 on Q and a 0 on Q bar? I've changed the state here, I've change the state down here; there's still no change. The beauty of putting the 1 and the 1 into the S bar and the R bar inputs is we store, we save, whatever state we're in and we don't allow it to change. Let me alter the truth table to read that the state in the case of a 1-1 input, Q is still whatever it was, and Q bar is still whatever it was. That's what that designation means. We have a set-reset flip-flop.

Let's take a look at what a set-reset flip-flop looks like, and then let's show you at least one interesting application of this device. Then we'll go on to build it into more sophisticated circuits. Let's move over to our board a moment. What I have here, again—there's a lot going on this board, and you can ignore most of it—I want you to focus on this particular chip. It's a 7,400 quad two-input NAND gate. There it is. It's connected in this cross-connection, the RS flip-flop I just described, and two push buttons are over here. I'm using one of them to be the set button, or the not set button, and I'm using the other one to be the reset button. Here in this light over here is the output of this circuit, the Q, or actually, it's the Q itself; I haven't bothered to connect the Not Q. Right now, Q is 0. I press the push button, and Q goes to 1. I press the push button again, and nothing happens. I press the other push button, which is the reset push button, and it goes back to 0. With these two push buttons, I can put that circuit in the Q equals 1 set state, or in the Q equals 0 reset state. That's an SR flip-flop. You obviously can't see the gates and all that kind of stuff, but I've wired exactly the circuit I just showed you on the big screen.

What's a circuit like that good? Before we move on to some of the more sophisticated circuits we're going to be building, let me talk about what happens when you turn a light switch on. You turn a light switch on, and the light comes on; great. You don't think a whole lot about that. But what's really happening when you turn a light switch on is two pieces of metal are coming together and making contact. This doesn't bother you when you turn a light on, but typically those pieces of metal come together, and they bounce apart a little bit, come together, bounce apart, come together, bounce apart a few times before they settle in to being fully on. With a lightbulb that doesn't matter; you don't notice the lightbulb turning on and off in a few thousandths of a second. In fact, if it's an old fashioned incandescent lightbulb that works by just heating up a filament, it doesn't even do that; it just interrupts the current a little bit and doesn't warm up quite as fast perhaps. But if you're dealing with digital logic circuits, particularly circuits that have to count pulses or something, then that can wreak havoc with what's going on. One of the simplest uses of an SR flip-flop, also call a latch in this instance, is to stop that bouncing of a switch from occurring.

I'm going to do a demonstration now with the oscilloscope. I'll put the oscilloscope on the big screen, in which I'm going to show you this process of switch bouncing. What I have over here now is a switch, a push button. It happens to be a push button that, when I push it, is going to bring to the oscilloscope screen up to a voltage of 5 volts; it's going to bring the voltage at a certain point here that I'm connecting to the oscilloscope up to 5 volts. I'm going to press this thing, and we're going to see what happens. There it is; ugly, ugly, ugly. Let's take a look. We were down in the low state; there's a bunch of noise in here, too. Then I pushed the button, and up went the switch. The switch changed state, and then it dropped down again—up and down, up and down, what a mess—and finally, over here, it probably was on continuously after that. No problem if that were turning on a lightbulb, but a terrible problem if that were sending pulses into a digital counting circuit—which we'll be building in a few lectures—because it would count every one of those pulses and we only wanted there to be one.

How do we handle that situation? Here's an example: Here's exactly the circuit I had. I had 5-volt power source connected through a resistor to that push button switch that then grounded that point between the resistor and the top of the switch. That's where the bouncing was occurring. There's typical what the bounce might look like in that case, that on and off rapid flipping. What do I do? I connect an SR flip-flop to that thing. As soon as the S bar goes low, that sets the flip-flop and sets Q to 1. There's the truth table we just worked out for that thing. If the state bounces back up again, that's the 1-1 state—notice I've got the R bar to set to 1, I'm just holding it there—that's the state where there's no change. Once this thing goes down for the first time, it never goes back up again. The same thing is true when we go in the opposite direction, which is what our oscilloscope happened randomly to catch in this particular instance. This circuit latches up. It holds the state the first time there's a change, and it completely eliminates that bouncing.

In this simple circuit I've just shown you here, I can't ever change it back again. What I'd typically do is modify that circuit a little bit so that I have a switch that actually can switch to two possible states. If I push it all the way to the right, that's the reset state, and that would get it ready to undergo another state change. But the point is, the so-called latch or SR flip-flops latches up and holds the state the first time there's a change, and that eliminates the bouncing of the switch.

Let me demonstrate that to you with the board also. This particular board actually has built into it two debounce push buttons. These two blue buttons say "debounce push buttons." They're push buttons that are connected in exactly a circuit like I've showed you. Now I'm going to change where I've got the oscilloscope connected. Right now, the oscilloscope is connected to the bad push button, the one that was bouncing, and I'm now going to connect it to the good push button, the one that's over here that's the debounced one. I'm going to hit the button. Look at that clean transition. Perfectly clean, because the SR flip-flop has latched up and prevented there being another change of state until the flip-flop resets, and we're able to do another change of state. There it is. We get these beautifully clean changes of state. That's one application of a simple SR flip-flop, an application called switch debouncing.

I'll give you one other application of this thing before we move on. Here's a circuit in which, very similar, I'm worried about—and I actually talked about a circle like this earlier with analog electronics—I'm going away on vacation, I want to know if my freezer has had a power failure and gotten above a certain temperature. What I've got here is a string of two resistors from 5 volts to ground, and the lower resistor is a thermistor, a resistor whose resistance decreases with increasing temperature. If the temperature goes up, eventually that decreasing resistance pulls the S bar input down into the state where it's the 0 state, and that's the state that puts that flip-flop into the situation where Q bar is equal to 1. When Q bar is equal to 1, you can see there's a warning light connected there. That warning light turns on, and then we've latched up, just like we did with the debounce push button. That warning light stays on even if the temperature then goes back down. You come back home, you think your food is all good, but actually the freezer has been above freezing and the warning light is on to tell you that. I've put a little reset button there on the R bar input, which is normally held there at 5 volts. If I push that, then I reset the thing and then it goes on to work just fine.

Those are two examples of the simple SR flip-flop. Now we're going to move on and get a little more sophisticated. That's great for applications like freeze alarms and debouncing push buttons, but in computer circuits we have to get more sophisticated with our flip-flops. The simple SR flip-flop I just showed you is called asynchronous logic. The output changes as soon as the input changes. If you have large scale circuits with lots of flip-flops, that's not good. That's what a computer is, for example. We're going to start talking about something more sophisticated called synchronous or clocked logic. In clocked logic, a central clock governs all the logic switching. Don't think of the thing with hands; that's not what I'm talking about for a clock. A clock is simply a square-wave generator that generates regular pulses. In today's computers, for example, the clock frequency is around 2–3 gigahertz; 2–3 billion clock cycles every second.

We're going to look now at how we make flip-flops clocked, how we make them synchronous circuits. Here's our simple SR flip-flop, and all I'm going to do is add some NAND gates in front of it. Now I'm going to call those inputs S and R. You can kind of see why that is, because to go from S to S bar at that top one, there's an invert, a little circle at the end of the upper NAND gate and similarly at the bottom one. Then there's this extra pair of inputs to these NAND gates, and that's the clock.

Let's think about what happens. Whenever I have a circuit, there's a NAND gate. Whenever I have a NAND gate, if both inputs are high, the output is low, and if either input is low, the output is high. If the clock is low—the clock constitutes one input to those NAND gates—if the clock is in its low state, the outputs of both NAND gates are high, and that's the state in which the SR flip-flop doesn't change. This flip-flop can't change state unless the clock is high. That's what gives us the synchronism. This pulse generator is coming along at some fixed rate and it's making these pulses, and the clock is going high and allowing this SR flip-flop to change state. When the clock as low, S bar, R bar are both 1, and there's no change. When the clock is high, the other NAND gate inputs get through. Again, this is an example I showed you in an earlier lecture about how these things are actually gates. Here we have to put those two lower input of the upper NAND gate, upper input of the lower NAND gate, high in order for the S and the R to get through. Then they get through and they're inverted, and they go in and they run the SR flip-flop just as we'd done before.

That's a clocked SR flip-flop. The clocked SR flip-flop is going to allow us to change state only when the clock comes along. I set S to 1, for example, and R to 0, and that doesn't automatically change the output of the flip-flop. It does so only when the clock comes along. If you have a billion flip-flops in the circuit—as you might, a billion literally—then they will all change state together and we have synchronous logic.

There are still some issues with this flip-flop that I've just described. As soon as the clock goes high, the whole time the clock is high, the flip-flop is subject to change. If there's any noise in the system or something else strange going on, it's possible that during the high phase of the clock, the flip-flop could flip back and forth several times. A more orderly situation would be if it would change only once per cycle. There's a much more subtle issue here, too. In a few lectures, we're going to build a flip-flop that's going to involve feedback from the Q bar back to the set input. With the simple clocked flip-flop I just described, that one would go into a wild oscillation if I tried to make this happen.

What we'd really like to do is build a flip-flop that can change state only once per clock cycle. There are two approaches to that system. One is called an edge-triggered flip-flop, in which literally the rise or the fall—and flip-flops are built both ways—of the clock is what causes the state change. That's called an edge-triggered flip-flop. An alternative that I'm going to describe—it's a little bit easier to understand; it's the one I have my students build—is called a master-slave flip-flop, and we want to look at how this master-slave flip-flop works.

Here we go. Here's our SR flip-flop now, our clocked SR flip-flop. Look how far we've gotten. We started with the basic bistable circuit. That's the pair of NAND gates on the right; actually, we started with a pair of inverters but now we've replaced them with NAND gates so we could set and reset. Now we've added another pair of NAND gates for the clocking, and we have this clock input. What do we do to make a master-slave flip-flop? We build the exact thing all over again. Now we're getting a lot of gates; one, two, three, four, five, six, seven, eight gates already. We've got a clock input there. I represent the clock again as what it is: It's a square wave that's coming up and down, up and down, meaning it's going from 0 to 5 volts if we're in TTL logic or it's swinging between whatever the two logic level voltages are in whatever circuit we're using.

How are we going to connect these things? Let's do the following. Let's call the first thing on the left, the first clocked flip-flop, the master, and let's call the one on the right the slave. There are the S and the R inputs to the slave flip-flop, and I'm going to connect them to the Q and the Q bar outputs of the left hand, the master flip-flop. Why am I doing that? Here's why: I'm now going to add one more gate; so we've got nine gates, the ninth gate is an inverter. It's going from the clock input to the master flip-flop into the clock input of the slave flip-flop, but there's an inversion in between. What does that inversion mean? It means when the clock is high—the actual clock signal that's coming in at the left is high—the clock input to the slave flip-flop on the right there will be low. I call that not clock; clock with a bar over it. Again, the bar signifies an inversion, a change in the logic state from 0 to 1, or 1 to 0. If the clock on the left is high, not clock is low. I've illustrated that in this diagram by showing the square wave going just the opposite: When the clock square wave is high, than not clock square wave is low.

What does all that mean? It means when the clock is high, which was the situation in which one of these clocked flip-flops was allowed to change state, then the master flip-flop can indeed change state. But at the same time the clock is high, the not clock—the clock input to the slave flip-flop—is low and it can't change state. While the master can change state, the slave can't. Here's what you do with this flip-flop: You run the clock, you put the S and R inputs to wherever you want to set the master flip-flop, and the Q goes to whatever it wants to be. But the slave's output doesn't change at that point; the Q of the master. The Q of the slave stays whatever it was. Then when the clock goes low, the not clock goes high, and that allows the slave to change state. Since the set input of the slave is connected to the Q of the master, and the reset is connected to the Q bar of the master, what happens is the contents of the master flip-flop; that is, the value of Q gets loaded into the Q of the slave. That can happen only once in the clock cycle. It happens in this particular circuit when the clock falls to low, and at that point the state of the slave changes and then it can't change again until the clock cycle comes along again and the state falls to low.

The master-slave flip-flop is a device that gives us increasing order in a complicated electronic system. It allows the clock to control the states of when those flip-flops can change state. Furthermore, with the master-slave arrangement, it allows that change to occur only once per clock cycle. That's really crucial to keeping order in a complicated circuit like a computer. Notice we've gained this order at the expense of now going from four NAND gates to nine total logic gates, eight NAND gates and one inverter.

Of course, these things are built in today's integrated circuits not by going out and buying 7,400 NAND gates and wiring them all up—although that's how I have my students do these when they're studying them—but they're all just built etched onto the silicon circuits with the transistors needed to make this. Nevertheless, this is a rather now complicated circuit. It's a little bit expensive to make, and it consumes a good bit of power, and maybe it isn't the best way to make, for instance, computer memory, as we'll see in a subsequent lecture. But it's a very good way to make a system that's very orderly because each flip-flop can change only once in a clock cycle.

Now we're going to take this whole complicated circuit consisting of eight NAND gates and one invert, and we're going to shrink it down and forget about what's going on inside it. We'll remember that, but we're going to build it as one single device. We'll shrink it down, and we'll give it a name and a symbol. The symbol is this rectangle; that's going to be our symbol for different kinds of flip-flops. On the left we have the S input, the set input to the master flip-flop, and the reset input to the master flip-flop, and the clock input to the master flip-flop. On the right, we have the Q and Q bar of the slave. We have this system where we could put whatever values we want on set and reset. When the clock comes along, it'll load set into the master. When the clock goes low, the slave clock will go high because it's a not clock, and will load those things into Q and Q bar. This is a master-slave flip-flop, and we'll use this symbol as we go on and build additional flip-flops. We'll have similar symbols, and we'll take the master-slave and build other flip-flops from them.

Again, I want to say something I did at the beginning: Some electrical engineers would not grace this object with the name "flip-flop," because it's still a little hard to make it flip and flop all by itself. It would be called a latch, a master-slave latch, a clocked latch, but it could also be called a flip-flop. I'm going to use those terms interchangeably.

Let me wrap up and then give you a little project to work on. We've seen the basic bistable circuit; we started with that circuit with just two inverters. We grew it into a more sophisticated circuit where we could control the setting and resetting with those not set and not reset inputs. But still, chaos reigned, although we could use that in cases—in fact, we had to use that in cases—like the debouncing push button or the freezer alarm where there wasn't any synchronism throughout a whole complicated circuit. But once we get to computers and more complicated circuits, we needed the clocked flip-flop, so we built the clocked flip-flop. The clocked flip-flop was much better, except that it still had this problem that the state could change many times during a clock cycle as long as the clock was high. So we added the master-slave flip-flop, which allowed the system to change only once in a clock cycle. Once again, when you have tens, hundreds, millions, and literally today billions of these things in a single circuit, you really need that everybody changing only once a clock cycle and everybody changing together. That's the first of many sophisticated—three or four sophisticated—flip-flops that we're going to develop in subsequent lectures.

If you'd like to look a little bit more at flip-flops, do the project. In the project, I'd like you to design an unclocked SR flip-flop—like the early one I designed after the two-inverter bistable—but make it using NOR gates. Be sure to label your inputs correctly and determine the truth table. Ask yourself the question: How does it differ from the NAND-based flip-flop that I presented in this lecture? Then, if you want to use one of your circuit simulators, go into the circuit simulator and simulate either the design you came up with in part one or the NAND-based flip-flop from this lecture. Go through it, putting different inputs, and verify that it works. Then you'll really understand flip-flops.

Shift and Divide—Your USB and Your Watch
Lecture 19

The simple SR flip-flop stores 1 bit of information, and that information is available at its Q output, whose value is either logic 1 or logic 0. Any piece of information that can be addressed with a yes/no, true/false response can be stored as 1 bit. But we've seen that computers and other circuits use many bits. For example, the ASCII system represents each character in the alphabet by an 8-bit sequence, and computers in general store information in strings of many bits (called *words*). Further, as computers have evolved, those words have gotten longer. For all these reasons, we need more than just one flip-flop storing 1 bit. In this lecture, we'll begin to learn how to link flip-flops to store and process multiple bits of information. Key topics include the following:

- Organizing data: The computer word
- Meet the D flip-flop and shift registers
- Serial-to-parallel conversion
- Parallel-to-serial conversion

- A computer communicates
- Inside your USB
- Meet the T flip-flop
- Frequency dividers: what's inside your quartz watch.

Organizing Data: The Computer Word

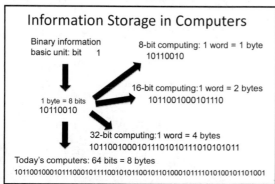

The basic information unit in binary is the bit, which can be either a 1 or a 0. The next largest unit is the *byte*, containing 8 bits. Early computers used 8-bit words; each word was 1 byte. Today's computers are mostly 64-bit units; smaller microcomputers in smart devices may still use 8- or 16-bit words.

Two Approaches to Digital Communications

- Parallel
  - Send a whole word at once
  - Requires *N* wires, with *N* the word length
  - Increasingly impractical as *N* has grown

- Serial
  - Send one bit at a time
  - Requires parallel-to-serial conversion at sender...
  - ...and serial-to-parallel conversion at receiver

There are two approaches to computer communication: *parallel* and *serial*. In parallel, an entire word is sent at once down a cable that contains one wire for each bit. Parallel transmission occurs almost instantaneously, but as the length of the word has grown, parallel communication has become impractical. In serial communication, 1 bit is sent at a time. This method requires far fewer wires, reducing the cost and bulkiness of the cable, but it comes at the expense of slower transmission and the need to convert from parallel to serial on the sending end and back to parallel on the receiving end. Flip-flops can be used for that purpose.

Meet the D Flip-Flop and Shift Registers

For a data (D) flip-flop, we start with an SR flip-flop, and we connect the set input to the reset input through an inverter. If we put a 1 at the D input (the set input), we get a 0 at the reset input, and Q goes to 1. If we put D to 0, we get 1 at the reset, and Q goes to 0. One of the advantages of the D flip-

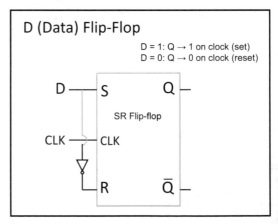

D (Data) Flip-Flop

D = 1: Q → 1 on clock (set)
D = 0: Q → 0 on clock (reset)

flop is that it eliminates the state with 1 on both inputs, which we've seen is to be avoided.

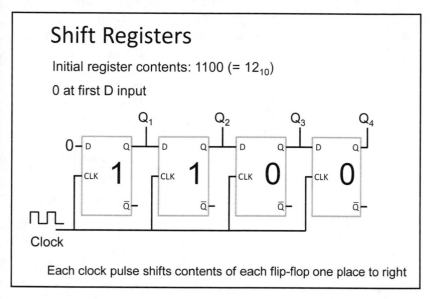

Shift Registers

Initial register contents: 1100 (= 12_{10})

0 at first D input

Each clock pulse shifts contents of each flip-flop one place to right

A *shift register* is a number of D flip-flops connected so the Q output of each flip-flop feeds the D input of the next flip-flop. All flip-flops are connected to the same clock, and when the clock pulse occurs, each flip-flop delivers the contents of its D input to its Q output. As a result, each clock pulse shifts the contents of the entire register one place to the right.

Suppose the initial register content is 1100 (12 in base 10). If we put a 0 to the first input, after the first shift, 0 will load into the leftmost D flip-flop, leaving us with 0110 (6 in base 10) after the first clock pulse. After the second clock pulse, we get 0011 (3 in base 10).

Shift Register: A Divide-by-2^n Circuit

Initial register contents: 1100 (= 12_{10})

After first clock pulse: 0110 (= 6_{10} = 12 ÷ 2)

After second clock pulse: 0011 (= 3_{10} = 6 ÷ 2 = 12 ÷ 4)

What's this do?

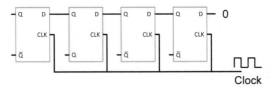

Clock

Each clock pulse shifts contents of each flip-flop one place to left: multiplies by 2

If we interpret the contents of the shift register as a binary number, notice that we are dividing that binary number by 2 with each shift. We can imagine taking as many flip-flops as we want and dividing by any power of 2. Thus, a shift register is one way of doing arithmetic in a computer. If we reverse the directions of the flip-flops, as shown in the bottom of the figure above, then we get a circuit that multiplies by 2 with each clock pulse.

Serial-to-Parallel Conversion

Serial-to-Parallel Conversion

Each clock pulse shifts contents of each flip-flop one place to right

After four clock pulses: 4 bits available in parallel at outputs

Input: 4 binary bits, appearing sequentially 1100

$Q_1 = 1$ $Q_2 = 1$ $Q_3 = 0$ $Q_4 = 0$

In

Clock

To accomplish serial-to-parallel conversion, we start with a register consisting of four D flip-flops. If we want to load 1100, we bring in each digit sequentially to the left-hand flip-flop. As we do that, on the clock, everything shifts to the right. After four clock pulses, we end up with 1100 stored in the register.

Parallel-to-Serial Conversion

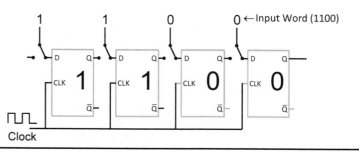

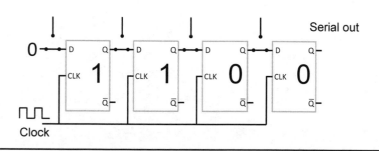

Parallel-to-serial conversion is a little more complicated, requiring some switches. Here, the switches at the top of the diagram are first connected to the information we want to bring in. Again, our input is 1100. With the switches in their upper position, we load the 4-bit word into the register. Then, we move the switches to the other position so the Ds are connected to the Qs again, as we had earlier in our shift register. This allows us to start the process of shifting the word to the right. The bits appear in sequence, one at a time, at the serial output.

A Computer Communicates

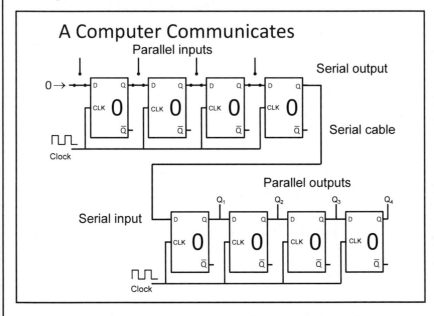

We can now see how a computer communicates with external devices via a serial connection. The information to be communicated (here, the 4-bit sequence 1100) is loaded in from the computer in parallel, sent serially over a cable, and converted back to parallel for use in another device, such as a printer or another computer.

Inside Your USB

- **USB 2**
  - Four wires; two for signal, two for 5-V power
  - Max data rate: 480 megabits/second
  - Unidirectional data transfer
    - Send/receive happen separately
  - Max current supplied: 500 mA
- **USB 3**
  - Nine wires; seven for signal, two for 5-V power
  - Max data rate: ~5 gigabits/second
  - Simultaneous send/receive
  - Max current supplied: 900 mA

USB stands for *universal serial bus*; a *bus* is something that carries signals typically over substantial distances in electronic circuits.

Meet the T Flip-Flop

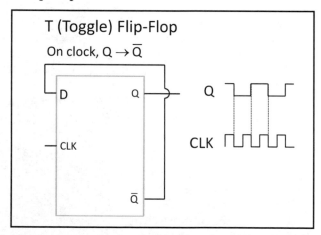

The T flip-flop represents an interesting use of negative feedback. We start with a D flip-flop, then connect the $\overline{Q}$ output to the D input. If Q = 1, then

$\overline{Q} = 0$ and so the D input will have 0. When the clock pulse arrives, that 0 is loaded into Q, and Q changes to 0. On the other hand, if Q = 0, then $\overline{Q} = 1$, because those two are always opposites. So if Q = 0, there's a 1 at the D input. When the clock pulse arrives, that 1 is loaded into Q. The point is that this flip-flop toggles between its two possible states. Whatever state Q is in, on the next clock pulse, Q goes to $\overline{Q}$. T flip-flops are the building blocks of many circuits, including frequency dividers and digital counters.

Frequency Dividers: What's inside Your Quartz Watch

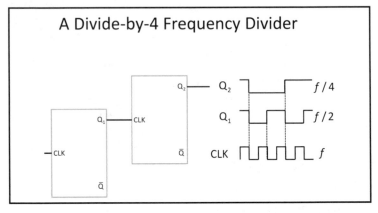

Earlier, we saw a divide-by-2 frequency divider with one T flip-flop. If we take the Q output of that T flip-flop and put it into the clock input on the next T flip-flop, the second flip-flop acts exactly like the first, except now, it's working on a signal that's coming in at half the clock frequency. That gives us a divide-by-4 frequency divider.

Inside a quartz timepiece, there is a small crystal of quartz that can be set vibrating by applying a voltage across it. Typically, the crystal is engineered to vibrate at a frequency of 32,768 Hz (cycles per second)—which is 2^{15} Hz. That frequency is chosen because when it's divided by 15, the result is 1 Hz, and that 1 Hz drives the second hand or the digital display of the timepiece.

Introductory

Brindley, *Starting Electronics*, 4th ed., chapter 11, pp. 194 to end.

Lowe, *Electronics All-in-One for Dummies*, book VI, chapter 5, pp. 596 to end.

Advanced

Horowitz and Hill, *The Art of Electronics*, 3rd ed., chapter 10, sections 10.4.2, 10.5.1, 10.5.3; chapter 14, section 14.2.4.

Scherz and Monk, *Practical Electronics for Inventors*, 3rd ed., chapter 12, sections 12.6.3, 12.6.4, 12.6.5 through p. 780; section 12.8.

Projects

T Flip-Flop Simulation

Simulate a T flip-flop using your circuit simulator's built-in $\underline{D}$ flip-flop and verify that it works. Note: CircuitLab's D flip-flop has no $\overline{Q}$ output. How will you deal with that?

Frequency Divider

Make a frequency divider that divides the input frequency by 8 and verify that it works. Use your flip-flop(s) from the project above, CL's built-in T flip-flop with a 1 at the T input or DC's JK flip-flop with all J, K inputs to 1.

Questions to Consider

1. Why has parallel data transmission become increasingly impractical as computers have evolved?

2. What's a T flip-flop do? And what does the *T* stand for?

3. What's special about the number 32,768, which is the frequency of the quartz crystal vibrations in a quartz watch?

Shift and Divide—Your USB and Your Watch
Lecture 19—Transcript

Welcome to the next lecture in exploring the magic of flip-flops. In this lecture, we're going to get both divisive and shifty, and those are good things with flip-flops.

So far, we've seen the simple SR flip-flop, although we've embellished it to the point where we have the clock-to-master slave version of it that could act synchronously in a complicated circuit like a computer. But what that flip-flop did for us basically was to store one bit of information: It had a Q output that was either a 1 or a 0. You may think that's trivial; what good is that? But that could stand for a yes or no decision on almost anything. Is the weather today going to be clear or cloudy? One, it's going to be clear; zero, it's going to be cloudy. Any piece of information that could be answered with a yes/no, true/false answer can be stored as 1 bit in a flip-flop.

But we've already seen that computers and other circuits use many bits. For example, I showed you a few lectures ago how the ASCII system of information storage represents each character in the alphabet by an 8-bit sequence. That's 8 bits. Computers in general store information in strings of many bits. They're called words. As computers have evolved, the lengths of those words have gotten bigger and bigger, and computers typically manipulate a whole word at a time. We're going to need more than just one flip-flop storing 1 bit. We're going to see how to link flip-flops together to store lots of information. That will be the subject of the next couple of lectures.

Let's look at the situation in terms of information storage in computers. The basic information unit in binary is 1 bit, and the bit could be either a 1 or a 0. Usually, the next thing up that people talk about is the byte, and many of the quantities about your computer are measured not in bits but in bytes. For instance, the amount of memory you have in your computer is measured in bytes, or probably in gigabytes, or hundreds of megabytes. Similarly, for the amount of information that's stored on your hard drive or whatever information storage you have, in bytes. A byte is 8 bits. We've gone from the basic unit, a 0 or a 1, a single bit, to a byte, which is 8 bits. I remember my very first computer back quite a few decades ago, and that computer used 8-bit words. In other words, each word was 1 byte. Computers eventually graduated to 16-bit computing in which one word was 2 bytes. Among many other things they could do with 16 bytes as opposed to 8 was to process, do arithmetic, and do mathematics to a higher level of accuracy because they had more digits. Then came 32-bit computers in which the word is 4 bytes, 32 bits. Today's computers are mostly 64-bit units, and 64-bit units for personal computers began to be developed in the early 2000s. Most computers today are 64-bit, although many of you'll remember, if you're into computers, that that transition didn't occur that long ago, and there were some pieces of software that maybe took a few more years to get caught up with the 64-bit processing.

Having said that, there are plenty of computers that still use 8 or 16 bits. They probably aren't your personal computer or a big computer that scientists are using for computations of climate models or something, but they may be small microcomputers that are built into many places in your car or into any smart device these days, and they use fewer bits because they don't need those big numbers of bits.

We store information in computers in words; that's the word for it. Whether we're dealing with it as mathematics, or a string of symbols, or whatever else, it doesn't matter; the unit of information storage is called the word. Today's words in modern computers are about 64 bits long.

Computers have an issue. They need to communicate with other computers. They need to communicate with peripheral devices like monitors, like printers. In the old days, they had to communicate with a modem that then communicated out over the telephone line. Today, they need to communicate with perhaps a wireless device that sends signals off into the ether as radio signals. They need to communicate. How do they communicate? There are two approaches, one of which is becoming rapidly obsolete as the word length grows in computers, and that's called parallel. In parallel, you send an entire word all at once; all 8 bits, 16 bits, 32 bits, 64 bits, or whatever. You send them down a cable, and they go all at once. There's a virtue to that: The virtue is that transmission happens very fast. You put 64 bits on one end of the cable, and speed of light later 64 bits come out the other end instantaneously. But as the length of the word has grown, parallel cables have become impractical; parallel communication. Just to remind you, if you've been into computers for a while, you probably 10 or 15 years ago had a computer that connected to its printer with an ugly, massive connector like this thing and a huge, thick cable that was very hard to bend. If you look at the end of this cable, you'll see there are many, many, many wires in here. This is a parallel cable I took from an old computer-to-printer connection. It sent a single word all at once to the printer, so it was fast.

On the other hand, electronics have gotten a lot faster—think Moore's law— and we don't necessarily need parallel for good speed anymore. But you look at this cable, many of you have used a cable like this, and that's a parallel cable. All the bits are going down there simultaneously; each one has its own wire. Increasingly impractical as the number of bits has grown.

The second approach is what's called serial. In serial, you send one bit at a time. They can all share the same wire. Bit 1 goes down, bit 2 goes down, bit 4 goes down, bit 8 goes down, bit 16 goes down. It's slower; although, again, with modern electronics it's not prohibitively slow. It requires in principle one wire instead of multiple wires. In practice, it requires two because you've got to establish a return or a ground common to the two devices that are communicating. But the big deal is it requires that you somehow convert parallel, which is what the computer is using internally. The computer would be far too slow if it processed things one bit at a time. That's the whole point of having words. It processes 64 bits all at once. It adds two 64-bit words representing decimal numbers, or it does something to compare two 64-bit words representing strings of letters in a plagiarism-searching routine, or whatever. You need to convert that parallel information that's handled inside the computer to serial when you're going to send it out to another device, or put it over Wi-Fi, or whatever. Then at the other end, you need to convert serial back to parallel. How on earth are we going to do that?

Before we do, let me just give you an analogy and let me show you what happens when we do use serial cables. Here's a serial cable. This is a very familiar cable to anyone who has an iPhone; it's a USB cable for the iPhone. One end plugs into the phone, the other end plugs into any USB port. There it is, a USB cable. It's very thin. It's only got four wires in it, and that's more than it really needs to just do USB communications. It's very easy to handle compared to this massive cable we had in the old days. USB. You may have a printer that uses a USB cable, or maybe the printer's completely wireless. There's still serial data transmission involved.

We've gone from parallel to serial, and the reduction in the cost and the bulkiness and so one of the cable that's doing the communication is huge. That's coming at the expense of slower transmission and also the need for this conversion from parallel to serial and back to parallel at the other end. That's what I want to explore now, and that's where flip-flops are going to be able to help us.

But before I do that, again, an analogy: Here's a highway carrying lots of cars, a wide, divided highway; many cars going down at once, side by side in different lanes. That's like parallel data transmission. On the right, you see a one-lane road carrying cars one at a time. They're all sharing the same lane; it's obviously slower. On the other hand, the infrastructure needed is a lot cheaper. A one-lane road is easier and cheaper to build than a multi-lane highway. There's an analogy with parallel and serial data conversion. What we want to now understand is how to do the conversion from one to the other, parallel to serial at the sending end and serial to parallel at the receiving end. Flip-flops are going to come to the rescue for that purpose.

Before we do that, we need to develop one more kind of flip-flop. In fact, it's not the last flip-flop; we'll be developing at least two more kinds after this. Here's the flip-flop I showed you in the preceding lecture; it's the SR clocked flip-flop. Probably the master-slave variety, so it switches only once on the clock. But we're not worrying about those details anymore, it's just a flip-flop. It can be set, and it can be reset. Q can be set to 1, and Q can be set back to 0. What are we going to do with this flip-flop?

We're going to make it into a D flip-flop. "D" stands for "Data." In a D flip-flop, we take a set-reset flip-flop, and we take the set input, and we take whatever's coming in to the set input, and we put it also to the reset input, but we put it through an inverter. If you put a 1 at the D input, as it's now called, at the set input, you get a 0 at the reset input. If you put D to 1, when the clock comes along Q goes to 1. If you put D to 0, when the clock

377

comes along, you've got a 1 at the reset because of that inversion, and Q goes to 0. This is a data flip-flop, and one of its advantages is it eliminates that redundant state that we had, the possible state of putting a 1 on both inputs, which really didn't make it work like a flip-flop. That's not possible because whatever you put on S, the reset has the opposite. It's actually a simpler flip-flop in a sense. It has one input, not two. It has the D input for the data. The data is what you're going to store in this thing. Again, even this more sophisticated device is often not graced with the name flip-flop. It's called a latch or a register; it's a storage register. In fact, we'll see in another lecture that it's going to be the fundamental unit of some kinds of computer memory, not the kind that's in your everyday computer, but a more sophisticated and faster kind of computer memory.

Now we have a D or data flip-flop. We've eliminated the separate S and R's, Set and Reset, for our flip-flop, and replaced them by just one input D, and then this line that goes down to the reset and inverts whatever's at the D. We can set or reset Q with a single input at the D. What we're going to do is take this flip-flop and do what we've done before: We're going to reduce it down and give it its own name. It's a slightly simpler rectangle, which represents the master-slave SR flip-flop inside there with this extra connection from the S to the R through an inversion, giving us only one input, the D input. That's the symbol for the D flip-flop, and we'll be working at some length now with D flip-flops.

Let's see what we can do with them. The first thing we need to do is this business of serial-to-parallel and parallel-to-serial data conversion. We're going to build something called a shift register, and this is why I said we're going to be getting shifty here. What's a shift register? It's a bunch of D flip-flops connected together, and notice the connection here at the top of the flip-flops. The Q of the leftmost flip-flop—that's the stored bit that's either a 1 or 0—goes to the D input of the second flip-flop. I've marked above there, Q sub 1 I'm going to call it. The next flip-flop has its Q output; I'll call it Q2 in this case. Later I may begin to modify some of these numbering schemes a little, so bear with me. But I'm going to call that Q2 now; in fact, I'll call it Q2 later. It goes from Q to the D input of the third flip-flop. The third flip-flop's output is Q3 and goes to the fourth flip-flop whose output is Q4.

What happens? All these flip-flops are connected to the same clock, so remember what happens on the clock pulse. The clock pulse takes the SR flip-flop and delivers the contents of the set input, which is now the contents of the D input on this D flip-flop, delivers it to Q. The first flip-flop has its D input connected to whatever I want to send into it, but the remaining flip-flops are connected to the previous one. What happens when the clock comes along is that the contents of the flip-flops shift one place to the right.

Let me give you an example. Suppose initially this register—and it's a shift register, as it's called; it's called a shift register because it shifts the contents—suppose it initially contains 1, 1, 0, 0, which happens, if you want to interpret it as a number, to be the binary number 12 because we've got one 8 and one 4 and no 2s and no 1s. But I don't have to interpret it that way; it's just a piece of information representing true, true, false, false, 1, 1, 0, 0, high, high, low, low, whatever it represents. There are the initial register contents. Let me put a 0 to the first input. That means after the first shift, 0 is going to load into the leftmost D flip-flop in the register. After the first clock pulse, what do I have? Everything's shifted to the right, and I've got 0, 1, 1, 0 in the register. Then after the second clock pulse, I've got 0, 0, 1, 1, which happens to be the number 3, 1, 1, and 1, 2. The intermediate state was the number 6. Here's what happens with this shift register: On every clock, the contents shift to the right.

What good is this? I'm going to give you several goods to this, one of which is going to involve our problem that we're trying to deal with here of serial-parallel conversion. But let me pause and show you another use for this thing. Let's just rehash what that circuit did. We set the initial register contents of 1, 1, 0, 0. That translated into 12 in base 10; again, one 8, one 4, no 2s, no 1s. After the first clock pulse, we had 0, 1, 1, 0. That translated into 6 in base 10; that is, no 8s, one 4, one 2, no 1s, 4 and 2 is 6. Notice what that is: That's 12 divided by 2. After the second clock pulse, both the 1's have moved as far right as they can and still be in the register. We have 0, 0, 1, 1. That's 3 in base 10; 1, 2 and 1, 1 is 3, that's 6. The previous number we had divided by 2 is the same as 12 divided by 4. What we've done with each successive clock pulse is shifted the contents to the right, and if we interpret the contents as a binary number, we've divided that binary number by 2 each time. You could imagine taking as many flip-flops as you want, and you would divide by any power of 2. This is a way of doing arithmetic by computer with a shift register. It's only a way of dividing by powers of 2, however, but it's a way of doing important arithmetic that we'd want to do in a computer.

That's sort of an aside; and if you like that aside, let me give you a little challenge: Here's another shift register configuration, and if you want to pause and ask yourself what does this do, you can do that. Let me just point out a few things: I've turned the flip-flops around, the D flip-flops around, so the D input is on the right. I'm bringing in a 0 from the right. The clocks are all connected together again. Ask yourself, what does this circuit do?

Here's what it does: Each clock pulse now shifts the contents one place to the left, so this is a multiplier by 2^n. In this case, I've got 4, so it multiplies by 2^4 after I go through the entire shifting. What this circuit does is to multiply. Because I'm shifting left and making the numbers bigger, it's like adding a 0 on a decimal number. What's 10x15? It's 150; you just tack a 0 on. This circuit is doing the same thing, except it's in binary so it's only multiplying by 2 rather than by 10.

That was a little aside on what shift registers do. We want to move back to our original challenge, which is the problem of parallel and serial communications. Let's move on and look at first serial to parallel; that's actually a little tiny bit easier. Then we'll put the whole thing together and get computer communications working.

Here, again, is our system in which we have a register consisting of four D flip-flops connected as I had them originally. Suppose I want to load this with the number 1, 1, 0, 0, the number 12, the same number I've been talking about (although you don't have to interpret it as a number; it's a string of four binary digits, four bits). I'm going to bring the 0, and the 0, and the 1, and the 1 in sequentially to the left-hand flip-flop, and I'm going to sequentially on the clock shift everything to the right. What happens? After the four clock pulses, I brought that whole thing in. The first 0 came in and shifted all the way to the rightmost register; the second one came in and shifted. After four clock pulses, I end up with 1, 1, 0, 0 stored in that register. Available at the Q outputs at the top there is, now in parallel form, that binary quantity, binary number, string of bits, whatever you want to call it. That's how you do serial-to-parallel communications.

If you want to do parallel-to-serial conversion at the other end, it's a little bit more complicated. You've got to have some switches. Here I again have the flip-flops. I show you these switches at the top, and those switches are connected to whatever information I want to bring in. That information is now in parallel. Here we see our 4-bit sequence, 1, 1, 0, 0—we could call it a word, a 4-bit word, I suppose—that's the input sequence 1, 1, 0, 0. We're going to have the switches connected so those things individually go into the four registers, and so we load them up. First thing we do, we set the switches to their upper position. They're loading in the 4-bit word. Then we have 1, 1, 0, 0 in the register. Then we switch the switches to the other position, to the left, so the D's are connected to the Q's again, as we had before in our shift register. When I say switches, nobody's in there throwing mechanical switches. These are CMOS transistors or some other kind of electronic devices that are acting as switches. Now I can start the process I described before and shift that word to the right. Out on the rightmost one come, sequentially, one of the bits after the other. After four shifts, the 1, 1, 0, 0 are heading down the line. That's parallel-to-serial conversion.

Let's talk about how we use that in computer communications. Here's how a computer communicates: At the top, I've got a parallel-to-serial converter. At the bottom, I've got a serial-to-parallel converter. What I'm going to do is load an input word in those parallel inputs. I'm going to have a serial output, just as I described, as the clock comes along. I'm going to connect by means of a serial cable; one wire or maybe two is all we need here. Those bits are going to go down that cable, and they're going to load themselves with each successive clock pulse at the bottom end of the receiving apparatus into that bottom register. I will have transferred the 1, 1, 0, 0 that was loaded in from the computer in parallel, send it serially over the cable, and it will load then in to the other device, where it becomes available in parallel for that device to use for processing.

By the way, that cable need not be a cable; it could just equally well be a Wi-Fi connection or whatever. But the point is we're sending the information one bit at a time. That's how computers communicate.

If we want to talk a little bit more about that, here's what's going on inside the USB cable that I talked about. This happens to be a USB 2.0 cable, the older kind. We introduced USB 3.0 in 2008, but there are plenty of USB 2.0 cables out there. There are four wires in a USB 2.0 cable. Two are for the signal—the main signal wire and then a return ground—and two actually supply 5-volt power, and that's why you often get a little device like this with your USB cable, because you can plug that into a wall and power other devices. It's also why you can charge your phone by plugging it into the USB port of your cable, because the USB ports also supply power. They can supply up to 500 milliamps. They can transfer data at about almost 500 megabits, 500 million bits a second. But they can only transfer it in one direction; so they have to send for a while, and then receive for a while. USB 3.0, which, as I said, was introduced in 2008, has eight wires, six are for the signal. It carries signals in opposite directions at the same time. It's got two for 5-volt power. It can have a data rate maximum of about 5 billion bits per second, much faster, a factor of 10 faster. It simultaneously sends and receives, as I said. It can supply almost an amp of current. That's what USB is about. What's the "S" in USB? Serial; it's a serial cable. What's the "U"? Universal; it's a serial cable used universally for all kinds of devices. "B" stands for bus, which is a computerese term for something that carries signals typically large distances or substantial distances in electronic circuits. That's how a computer communicates.

I promised you a watch, and let's end with the watch. Before we can get to the watch, we have to introduce still one other flip-flop, and that's a T or Toggle flip-flop. This really is a flip-flop; nobody's going to object to my using that term because it flips and flops from one state to the other. It toggles states every time the clock comes on. How do we make a T flip-flop? Again, it's got a kind of feedback, and this is a sort of interesting negative feedback. We take the Q bar output and we put it to the D input. Think about that. If Q is 1, then the D input is going to have a 0 on it. When the clock comes along, what that's going to do is load that 0 into Q, so Q is going to change to 0. On the other hand, if Q is 0, then Q bar is 1, because those two are always opposites. That means if Q is 0, there's a 1 at the D input, and when the clock comes along, that 1 is going to load into Q. The point is, it's going to toggle; it's going to change state. Whatever state Q is in, on the next clock Q goes to Q bar, is a quick way of talking about what the toggle or T flip-flop does.

If we sort of looked at a picture of what went on with the toggle flip-flop, here's Q, and here's the clock. The clock is flipping twice as fast as Q is, and I'm assuming this is a flip-flop that changes state when the clock falls. The first time the clock falls, that's when the D input is accepted. Q might have been high, and now it goes low, and it stays low until the next time the clock falls, which is a full clock cycle. So half of a cycle of the flip-flop, of the Q, is a full cycle of the clock, and so the Q is flipping at half the rate that the clock is.

Let's see that working in an actual system. Let me go over to my demo table and look at that situation. I actually have that situation happening right now. I have a T flip-flop up here. It's part of a 7476 dual T flip-flop. We don't need to worry about that, but that's what it is, and it's flipping and flopping in there. I've got its Q output and the clock connected to these first two lights. The first light is the clock, and the clock is flipping back and forth at a certain rate. If you watch the second light, that's the output of the flip-flop, you'll see it's flipping at half the rate. If I speed things up a little bit, perhaps you can see it a little more clearly. There it goes flipping. The second red light is coming on only half as often as the first red light. It's a toggle flip-flop. It's producing an output frequency, if you will, that's half what the input was. If I crank that up really fast and look at it on the oscilloscope, the frequency up by quite a bit—that's going to go about there—you can see that the clock, which is the yellow signal, and the Q output, which is the green signal, are happening at a factor of 2, different in frequency. Their one full cycle of the clock is only half a cycle of the Q. This has acted like a frequency divider. It's divided the frequency by a factor of 2.

That's a T flip-flop, and the T flip-flop is going to be a building block of many circuits. In this lecture, it's going to be frequency dividers. In subsequent lectures, it's going to be circuits that do digital counting.

But I promised you a watch, so let's talk about a watch. We'll leave the oscilloscope on because I'm going to demonstrate some more with it in just a moment. As we move on, we're going to shrink that T flip-flop down. We'll simply make a flip-flop that looks like it has no input except a clock now, and its Q flips every time the clock comes on. There's our symbol for a T or toggle flip-flop. Now let's think about, say, how we might make a divide-by-4 frequency divider. I just showed you a divide-by-2 frequency divider with one T flip-flop. If I take the Q output of the T flip-flop, put it into the clock input on the next T flip-flop, it acts exactly the same way as the first one did, except now it's working on a signal that's coming in at half the frequency. I have a clock, and then I have Q1, and then I have Q2 at a quarter of that frequency. That's a divide-by-4 frequency divider.

Before we move on, let me talk about another circuit that has been sitting here all through this lecture with these lights blinking rapidly, and let me just adjust the frequency. This is a system consisting of four flip-flops, four T flip-flops. There are 16 possible states, so it's a divide-by-16 frequency divider. If I turn it down, you watch: This is the clock, and these are the subsequent flip-flop's Q outputs. You'll see each one changes less rapidly than the one before, and the less rapid changing is by a factor of 2. If you don't like watching those lights go slowly changing, I can turn it up a little bit. But more interestingly, I can turn it up a lot and look at that from the oscilloscope. Let's take a moment and rewire the circuit and do that.

Here we have exactly the same circuit I was talking about with the lights, and you can still see that the lights are blinking. The lights at the later flip-flops at least you can see blinking. The other ones are blinking way too fast, about 100 times a second, and you really can't see that one blinking; so the clock rate here is pretty high. I've put the output of the last flip-flop, the fourth one, the one that ought to have divided by 16, on the oscilloscope and I've also put the clock on there. Let's look at what happens. By the way, I didn't build this out of individual flip-flops, I got a circuit that has four flip-flops in it already connected together to do this. Here we go on the oscilloscope, and here's the clock, the yellow trace. Let's just see what's happening. There's one full cycle of the clock: down, up, up, down, up, up, down, up, up, down, up, up, down, up; 4, 5, 6, 7, 8 cycles of the clock in 1/2 a cycle of that final output. There's the full cycle of the final output, down half the time, up half the time; and that would be 16 clock cycles. This has been a division of the frequency by a factor of 16. What we have in these cascaded D flip-flops are frequency dividers, and if you have n flip-flops, you'll divide by 2^n. Here I've got four flip-flops; 2^4 is 16, so I've divided by 16. One flip-flop, you divide by 2. Two flip-flops, you divide by 2^2 or 4. Three, you divide by 8, and so on.

I promised you a watch. What does this have to do with watch or for that matter with clocks? If you have a quartz watch or a quartz clock—and chances are you do, because this is the technology that's simple, cheap, and accurate used in today's timepieces—what's going on in there? Your watch or clock probably says quartz; what's going on? Inside the watch or clock is a tiny little sliver of quartz crystal. The interesting thing about quartz is you apply a voltage to it and the quartz kind of deforms. If you apply an alternating voltage, it deforms at a particular frequency. In most quartz timepieces, that frequency is 32,768 Hertz, 32,768 cycles per second. Why that number? Because that number is 2^{15}. If you take that number and you divide it by 2, 15 times—in other words, you run it through 15 D flip-flops cascaded together to make a binary divider; a 15-stage, 2^{15} binary divider—you'll get out a pulse every 1 second. If you look closely at a timepiece, you'll sometimes see the hand moving, taking a little jump every second, and that little jump is caused by the 1 Hertz pulse that's coming out after the 15 stage frequency divider's done its thing. That's quite an interesting use for flip-flops.

Let's wrap up. We've seen two new flip-flops in this lecture: We've seen the D flip-flop and the T flip-flop. We'll use the D flip-flop more in the next lecture for memory, and then we'll introduce one more flip-flop as we move on in subsequent lectures.

If you want to know more about flip-flops, try the project; just a couple of projects for you to do. If you want to, simulate a T flip-flop using your circuit simulator's D flip-flop if it has a built in one—which it does—and verify that it works. If you're working with CircuitLab, the D flip-flop doesn't have a Q bar output. That's going to be a little bit of problem for you when you try to make a T flip-flop, but you can overcome that. Once you have your T flip-flop, make a frequency divider that divides the input by 8 and verify that it works. You can do that either using flip-flops from part 1 that you built yourself, or you could use CircuitLab's built-in T flip-flop, which it has; but if you do that, it has a separate T input and you have to put that to 1 to enable it to flip at all. Or you could use DoCircuit's J-K flip-flop, which we'll get to in a subsequent lecture. They don't have a T flip-flop, but if you put all the J and K inputs of that thing to 1 it functions as a T flip-flop. Enjoy playing with flip-flops.

Digital Memory
Lecture 20

We've already discussed flip-flops as circuits that can "sort of" remember, but in this lecture, we'll build them into substantial memories, such as would be used in computers and other kinds of devices to remember both data and program instructions. We will look in some detail at different kinds of memory, how memory is used, and how memory can be made using flip-flops. We'll also discuss other kinds of memory that aren't based on flip-flops. Key topics in this lecture include the following:

- Volatile versus nonvolatile memory
- Random-access versus sequential memory
- Memory addressing

- A simple flip-flop–based memory
- Static versus dynamic memory
- Flash memory.

Distinctions in Electronic Memory

We need to make two important distinctions concerning memory. Memory can be volatile, meaning that information is lost if the power is lost, or it can be nonvolatile, meaning that information is retained even without power. Second, memory comes in two forms: random-access memory (RAM) and sequential memory. RAM is memory that can be accessed equally quickly, no matter what piece of data is being accessed. Most semiconductor-based memory is RAM. Sequential memory requires passing through a sequence of memory items before you can access the one you want. Examples of sequential memory devices include magnetic tape, hard disks, CDs, and DVDs.

Memory Addressing

Another issue with memory is how we address it. Think of a single item of memory storing a bit, a 1 or a 0. If you want that bit, you must know where that memory is—its address. The amount of memory available is usually expressed in bytes (1 byte = 8 bits). The maximum amount of memory that can be addressed is ultimately set by word length; it's 2 raised to the number of bits. With a 32-bit word, we can address up to 2^{32} individual memory locations. That's only about 4 billion addresses—a rather small amount of memory for a modern computer. With 64-bit words, we can address about 20 trillion billion locations. In dealing with memory, we also need to know whether the memory location is being read or written to.

A Simple Flip-Flop–Based Memory

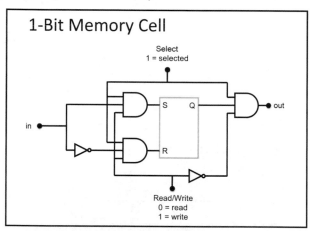

We can build a single memory cell that stores 1 bit based on the SR flip-flop. To this SR flip-flop, we add a pair of three-input AND gates. Remember, with this type of gate, the only way to get a 1 at the output is to have a 1 in all three inputs. The output circuitry consists of another three-input AND gate. The information in the memory cell is stored as the value of Q.

A 1 at the select line means that this memory cell is selected to be read or written to. A 0 at the select line means that the memory cell is inert. The read/write line tells us whether we want to read the information that's in the memory or write new information to the memory.

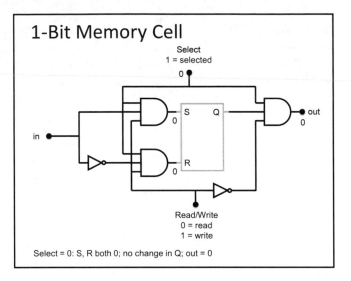

A 1 is needed to select this memory cell. If there is a 0 on the select input, the AND gate outputs (the S, R inputs to the flip-flop) are both 0, and Q can't change. Further, the output is 0 regardless of what Q is because the output AND gate has a 0 at one of its inputs.

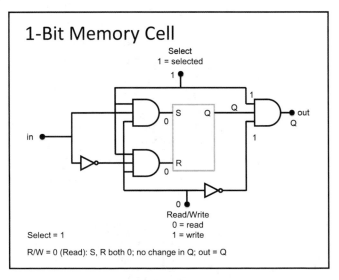

What if we select this cell and set the read/write to 0 (meaning *read* in this configuration)? Now, the output AND gate is open, in the sense that the value of Q "gets through" to the output of the memory cell. If Q is 1, we get 1 at the output because all three inputs to the AND gate are 1. If Q is 0, we get 0 at the output. Here, we're reading the memory cell. We can't change Q in this configuration, but we can find out what Q is. On the other hand, if we set the read/write input to 1, the output gate is closed, and we can't tell the value of Q from the output. But in this state, the input to the memory cell "gets through" the input AND gates, and we set or reset the flip-flop, loading the input into Q. Thus, we've written information to the memory cell.

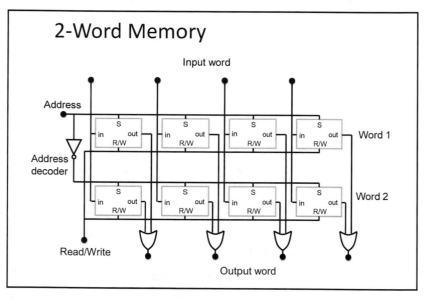

Computers have many (billions!) of words, usually of 64 bits each. Above is a simple example of a multiple-word memory, in this case containing two words of 4 bits each. Each word has a select line, and the select lines of all the memory cells in each word are tied together. The output lines of the corresponding bits from each word are tied together through OR gates. Thus, whichever word is selected, we can read or write from/to that word only.

Here, the address requires just a 1 or a 0 because there are two input words. If it's a 1, we address the upper word because the address line is connected directly to all the selects of Word 1. If it's a 0, the inverter causes us to select Word 2. The address decoding logic gets much more complicated with billion-word memories, but the idea is the same.

Static versus Dynamic Memory

In static memory, the basic information storage unit is a flip-flop. It's a volatile memory, but it's energy efficient and fast. It also takes up space, however, and it's a bit expensive. The memory used in most computers, dynamic RAM (DRAM), is much simpler. Its basic information storage unit is the capacitor.

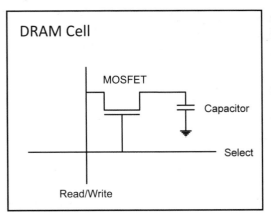

DRAM has two lines—a select line and a read/write line. It has a capacitor that is built onto the integrated circuit chips. A MOSFET turns on or off, in this case, depending on whether we put a 1 or a 0 on its gate.

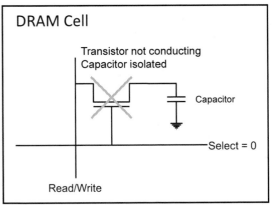

If we put a 0 on the select line that is connected to the gate, then the capacitor is isolated because the transistor is off. Putting a 1 on the select line turns on the MOSFET transistor. We can then write to the memory cell by putting charge through the MOSFET and onto the capacitor. Or we can read the memory by sensing whether or not there's a voltage on the capacitor.

But capacitors gradually lose their charge, and so we need to go through the entire memory periodically and read the state of each capacitor. If a capacitor is in the 1 state, we boost it back up to maximum charge. This is typically done about 15 times each second. It's this "dynamic refreshing" of the memory that accounts for the *D*, for *dynamic*, in the term *DRAM*.

Types of Memory

Type	Speed	Volatile	Access	Uses/Comments
SRAM	Fastest	Yes	Random	Small, high-speed memory
DRAM	Fast	Yes	Random	Main memory in computers, etc.
ROM **PROM** EPROM	Slower	No	Random	Storing permanent information, startup sequences, etc.
Flash	Slower	No	Random	Long-term storage in smart devices, tablets, "flash drives," increasingly in computers
Magnetic disk	Slow	No	Sequential	Long-term storage in computers Uses spinning magnetic disk Rapidly giving way to flash storage
Optical disc	Slow	No	Sequential	CDs, DVDs for music, video, computer data Data read/written by laser
Tape	Very slow	No	Sequential	Backup and archives
Magnetic core	Slow	No	Random	Obsolete; used tiny magnetic rings on intersecting wires

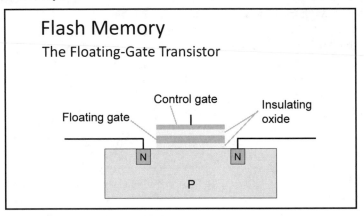

Flash memory uses transistors that look similar to MOSFETs, except they have an extra gate, called a *floating gate*, sandwiched between two layers of insulation.

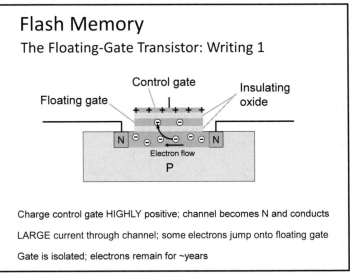

If we want to write a 1 to this floating-gate transistor, we put a highly positive charge on the control gate. That voltage pulls electrons into the channel, which becomes N-type and conducts. Some of the electrons that are conducted through the channel undergo a process called *quantum tunneling*, in which they leap onto the insulated gate. The presence of those electrons tells us that we have a 1 stored in that memory unit. If we want to write a 0, we put a highly negative charge on the control, and if there are any electrons in the floating gate, they're pushed out through the insulation and into the substrate.

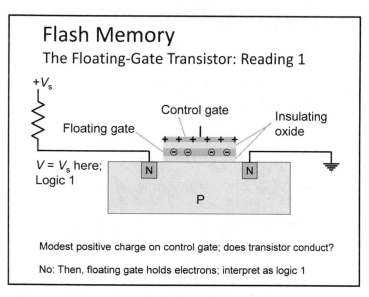

Flash Memory
The Floating-Gate Transistor: Reading 1

$+V_s$

Control gate

Floating gate

Insulating oxide

$V = V_s$ here; Logic 1

N

N

P

Modest positive charge on control gate; does transistor conduct?

No: Then, floating gate holds electrons; interpret as logic 1

If we want to read this transistor, we put a modest charge on the control gate. If the transistor doesn't conduct, that means the floating gate has some electrons on it. That situation is interpreted as a logic 1. If the transistor does conduct, then the floating gate doesn't have any electrons; that's a logic 0.

Suggested Reading

Introductory

Brindley, *Starting Electronics*, 4th ed., chapter 11, pp. 194 to end.

Lowe, *Electronics All-in-One for Dummies*, book VI, chapter 5, pp. 596 to end.

Advanced

Horowitz and Hill, *The Art of Electronics*, 3rd ed., chapter 14, sections 14.2.2–14.2.3.

Scherz and Monk, *Practical Electronics for Inventors*, 3rd ed., chapter 12, section 12.12.

Projects

Memory Cell Simulation
Simulate the 1-bit memory cell described in this lecture. In CL, you'll need to make three-input AND gates; DC allows you to select the number of gate inputs. Clock the read/write input and the data input at a slower rate (1/4 as fast is good).

4-Word, 4-Bit Memory
Consider a 4-word by 4-bit memory, similar to the 2-word memory described in this lecture. Design (but don't build) an address decoder for this memory. You'll need two address lines to address four words.

Questions to Consider

1. Both RAM and your hard drive or solid-state drive (SSD) are forms of memory. How are they different, and what's the purpose of each in your computer?

2. What's the difference between volatile and nonvolatile memory?

3. Find out how much RAM is in your computer, as well as the capacity of its SSD or hard drive.

4. Your phone, tablet, flash or USB drive, and many other devices use flash memory for long-term storage. Describe the physical principle behind flash memory.

5. How is a computer's word length related to the maximum amount of RAM that can be installed in the computer?

Digital Memory
Lecture 20—Transcript

Welcome. Here's a lecture I hope you're going to remember, because this lecture is about memory. Digital devices need to remember. We've already talked about flip flops as devices that can sort of remember, but here we want to build them into real substantial memories such as would be used in computers and all kinds of other devices.

What do these things need to remember? Your smartphone needs to remember your contacts. It needs to remember your pictures that you've taken with it. It needs to remember the apps you've downloaded to it. It needs to remember all these things. Your camera needs to remember photos. Instruments that scientists might use or that you might use need to remember. You take a temperature with your digital fever thermometer and it beeps when it's done, and then it locks that temperature into its memory and it remembers it. Have you got a maximum/minimum thermometer? If you've got a maximum/minimum thermometer, it has memory in it that remembers the maximum temperature and remembers the minimum temperature, and it only replaces them if the temperatures exceed those limits.

Computers need to remember lots of things. You're typing at your computer or maybe texting on your phone, and every keystroke you make, every letter you enter in is converted into one of those 8-bit strings—those strings of digits, binary digits, that use the ASCII code to represent a particular character—and then those things are stored in memory. Maybe temporarily with the text message before it gets sent out by some kind of parallel-to-serial conversion scheme like we've already talked about, or maybe more permanently. Maybe you're typing your term paper or an email message that you're going to save, and it gets saved for long term to memory.

Finally, one thing we tend not to think about as much, but this is really important to computers in particular and was important in the development of digital computers, not only do they store information—that is, data; the data they're going to work with, the information on the pixels that make up the photograph, the information on the phone numbers, whatever it is—but they also store the instructions that are used to tell the computer what to do. When a programmer writes a computer program, that program is also a set of binary numbers. In fact, think of your favorite spreadsheet or word processing program. In some sense that program is really nothing but an enormous, long binary number. Pieces of that program are loaded into memory and executed as the computer does what it's supposed to be doing.

In this lecture, we want to look in some detail at memories and different kinds of memory and how memory is used and how memory is made using flip flops. Then we'll talk about some other kinds of memory that aren't made using flip flops. There are several distinctions; many, many different kinds of memory. Later in this lecture, I'm going to show you a table of many of these kinds; but I want to make two big, important distinctions right at the outset. Memory can be either volatile or non-volatile. What does that mean? Volatile memory is memory that forgets if the power is lost. When you're working on some important project like that term paper, you're always told, "Hey, you ought to save frequently," because if the power fails and it's not a laptop with its own battery built in, every keystroke you typed in is gone. Most modern programs have auto save every 10 minutes or every 5 minutes, or maybe you can set the timing; they'll save what's in there to a more permanent, non-volatile memory. But the basic memory that's there in your computer when you're first working on it, that's volatile. If you quit a program and it's giving you the option to save or not save and you say "Don't save," gone, because that was in volatile memory. Volatile memory forgets if the power is lost. Why would we ever want volatile memory? It's fast and it's cheap, but it's volatile.

Non-volatile memory is memory that remembers without needing power. It doesn't need the power to keep on. You turn the power off, the non-volatile memory is still there; it remembers. Many, many, many devices have non-volatile memory. Why can you turn your smartphone off and then summon up the picture you took three months ago? Because it's in non-volatile memory. How does your computer know to get itself started in the first place, to execute the few basic instructions that get the computer booted up, the term we use? ("Bring it up by its own bootstraps" is what that term comes from.) It's got some nonvolatile memory in there that can remember.

There are actually gradations between volatile and non-volatile. There's, for instance, memory that can be read-only memory, it's called, but it can be rewritten. It's non-volatile, and so I'm not going to go into those details. But the distinction between volatile and non-volatile is just a little bit less clear than you might think. Volatile memory is basically memory that forgets if the power is lost, and non-volatile memory remains there even when the power is gone off.

That's one distinction; the second distinction is what's called random versus sequential memory, random-access versus sequential. Random-access memory, RAM; and you probably have thought about your computer and maybe you paid an extra few hundred dollars to get more RAM in your memory, more random-access memory. What's random-access memory? Random-access memory is memory that can be accessed equally quickly no matter which piece of memory it is. It's like looking on a bookshelf and saying, "I can grab that book or I can grab that book with equal ease." That's what random-access memory is. Most semiconductor-based memory, made of flip flops or other devices on integrated circuits, is random-access memory. The virtue of random-access is it's fast. You want a piece of memory; you want a piece of information that's in memory? You can go grab it. It doesn't matter which piece it is, it's equally easy to grab,

On the other hand, there's sequential memory. Suppose you couldn't pull the book off the shelf without first pulling out all the books to the left of it. That would be a pain. That's sequential memory; memory that you need to pass through a sequence of memory items before you can access the one you desire. There are some memory devices that are inherently sequential. A very old-fashioned one is magnetic tape. You have to run the magnetic tape past the read head on the tape machine, and if the information you want is a mile down the tape you've got a spin a mile of tape past there before you can read it. Maybe that's not a problem, but maybe it is if you want to get that information quickly. Hard disks, spinning hard disks, which are still widely used in computers—although by the end of this lecture, you'll appreciate that they aren't going to be widely used for much longer—are sequential devices. There's a spinning magnetic disk with information stored on it, and that magnetic disk spins past a so-called read head, and you have to wait till the information you want comes by. CDs and DVDs are other examples of sequential access devices. They're spinning disks, and you have to wait until the part you want to read from gets under the laser that reads the CD or DVD. These are called optical discs. Those are sequential memory.

Why would you want sequential memory? Because it's cheap and it can store a lot. The disadvantage is it's slower to access. Two important distinctions: volatile versus non-volatile memory; random-access versus sequential memory.

Another issue for memory is how we address it. Think of a single item of memory storing a bit, a 1 or a 0, as like how. It's got an address; where is it? If you want that bit, and find out whether that bit is a 1 or a 0, you've got to know where that memory is. You've got to know its address; you've got to be able to tell the circuitry to go to that bit. Maybe you want to put some new information in that particular memory location; you've got to know the address to put it in. There's an issue about addressing memory. The address of a memory, the amount of memory that you have available, is usually expressed in bytes; one byte, again, being eight bits.

The question arises, then, what's the maximum amount of memory that you can address? That's ultimately set by the word length; remember we talked about word length. Computer word lengths have grown from typically one byte, 8 bits, to 64 bits. What determines the amount of memory you can address? The answer is it's 2 to the power of the number of bytes, a certain number of bits. If you have a 32-bit word, you can address up to up to 2^{32} individual memory locations, and those locations might each involve bytes of information. By the way, 2^{32} comes out to be about four billion. It sounds like a huge number, but when we had 32-bit computers, which wasn't that many years ago, they could address at most four gigabytes of memory. Today, four gigabytes is a pretty puny amount of memory in a computer. With 64-bit words we can address 2^{64} different bytes of memory, and that's about 20 trillion gigabytes, 20 trillion billion bytes, and we're not likely to exceed that limit anytime soon.

Another issue when we talk about addressing memory is: Are we addressing that memory because we want to find out what's in it or because we want to write some new information to it? We're going to use the words *read* and *write*; *read* means extract the information from memory, *write* means put new information into that memory. The question is about addressing memory; and again, the amount of memory we can address depends on how big the words are in our computer or other device.

Let's go to our big screen and look at a particular implementation of memory. This is, after all, a course in electronics, not just in computers and memory, and we want to know how you'd build a memory. We're going to build a single memory cell that stores one bit, and we're going to base it on the kinds of flip-flops we've been studying already, but we're going to really embellish it to do everything you need to do with memory. I'm going to start my one-bit memory cell with that SR; that set/reset flip flop. By the way, I'm not going to show the clock just to avoid complication. This memory cell would certainly be clocked, so it would change in sychronism with the other memory cells in whatever device it's in. Here's going to be a one-bit memory cell. The basic essence of it, the place where the bit is stored, is in this flip-flop, and the stored bit exists as a 1 or a 0 at the Q of that flip-flop.

I'm going to add different circuitry to this. I'm going to start with an input circuit, and the input circuit consists of an input. That's where you send the information into this memory cell. Two, three input AND gates. We haven't built three input gates, although you had a project where you could've built some multiple input gates if you'd wanted. They work just like two input gates. The only way the output of an AND gate can be one is if all its inputs are one. These are three input AND gates; you've got to have a one in all three inputs to get a one at the output. This is the input circuitry. Notice the only difference here is the input comes in, and it goes to the middle input of the upper AND gate, and it goes to the middle input of the lower AND gate—the one that's connected to the flip-flops reset—but it does so through an inversion. Whatever you put in here appears at the input to that gate, and the opposite of it appears at the input to the lower gate. That's the input circuitry.

Here's the output circuitry. The output circuitry consists of a three-input gate connected to the Q. Not surprisingly, the Q is what contains the information that's in the flip flop. After we go through that output gate, we have what we call the output terminal of the memory. Whether or not that output tells us what's in Q depends, and we'll see what it depends on. Here's the select line. The select line says if you make that a one, you've selected this memory cell to work on; to read from or to write information to it. If that select line is a 0, this memory cell is inert, nothing happens, because you can't get information in and you can't find out what the information is that's in it. Finally, we have a read/write line that tells us whether we want to read the information that's in the memory or write new information to the memory. That complete system is a one-bit memory cell. Let's now see how that one-bit memory cell operates.

Suppose I have a 0 on the select input; I've argued that 1 is needed to select this memory cell. If there's a 0 there, there's a 0 at the input to those two gates and a 0 at the input to those gates. Those are all AND gates, and an AND gate's output is 1 only if all its inputs are 1. If you put a 0 to the input, any input of any of those gates, the outputs are 0. We got a 0 on S and a 0 on R and a 0 on Out. What does this zero do on S and R? We spent a lot of time with S bar R bar flip-flops, where a 1 and a 1 kept it from changing. A 0 and a 0, once we've got the S and the R, keeps it from changing. Q can't change in this configuration. Furthermore, the output is 0 regardless of what Q is, because this gate has a 0 to one of its inputs; and also, by the way, regardless of what the read/write is doing. These gates have zero, so the outputs of those gates are 0s and that keeps the Q from changing; so nothing happens. If you don't select this particular memory cell, you can't read it, you can't write it, you can't do anything with it.

What if we select it? Then we've got a 1 up here; that's encouraging. What if, in fact, in addition to selecting it, we set the read/write to 0? Read/write of 0 is read in this particular set up. You could've made it the other way; but the way I've designed this circuit, a 0 means I'm going to read it. Let's see what that means. If I have a 0 here, I've got this inverter. Then I've got a 1 here. We've got a 1 at the lower input of that gate, a 1 at the upper input of that gate. Now this thing is behaving like a gate. It's an open gate, because if Q is 1, we're going to get one at the output because all three inputs will be 1; if Q is 0, we'll get 0 at the output. What's that mean? It means the output is Q. Here I'm reading the memory cell; I get on the output whatever Q is. I can't change Q in this configuration, but I can find out what Q is. There's the one-bit memory cell selected; so we're going to operate on this particular memory cell, and we're going to read from it.

On the other hand, if the select is 1—still selected—but the read/write is 1, let's see what happens. Then I have a 0 at this output gate. I don't care what Q is; I don't care about that 1; 0 is at the output. When I've got the right selected, I can't tell what's in the memory cell; I just get a 0. But I've got a 1 coming in here; that 1 is going to these two gates. By the way, in drawing this rather complicated circuit where two wires cross and they're different colors, they're not connected, so I'm not drawing that little curvy thing for non-connection. There's no connection. There's a connection here, there's no connection here, and there's a connection here. That 1 is going to the bottom of that gate and going to the bottom of that gate. There's a 1 at the tops of these two gates because they've been selected with the 1 at the select input. What that means is what we get at the input here is coming through this gate. In that case, S becomes the input and R becomes not the input because it's been taken through this inversion. Q therefore goes to the input, because that's what happens in this as our flip-flop; it loads Q from the input to the S connection there. This is how I write to the flip-flop. When the clock comes along, I'll have loaded whatever's at the input into Q.

That's a complete one-bit memory cell. It took an SR flip-flop, three input gates, AND gates, and a couple of inversions. We have a complete memory cell that we can address; we can select; we can do whatever we want with. Now let's take that thing and shrink it down and make a one-bit memory cell; just give it its own symbol. It's got a read/write, an Out, a Select, and an In. That's our one-bit memory cell.

There's how we make a one-bit memory cell. That's great; one bit. We can store it, we can address it, we can read it, we can write it. What do we need to do to go further? We need to store lots of words of memory. Here's an example of a two-word memory. Each word in this case is four bits; this is a pretty simple memory. Each one has a select line; each of the words has a select line. The select lines of all the memory cells are tied together. Each one has an output line; the output lines of each corresponding bit are tied together through OR gates. Whichever cell is selected, I can read the output from that one only. There are the read/writes; the read/writes are all connected together so we know whether we want to read them or write them. Again, the one crucial thing is the so-called address decoder; the thing that says, "Which of these two do I want to select to work on?" The address decoder in this case is very simple; if you do the project, you'll have a more challenging address case. But here the address case requires just a one or a zero, because there are two input words. If it's a 1, we address the upper word because the address line is connected directly to all the Selects of that one. If it's a 0, we go through that inverter on the left there, and that inverter causes us to select word two. There's a two-four-bit word memory. Very simple, but you could scale this up billions of times and make the memories in modern computers.

That's an example of a memory built from flip-flops. There's plenty of memory out there built from flip-flops, and it's good memory; it's very fast. It tends to be a bit expensive; it tends to be a bit power consumptive; but it's good. It's a little too expensive and a little too bulky to use in our everyday computers, and so we go to a different kind of memory. I want to distinguish now another distinction between two types of memory.

We have static memory, in which the basic information storage unit is, as in the memory cell I just described, a flip-flop. It's a volatile memory, but it remembers as long as power's applied. It's energy efficient and it's fast, but it takes up space and it's also a bit expensive. The memory used in your computer is much simpler. It's called dynamic memory, DRAM. Its basic information storage unit is the capacitor. We talked about capacitors a lot in this course; they store charge and energy, and they have a voltage across them because they have two conductors separated by an insulator. The insulator is never perfect, so the capacitor always discharges. Dynamic memory would forget rather quickly, in a fraction of a second typically, if it weren't periodically refreshed. It's somewhat slower and it uses some power, but it's very high density and it's ubiquitous in personal computers.

Here's how DRAM works; dynamic random-access memory. DRAM has two lines, a select line and a read/write line. It has a capacitor. This isn't a capacitor you go buy from some store, it's a capacitor that's built onto the integrated circuit chips. Modern memory systems have literally billions of these capacitor/transistor combinations on them. Remember that a metal oxide semiconductor field-effect transistor turns on or off, in this case depending on whether you put a 1 or a 0 on the gate. If you put a 0 on the select line that's connected to the gate, then that capacitor is isolated from the rest of the world because of that transistor being off. If you put a 1 on the select line, you've selected. Then you can either write to the capacitor, put charge on it through the transistor, or read what charge is on it through the MOSFET and you can read or write as you wish. The transistor isn't conducting when the Select is 0; it's conducting when the Select is 1; and you can read or write.

There's one caveat with this: I mentioned that those capacitors tend to lose their charge fairly regularly. Typical DRAM, typical dynamic random-access memory, is refreshed. The rate at which it's refreshed varies, but it could be as slow as about 15 times a second. In your computer's memory—my laptop has 16 gigabytes of memory—every one of those 16 billion memory cells, about 15 times a second, you read what's on it. If it's a 1, you pump a little more charge on it to keep it at that 1, and if it's a 0, you don't pump more charge on it. That's called refreshing, and it's happening 15 times a second. You say, "Whoa, 15 times a second I have to do this?" Fifteen times a second is awfully slow compared to the roughly three gigahertz, three billion times a second, that the clock is going in my computer and refreshing things. It's actually a very rare thing, in the context of the computer's internal timing, that that you refresh the memory. But it's dynamic random-access memory; has to be refreshed.

I've got an old dynamic random-access memory card out of an older computer. It consists of eight integrated circuits, each of which is a chip containing several hundred million—probably 256 million in this case—individual memory cells of the type I just showed you with a capacitor. There are eight of them, so we represent one byte with this; or rather 256 billion bytes with this. Out of a modern computer, this thing would have billions of transistors in it. That's a memory.

Let's take a closer look at memory, different kinds of memory. I mentioned there are all kinds of different kinds, so let's just run down some of them. There's what's called SRAM, static RAM. That's the fastest; that's what we talked about. It's volatile, it's random-access, it's small, and it's high speed. By the way, if you're an electronics experimenter, you probably want to use SRAM even though it's a little more expensive because you don't have to mess with the circuitry to dynamically refresh the DRAM. DRAM is also fast. It's volatile, it's random, and it's the main memory in most everyday personal computers. There's memory called ROM, read-only memory; PROM, programmable read-only memory in which you can only read, except you can program it; and E-P-ROM, EPROM, which is erasable programmable read-only memory. They're slower, they're not volatile, they're used for things like the boot up sequence in computers, simple pieces of computer code that you have to store permanently. They're random-access, and you store permanent information; start up segments and so on. There's flash memory, which I'll talk more about in a moment; that's the memory in most of our modern electronic devices other than computers. It's not the random-access memory, but it's the long-term storage memory in many devices. There are magnetic disks. These are the spinning hard disks. There are optical discs, CDs and DVDs. There's magnetic tape. Finally, there's something called magnetic core, which is long since obsolete, which consisted of crossed wires that were woven together by hand, and a little ferrous core was put on them and it was either magnetized or not. We can look at an ancient memory items; magnetic core memory is here and it looks like this woven cross-hatch pattern. Magnetic tape, used in these giant tape machines. Magnetic tape is still in use, but you probably don't see it. We have floppy disks, and we have hard disks. I have some of these things here, which I'd like to show you.

This is an example of magnetic tape. This is like the tape you used in old-fashioned tape recorders. It's a predecessor of the little cassette tapes you may have used or the VHS tapes, all of them now obsolete, although this is still used in long-term compute storage. This happens to have my PhD thesis on it from a long time ago as the backup storage. We developed computers that had floppy disks; spinning magnetizable media. This is truly a floppy disk; that's why they're called floppy. If you came into the computer era little later, you probably used floppy disks that looked like this. They didn't look floppy, but if you got inside them, they had a floppy magnetic component. We're still using optical disks for storage, some of them rewritable, some of them not. Here's the hard drive such as might still be in your computer. There's the rapidly spinning disk, and here's the head that moves over that disk and picks off or writes to bits of magnetic information from that disk. There are some examples of different memory types.

Having talked about those types: That's kind a museum; everything here is either still in use but fading out. When was the last time you got a DVD instead of streaming something, for example? Even optical discs are fading out. Hard drives are going fast. They're going in favor of something else called flash memory. Let me just show you some examples of flash memory, and then we'll talk about how it works.

What's in a camera? No film; film's long gone. What's in here's a sensor, probably a CMOS-based sensor (complementary metal-oxide semiconductor) and the camera information—the pixels, whether it's red, how much red, how much green, how much of blue, and what intensity for every one of the 18 million pixels that this camera can take—is stored in flash memory. Here's a flash drive, flash, USB stick, whatever you want to call it. It's got a USB connection, serial connection, to whatever you plug it into and it stores information, 32 gigabytes of information in that case. Here's a smartphone. The smartphone's non-volatile storage memory is flash memory. Here's a tablet reader, which also stores the books or whatever you're doing in flash memory. Let me take a few minutes and talk about flash memory, and then we'll wrap up.

Flash memory uses transistors that look a lot like the transistors we've talked about before, the metal-oxide-semiconductor field-effect transistors, except they've got an extra gate. It's called a "floating gate," and it's sandwiched between two layers of insulation. This has to be really good insulation, because your term paper, your photograph collection, your contact information in your phone, all that stuff is stored in the form of electrons sitting on that control gate. If they could jump off—and this is why early flash memory, there were worries about would it last 20 years, for example; I think the jury's still a bit out on that, but it lasts a long time—those electrons on that gate are carrying the information.

How does it work? If you want to write information to this floating gate transistor and you want to write a 1, you charge that control gate very highly positive; you put a big voltage on it. That big voltage works just like any

MOSFET transistor: It pulls electrons into the channel, the channel becomes n-type and conducts, and some of the electrons that are conducted through that channel undergo a process called "quantum tunneling" and they leap onto that insulated gate, and the presence of those electrons tells you that you have a 1 stored in that transistor, in that memory unit. Those electrons will remain there for years. You want to write a 0, you charge the control gate highly negative, and if there are any electrons in that floating gate they're pushed out through the insulation and into the substrate. You want to read the thing, you put a modest charge on the control gate and you ask yourself, "Does the transistor conduct or doesn't it conduct?" If the transistor doesn't conduct, the reason it doesn't conduct is because the floating gate has some electrons on it. It's repelled the electrons out of the channel, and the transistor doesn't conduct. Then you know that the floating gate holds electrons, and you interpret that as a logic 1. On the other hand, if the transistor does conduct, then the floating gate doesn't have any electrons because the charge on the control gate has pulled electrons into the channel and made it conduct.

That's how flash memory works. Flash memory is with us today; it's almost everywhere. It's slower than RAM. It is not volatile. It's there when the power's turned off. It'll last for years and years.

Let me end by looking at your future in memory. Two examples of memory: Here's a mechanical hard drive like I just showed you. These are still in use in computers. You can still buy computers with them, and you'll be able to for a number of years. I priced a 500 gigabyte hard drive in 2013 and it was $50, and the same unit in 2014 was $50. By the way, that's remarkable because that's half a terabyte, and when I started in as newly-minted PhD we were talking about the world's best computers having terabytes of memory storage. Now you can buy $50, half a terabyte; pay a little more, you get a whole terabyte. The other device shown here is a solid state drive, an SSD, also 500 gigabytes. By the way, when we talk about numbers of gigabytes, it's really always a power of two because of the address ability of the device; how we have to address it with the power of two. So 500 really means 512, which it says on this picture. In 2013, that thing cost $350, seven times as much as the hard drive. In 2014, it was down to $250, five times as much. They're on a collision course; nobody knows exactly when. You can talk to the electronics economists, but probably before 2020 they'll be at parity. I paid some extra dollars for my current laptop. It has solid state drive entirely. It has 768 gigabytes worth of solid state drive, because it's much faster to load programs or load lots of data into memory from the much faster solid state drive, and there's nothing to break. If it drops ....

That's a lot about memory. I hope you remember that. If you want to know a little bit more about memory, do the project. You can simulate the one-bit memory cell described in this lecture. If you're going to do it in CircuitLab, which doesn't have three input AND gates; you'll need to make them yourself, but you can do that. If you go to Do Circuits, you'll find when you select the gate, you can choose how many inputs it's got, so you can use three, and put gates right there. I suggest to watch your memory work; that you clock the read/write input and the data input at a slower rate, maybe a quarter the rate of the rewrite input. You can do that with the built in clocks both these simulators have, and you'll be able to see your memory go through its paces. Then consider a four-word by four-bit memory. I showed you in the lecture a two word four-bit memory, so it's similar to that one. Design, but don't build it please, an address decoder for that memory. You're going to need two address lines, because you've got to address four different words, and two lines give you four possible combinations; that is, four addresses.

Digital Counters
Lecture 21

In the last lecture, we enabled digital circuits to remember information; in this lecture, we'll enable them to count. Of course, counting has many applications: in business (counting products on the assembly line), in the medical profession (counting blood cells), in physics, even in archaeology (counting atoms for carbon dating). Counters also have other, less obvious uses, such as making conversions between the analog and the digital realms. We'll begin our exploration of counters by looking at some of the questions and subtleties involved in building electronic counters. Important topics we'll cover include:

- Frequency dividers and counters
- A 2-bit binary counter
- n-bit binary counters
- Decade counters

- Asynchronous versus synchronous counters and the JK flip-flop
- Application: a practical counting device.

Frequency Dividers and Counters

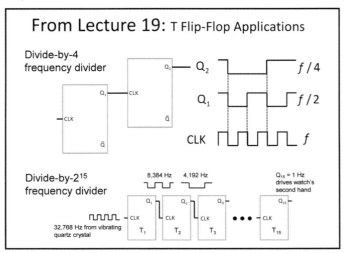

From Lecture 19: T Flip-Flop Applications

Divide-by-4 frequency divider

Divide-by-2^{15} frequency divider

8,384 Hz   4,192 Hz

Q_{15} = 1 Hz drives watch's second hand

32,768 Hz from vibrating quartz crystal

In Lecture 19, we saw the divide-by-4 frequency divider, which consisted of two toggle flip-flops connected so that the output of one went to the clock input of the other, and each one went at half the rate of the clock signal coming into it.

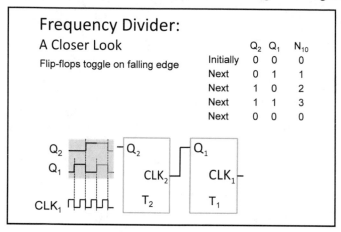

Frequency Divider:
A Closer Look

Flip-flops toggle on falling edge

	Q_2	Q_1	N_{10}
Initially	0	0	0
Next	0	1	1
Next	1	0	2
Next	1	1	3
Next	0	0	0

Here, the frequency divider is redrawn with the flip-flop on the right (T_1) receiving the clock input (CLK_1). We'll treat that as binary digit 1, the digit in the 1s column. The output of T_1, the Q_1 connection, is connected to the clock input of T_2; its output is Q_2. The 2 here stands for the 2s bit, the number we would multiply by 2 to find the decimal representation of the binary number in this flip-flop combination.

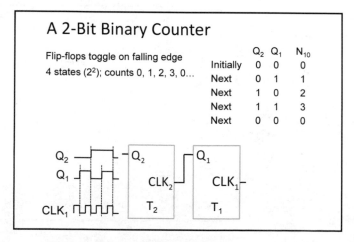

A 2-Bit Binary Counter

Flip-flops toggle on falling edge

4 states (2^2); counts 0, 1, 2, 3, 0...

	Q_2	Q_1	N_{10}
Initially	0	0	0
Next	0	1	1
Next	1	0	2
Next	1	1	3
Next	0	0	0

When we translate the intermediate outputs to decimal numbers, we can see that this device is not only a frequency divider but also a counter. With two flip-flops, we have two binary digits, which leads to only four possible combinations. These represent the decimal numbers 0 to 3. If we have n flip-flops, the number of combinations is 2^n.

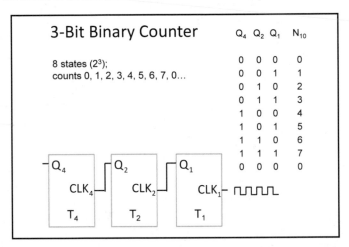

3-Bit Binary Counter

8 states (2^3);
counts 0, 1, 2, 3, 4, 5, 6, 7, 0...

Q_4	Q_2	Q_1	N_{10}
0	0	0	0
0	0	1	1
0	1	0	2
0	1	1	3
1	0	0	4
1	0	1	5
1	1	0	6
1	1	1	7
0	0	0	0

With a 3-bit binary counter, we can represent 2^3, or 8, states. These represent the decimal numbers 0 to 7.

Decade Counters

What if we want to count some number of states that isn't a power of 2? Logic gates help with that problem, along with a flip-flop that has a minor modification. On this toggle flip-flop with clear (CLR), if we put a 1 on CLR, no matter what else is happening, Q goes right to 0.

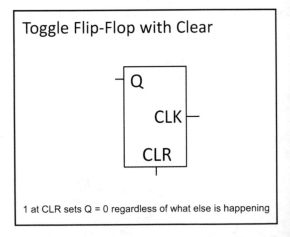

Toggle Flip-Flop with Clear

1 at CLR sets Q = 0 regardless of what else is happening

Decade Counter
(BCD Counter)

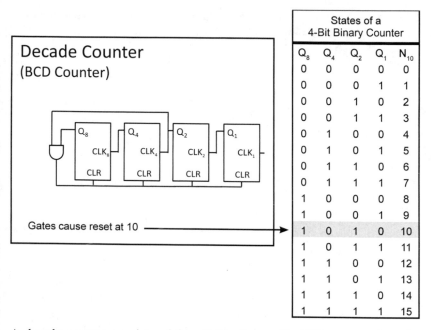

States of a 4-Bit Binary Counter				
Q_8	Q_4	Q_2	Q_1	N_{10}
0	0	0	0	0
0	0	0	1	1
0	0	1	0	2
0	0	1	1	3
0	1	0	0	4
0	1	0	1	5
0	1	1	0	6
0	1	1	1	7
1	0	0	0	8
1	0	0	1	9
1	0	1	0	10
1	0	1	1	11
1	1	0	0	12
1	1	0	1	13
1	1	1	0	14
1	1	1	1	15

Gates cause reset at 10

A decade counter consists of four T flip-flops with CLR connected to the output of an AND gate. If we encounter a condition in which both Q_8 and Q_2 go high, corresponding to the decimal number 10, the counter is set back to 0. So this counter counts from 0 to 9, but on the next clock pulse, it goes back to 0. A decade counter is also called a *binary coded decimal* (BCD) counter.

A 2-Bit Binary Down Counter

Flip-flops toggle on rising edge

4 states (2^2); counts 0, 3, 2, 1, 0...

	Q_2	Q_1	N_{10}
Initially	0	0	0
Next	1	1	3
Next	1	0	2
Next	0	1	1
Next	0	0	0

The easiest way to count down is to change the flip-flop to change state on the rising edge of the clock.

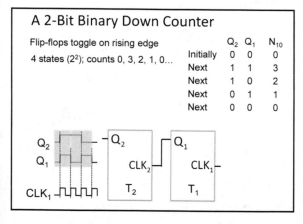

Asynchronous versus Synchronous Counters and the JK Flip-Flop

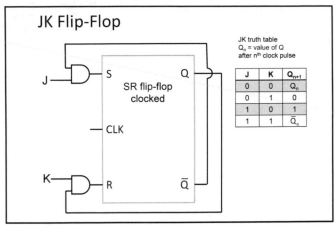

Digital gates have a *propagation time*, typically measured in nanoseconds, for the input to change the state of the flip-flop. Because of this, the count ripples through from one flip-flop to the next. Briefly, while it's rippling, the output lines have intermediate states that are not the correct count. A solution to this problem is to build a *synchronous counter* that clocks all the flip-flops together so that they all change at the same time. That requires either gating or more sophisticated flip-flop design or both. The JK flip-flop is a versatile flip-flop that can be used to build the synchronous counter.

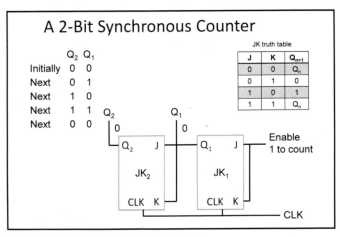

If the J and K inputs to a JK flip-flop are both 0, nothing happens. If J and K are both 1, then the device becomes a T flip-flop and toggles on the clock pulse. Here JK_1 is wired so it toggles on each clock pulse, but JK_2 toggles only when $Q_1 = 1$, making this a 2-bit synchronous counter.

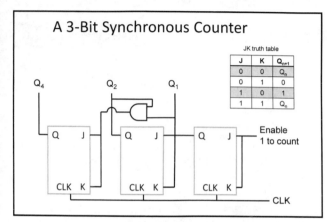

A 3-bit synchronous counter is more complicated because it also requires gating. The key point to remember with any synchronous counter is that all the flip-flops have their clocks connected, so they change state at once.

Application: A Practical Counting Device

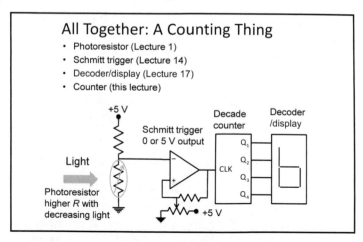

A circuit for a practical counting device consists of a photoresistor, a Schmitt trigger, a decoder/display, and a counter. It works by having the objects to be counted interrupt a light beam. This causes the Schmitt trigger to produce a single output pulse, which feeds the clock input of the counter. Thus, the device counts the number of interruptions of the light beam.

Suggested Reading

Introductory
Shamieh and McComb, *Electronics for Dummies*, 2nd ed., chapter 7, p. 165.

Advanced
Horowitz and Hill, *The Art of Electronics*, 3rd ed., chapter 10, section 10.5.2.

Scherz and Monk, *Practical Electronics for Inventors*, 3rd ed., chapter 12, pp. 781–802.

Projects

Exploring JK Flip-Flops
Explore your circuit simulator's built-in JK flip-flop. Is it rising- or falling-edge triggered? Verify the JK truth table.

Counter Simulation
Simulate the 3-bit synchronous up/down counter shown below. Verify that it counts both up and down.

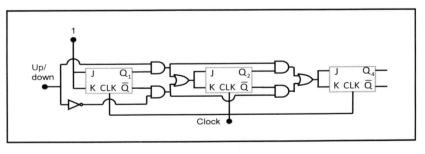

1. What does *BCD* stand for, and what distinguishes a BCD counter from a 4-bit binary counter?

2. What's the difference between ripple and synchronous counters? What's the advantage of the latter?

Digital Counters
Lecture 21—Transcript

In the previous lecture, we taught digital circuits to remember things; we built memory. Today, we're going to teach digital circuits to count.

Why would you want to count? I can think of lots of reasons for counting. You might want to count the number of people going in through a door. You might be working in a factory and you want to count the number of products going down an assembly line. You might be in the medical profession and you need to count red blood cells. What's my red blood cell count? You literally want to know how many red blood cells there are in a given volume. If you're a physicist or a health physicist or a radiation physicist, you might want to count radioactive decays. It's not called a Geiger counter for no reason; it counts radioactive decays. If you're an archaeologist, you want to count individual carbon-14 atoms for the most precise form of radiocarbon dating. You might want to count the pulses coming out of some electronic device as a very precise measure of the frequency. How many pulses occurred in one second? If it's 1,000, the frequency is 1 kilohertz, 1,000 hertz. In some very sophisticated physics experiments, we want to count the interference fringes, the light and dark regions when two waves interfere; the kind of thing that we dealt with, if you took one of the other physics courses, with the Michelson-Morley experiment and the advent of relativity. Today we do those with exquisite precision and part of the reason is we can count individual items. We want to build counters today.

You might think, "We're only going to use them to count," but that's not true. In the couple lectures hence, we'll understand some other uses for counters that really don't seem to have anything to do with counting individual objects but enable us to make conversions between the analog realm and the digital realm. We're going to be building counters today of all sorts and looking at some of the questions and some of the subtleties involved in building electronic counters.

Let's begin with a review from a couple of lectures ago when I introduced the divide-by-four frequency divider, which consisted of two toggle flip-flops connected so the output of one went to the clock input of the other, and each one went at half the rate of the clock signal coming into it. The first flip-flop went at half the rate of whatever clock I was bringing in; remember, a clock is a square wave generated, by the way, by a waveform generator of some sort, which could be like one of the ones we developed in the very last lecture on analog electronics. The second flip-flop sees the output of the first flip-flop, and so it halves the frequency again. I'd promised you a watch, and so at the end I gave you this example of the watch that has a clock, if you will, inside it, a quartz crystal oscillating at 32,768 hertz. That's 2^{15}, and so we have a string of 15 flip-flops that divides that down to make a output of 1 Hertz, which runs your watch.

I'd like to look in a little bit more detail at this frequency divider and see what it does. Notice first that I've redrawn the frequency divider with the flip-flop on the right that's going to get the clock. The reason I want to do that is because I want to treat that as binary digit one, the digit in the ones column. I'm calling that flip-flop one, T1, for toggle flip-flop one. Its input is clock one. The output of that flip-flop—the Q1 connection; the contents of that flip-flop whether it's a 1 or a 0—is connected to the clock input of the second flip-flop, which I'm calling T2. Its output is Q2; it's got clock 2. The 2, you might think stands for it being the second flip-flop; yes it does in this case, but it stands for being the 2s bit, the number you'd multiply by 2 to figure out what this final number is in this flip-flop combination.

Let's take a look at what happens now as we go through the clock. We begin with Q1 and Q2 both in the 0 state, so initially we have 0, 0. I'm making a table here in which I'm going to translate that binary digit, or that binary number 0, 0, consisting of two binary digits into a decimal number: 0, 0 is the decimal number 0. Next, the clock changes and that flips flip-flop Q1, and flip-flop Q1 goes to high. We now have 0, 1. By the way, these flip-flops, I've chosen devices that happen to be changing when the clock falls. That's similar to the master-slave flip-flop I introduced earlier in this course. But you can buy flip-flops that work either way, or design flip-flops that work either way. These work on the falling edge.

What happens? When the clock first falls, Q1 goes high, nothing happens to Q2. But then on the next clock cycle, Q1 goes low. That's a falling edge delivered to the clock input of Q2, and so Q2 goes high. Now we have the state 1, 0. Q1 is flipped back again to 0, Q2 has now gone to 1, and that's the binary equivalent of the number 2. Next time around, nothing happens to Q2 because Q1 doesn't fall, Q1 rises. Now we have 1, 1, and that's the binary equivalent of 3. Now we're in that 1, 1 state. The next cycle, Q1 falls again and that brings Q2 down also because the falling edge, as Q1 goes down, clocks the second flip-flop T2.

If we look at our table, n_{10}, these are the decimal numbers we're seeing out of this device, it's 0, 1, 2, 3, and 0 again. This frequency divider isn't only a frequency divider. If we just look at the input clock and the output of the final flip-flop, it just divides frequency. But if we look at the intermediate outputs and look at what they mean in terms of binary numbers, this is a counter. It counts 0, 1, 2, 3, 0, and keeps repeating that. That's the most we can do with two flip-flops because two binary digits have four possible combinations, and here they correspond to the numbers 0, 1, 2, and 3. We'll find that's always the case. If we have n flip-flops, we'll get 2^n states, and we'll go from 0 to 2^n minus 1. I'll show you some other examples of that.

But let's stop for a moment and actually demonstrate one of these devices. Over here on my board I have exactly the same 2-bit binary divider that I demonstrated a couple of lectures ago. The only difference is I'm now looking at the intermediate states, both Q1 and Q2. I've carried them over into this system of lights. You'll remember that the green means a 0 and the red means a 1. I've organized these the way I would a binary number, so the 1s digit is on the right and the 2s digit is on the left. I've connected the clock of this device—I'm going to call it a counter; that's what it is, a 2-bit binary counter—to one of my debounce push buttons; I talked earlier about how debounce push buttons are another application of flip-flops. It's really important here to use a debounced one because if I had a bouncing push button, I pressed at once, I might get 10 different pulses and I'd count all of them, and the counter would end up in some random state. But I'm going to get one nice clean pulse each time I press the push button.

Now I've done another thing. One of the early lectures in digital electronics, if you did the project—we had this funny project where I asked you to design a circuit that produced a 1 at its output when its input was the binary equivalent of 2 or 6 or 8, and that was a really strange request—you know that that was the decoder that turns on the bottom segment of one of these ubiquitous seven-segment displays that we use to display binary information in decimal form. So what I have here is one of those displays hooked up. I have the actual display there. I have some resistors that limit the current to the LEDs in that display. Down here I have a 7447 DTL binary to decimal decoder. It takes the binary input, a 4-digit binary input, and it produces 7 outputs, which do all the complicated logic to tell which of these segments to light.

We're going to be able to see the output of this counter in two places: one over here with these lights and also on this display right here. I'm going to press the push button once. The display goes to 1; a lot of complicated logic is happening to display that 1. But over here, we see very clearly that in the 1's place, the right-hand digit has gone to a 1, but the 2's place the digit is still a 0. Press the button again, now the 2's place has become a 1. Press the button again, now the 2's place is high and the 1's place is low, so we have the number 1,0. That corresponds to 2, and we indeed get a 2 on the display. Press it again. We go to 1, 1; that's the binary equivalent of 3. We indeed get a 3 on the seven-segment display. Press the button again and we're back to 0. This system cycles through these possible states: 0, 1, 2, 3, and back to 0. It has only 4 possible states that correspond to the numbers 0–3. It's a 2-bit binary counter.

Let me do one more thing with this before we move on. Let me turn on the oscilloscope; we'll put it on the big screen. Let me take what was the clock input to this circuit, and instead of connecting it to the push button let me connect it to my clock here, my square-wave generator. What do you see? You see the outputs of Q1, the 1's bit, and the output of Q2, the 2's bit. What do you see? Let's start anywhere. Here we are at 0 and 0, and the next time, we're at 0 in the 2's bit and 1 in the 1's bit. There's our 1. Next time, we're at 1 in the 2's bit, 0 in the 1's bit. That's 1, 0, or 2. Then they're both up. That's 1, 1, or 3. Then they go back down to 0 again. This pattern is repeating rapidly; the clock is running here at about 100 hertz. If, in fact, you look at the board, you can kind of see that this clock rate is slow enough that you can just kind of see the counter actually jittering. If I turn this down a bit, we can actually watch it count through its paces because now I'm feeding this with an input clock. You can see the numbers changing there; you can see the numbers changing there. Now the oscilloscope is really working in a regime where it isn't doing a very good job for us, but you can see things jumping up and down. There's a 2-bit binary counter.

We want to continue to look now at other examples of counting circuits. Now let me ask you what this is: Here I have three flip-flops, and notice something about them. I haven't called them flip-flop 1, flip-flop 2, and flip-flop 3. I could, and some conventions do that, but I much prefer to call them T1, T2, and T4. There's no T3, why? Because these are the places. We have the 1's place, the 2's place, the 4's place in the binary number sequence that these flip-flops can represent. Here we have three flip-flops, so we can represent 2^3 states; 2^3 is 8. We can't get all the way to 8 because we start at 0, so we have 0, 1, 2, 3, 4, 5, 6, 7. I've got the clock coming in on the right, just like I did before. The only difference between this and the previous circuit is I've added one more flip-flop. This is a 3-bit binary counter, it has 8 states.

If we look at the chart it goes through, we start at 0, 0, 0. The first clock pulse changes only T1. The second clock pulse changes T1 again. T1 always changes 0, 1, 0, 1, 0, 1. The second clock pulse changes T2 and we get 1, 0, the 2 state. Then we get 1, 1, the 3 state. In our previous counter, our 2-bit counter, we flipped back to 0 at that point. But now the change in Q2 flips Q4 and we have 1, 0, 0, and that's the number 4. Then we add the 1, 0, the 1, 1, and we go up to 7. Finally, we flip back down to 0 again. We can build a 3-bit counter that will count up to 7. We could build a 4-bit counter that would count up to, think about it, 15; 0–15, 16 states. You could build an n-bit counter that would count 2^n.

That's interesting. But we happen to come with 10 fingers, and 10 fingers is the basis of our decimal number system which is not a power of 2. What if you want to count something else? What if you want to count some number of states that isn't 2^n? Gating; logic gates can come to the rescue on this one. Before I show you how that's done, let me introduce yet another type of flip-flop.

Here's a toggle flip-flop, the kind we've been talking about. This is a flip-flop with only a minor modification, and most commercial-integrated circuit flip-flops have this modification. There's an additional input here called clear,

CLR, clear. What that does is if you put a 1 on that CLR, no matter what else is going on Q goes to 0. Sometimes that CLR is said to be asynchronous. You can really hit that with a 1 any time, and Q goes directly to 0. You clear it; you put it in the 0 state. Sometimes the CLR is so-called synchronous, and then when you put a 1 on the CLR at the next clock pulse Q will go to 0. I don't care which it is; the point is you have a flip-flop that you can have a different input to and you can force it to go to the 0 state by putting a 1 on that CLR input. We're going to use the toggle flip-flop with CLR in our next circuit.

Here's 4-bit counter, and I'd like you to pause a minute and see if you can figure out what this counter does and focus particularly on that AND gate at the left. Take a pause, think about it, I'll come back and describe its operation. Four T flip-flops with CLR, all connected together, those CLRs, and they're connected to the output of that AND gate. What does it do?

Let's take a look at it. Here I've made a list, now getting complicated, of all 16 states from 0, 0, 0, 0 to 1, 1, 1, 1. From 0–15, carries all the decimal numbers in between. But notice how I've connected that AND gate: The AND gate's inputs are connected to Q8 and to Q2. It's an AND gate, so its output will be high only when both Q8 and Q2 are high. The output of the gate goes to all the CLR inputs. If we ever encounter a condition in which both Q8 and Q2 go high, we'll set this counter back to 0. Now let's look down the table and see when that happens. If you scan down that table, you'll see the first time that both Q8 and Q2 are high is when we get the decimal number 10. What happens? This counter starts counting 0, 1, 2, 3, 4, 5, 6, 7, 8, 9, and then it flips just for an instant to 10, but it can't stay there. It immediately clears the flip-flops, brings us back to the 0, 0, 0 state. This counter actually counts 0, 1, 2, 3, 4, 5, 6, 7, 8, 9 and on the next clock pulse, it's back to 0. What is it? It's a decade counter. It counts the number of digits we have on our fingers. It's a decade counter. It's also called a BCD counter, which stands for binary coded decimal, because it does count decimal; it counts to 10. Not that it counts 10 states—it counts to 9, 0–9—but its numbers are still coded as binary numbers.

In a way, this counter is a sort of waste because we've wasted the capability of a 4-bit counter to go all the way up to the number 15, to have 16 states.

We've frustrated it by not allowing it ever to get into states corresponding to 11, 12, 13, 14, 15. But what we've gained, of course, is a counter that counts in our typical number system. There's a decade counter.

You don't want to just count to 9? Let's count to 99. Just take one of these BCD counters that goes from 0–9, take its Q8 output, put that into the clock of another BCD counter, and you count to 99. That's easy enough to do. Now we can understand how we could make counters that counted really to any amount we wanted. In base 10, in base 2, and in any other base we wanted, if we wanted some peculiar gating that would close the counter back to 0.

We might in some cases want to count down. In fact, these lectures I give you, I watch a clock that's counting down to tell me how much time I have left; there's a down counter involved in that clock. How do you count down? The easiest way to count down is to just change the flip-flop to change state on the rising edge. If you look at the situation when you have flip-flops that change on the rising edge, we start with the flip-flops in, say, the 0, 0 state. In comes the clock, the rising edge of the clock changes Q1 to a 1. Q1 goes up to 1, the rising edge of the clock changes Q2 to a 1; we're in the 1, 1 state. The next one flips Q1 only, and so we're in the 1, 0 state; that's the number 2. Next, we flip both of them and get into the 0, 1 state; that's 1. Then we get back to 0, 0 again. This is a 2-bit binary down counter. We could easily make a 4-bit binary down counter, or a 4-bit decade down counter, or whatever we wanted. In fact, we can make up/down counters. I'm not going to show you how to do that in this lecture, but if you do the project for this lecture, you'll build yourself a 3-bit up/down counter, which is a pretty sophisticated piece of electronics. You can choose which way it goes, and we're going to use that a couple lectures hence to build an analog to digital converter, but that's stuff to come.

Let me talk about a few other issues involving counters. There's a problem with these counters that have cascaded flip-flops, and the problem is the output of one flip-flop goes to the next flip-flop. It takes a little bit of time, something I haven't talked about, but digital gates have a so-called propagation time. It's typically measured in billionths of a second, nanoseconds, for the input here to change the state of the flip-flop. The electronics in there takes a little bit of time to do its thing. The first flip-flop changes, and only after it changes does the next one change, and so the count kind of ripples through from one flip-flop to the next. Briefly, while it's rippling through, the output lines have intermediate states that aren't the correct count. You could blank the outputs during those times if you wanted, but a much more elegant approach is to build what's called a synchronous counter that clocks all the flip-flops together, and they all change at the same time. That requires either gating or more sophisticated flip-flop design, or both.

Let's go there; and to go there, we have to look at one more flip-flop. This is the most versatile flip-flop, it's the final one I'm going to introduce. It's called the J-K flip-flop, and it can do anything any of the other flip-flops we talked about can do. You can make it into a D, you can make it into a T, you can make it into an RS; anything you want. We're going to start with our RS or SR, set reset, clock flip-flop, and we're going to add a couple of gates at the inputs, at the R and S inputs, a couple of AND gates. We're going to call the inputs to one of the AND gates J and the input to the other AND gate we're going to call K. We're going to do this kind of feedback thing where Q-bar goes up to the other input on the flip-flop at the S input and Q itself goes down to the other input on the gate at the K input.

What's this thing doing? Let's just take this thing before we see what it does, and shrink it down and give it, as we've always, a symbol of its own. J and K are its inputs, Q and Q-bar are its outputs, and it's got a clock. What does it do? I'm not going to go through the details, but I'll tell you; you can work this out. J and K, if they're both 0—and see the last column is Qn plus 1; that means what's Q after the next clock pulse, in terms of what it was after the previous clock pulse, Qn—if they're both 0, nothing changes and it's a storage unit; Q stays what it was before. If you put J to 0 and K to 1, you get a 0 on Q; that's a reset. If you put J to 1 and K to 0, you get a 1 for Q; that's a set. Finally, if you put J and K both to 1—and if you're working with TTL, sometimes you can get away with just leaving them unconnected— then it becomes a T flip-flop, and Qn plus 1 goes to not Qn. In other words, it changes state.

This is the most versatile flip-flop, and we're now going to use it to build the synchronous counter. Here's a 2-bit synchronous counter. It consists of two

J-K flip-flops. By the way, I'm using the J and K on the right most flip-flop, the 1's bit, to be a so-called enable input. I might under some occasions—and we'll see some of them in future lectures—want to be able to turn a counter on and off, let it start counting and let it stop counting, and so I have this enable. Remember, J and K, if they're both 0, nothing happens; so I've got J and K connected together. If I put them to 1, then this thing is able to count because it becomes a T flip-flop and the first flip-flop changes on the count just as on the clock, just as we've seen in the counters we've dealt with before. There's a J-K flip-flop. We're going to be initially in the 0, 0 state.

Let's ask what happens when the first clock comes along. The first clock comes along, and it flips the first flip-flop, which is acting as a toggle flip-flop, to the 1 state. However, the second flip-flop has its J and K inputs initially at 0 when that clock pulse came along because Q1 was initially 0, and so they aren't set up to change. The second flip-flop doesn't change on the clock, so we get the 0, 1 state. Now the next clock comes along. At this point, Q1 is a 1. That Q1 is connected into both the J and K of the second flip-flop, and so we've got that one in a state where it's going to toggle and we get the 1, 0. The first one always toggles—the flip-flop 2 toggles this time—and we continue that process. We have the same 2-bit counting we had before—0, 1, 2, 3, in decimal—but the point is the flip-flops changed at the same time; they changed synchronously.

Here is, if you want to get more complicated, a 3-bit synchronous counter. I'm not going to go through the details; you can work it out if you'd like to understand how it works. But it's more complicated because not only do we need the J and the K, but we also need this gating to decide what happens. The key to this counter, or any synchronous counter, is all the flip-flops have their clocks connected together so they change state at once and there are none of these spurious intermediate states. If you work through the gating you'll see that this flip-flop goes to 0, 0, 0, 0, 0, 1, 0, 1, 0, etc., counting the decimal numbers 0 through 7 and then restarting. It's just like the 3-bit counter I introduced before but all of the flip-flops change state at once. There's a 3-bit synchronous counter.

Don't go building these things, unless you do the project, in which case you're going to build a 3-bit synchronous up/down counter, which will be more sophisticated than this one I'm showing you. But you don't really build counters; you go out and buy counters. There are chips, semiconductor chips, integrated circuits, which do counting. Here are just a handful of them that I've found useful: 74XX160, the XX standing for low power shock LS or HC or HCT or whatever. It's a decade counter with asynchronous clear. That means you can always clear the counter to 0 whenever you want, and it's already got the gating in to make it a decade counter. The next one up is the same thing but binary. The next two are similar but with synchronous CLRs. Finally, the last one, which is getting obsolete but I really like it and we're going to use it in a circuit shortly in another lecture, is the 74190 decade counter. It's an up/down counter, and it has a companion, 74191, which is the same thing but without the gating to make it a decade counter so it's an up/down 4-bit binary counter. Don't build, buy; these things cost about $0.40 each. But it's nice to know how they work, and that's what this lecture is about.

Let's put a lot of electronics together now and build a real practical counting device. One nice way to count things—people going through a door, products going down an assembly line—is to have them interrupt a light beam and have that make the pulse that you're going to count. That's what I'm going to do in the circuit I'm showing you now. This circuit consists of a photoresistor. I showed you a photoresistor, a special kind of resistor, whose resistance increases with decreasing light. As it gets darker, it gets higher resistance. You'll notice on the left, I've got a voltage divider between 5 volts and this photoresistor. That means as the light goes down, the photoresistor's resistance is going to go up. Therefore, the voltage at the junction between those two resistors is going to rise. I've got that fed into an op-amp in comparator mode that says, "Give me an output that depends only on whether one input is greater or less than the other." However, because the change in light intensity could be gradual and there could be some noise on it, we really need that Schmitt trigger idea that I introduced in lecture 14 that has that hysteresis that says, "Once you change state, don't change back again until the threshold alters a little bit." In fact, I built this circuit first without the Schmitt trigger and it went haywire, and I realized that's what I needed. The second thing is the Schmitt trigger, and this op-amp is pretty much like the ones we used before, except it has an output that swings between zero and five volts, rather than plus 15 and minus 15, to make it compatible with logic. The output of that Schmitt trigger, that op-amp, is going into the clock input of a decade counter like the one I just demonstrated, and the decade counter goes through the decoder and display system.

There's the circuit for what I'm about to show you, the counting thing. We've got the photoresistor, the Schmitt trigger, the decoder display that you might've built part of in lecture 17's project, and the counter that we've introduced in this lecture. Let's take a look and see how it works.

What I've got here: The photoresistor is that little cylindrical object right there. The lower integrated circuit is the comparator, the op-amp. It's actually a chip that has four comparators on it, but I'm using just one of them. You can see the resistors that are involved in the Schmitt trigger and also the resistor going down to that photoconductive cell. I do have a volt meter hooked up, and the reason I do that is because this thing doesn't work the same way it did in my lab back at Middlebury because it's very sensitive to the ambient light levels. What I have to do is set the comparison voltage on that Schmitt trigger—the voltage at the plus input of the op-amp—to somewhere between the voltage that's at the junction of the photoresistor and the other resistor when it's in full studio light versus when it's obscured. I had to do that just before I got this thing hooked up or it wouldn't work well in this environment. What I then have is the output of that comparator is going into the clock input to this 74190 decade counter, and that's feeding in again to my display system.

I'm going to take this black eraser, it makes a good blocker, and I'm going to move it across the photoconductive cell. The way this circuit is set up, the change of state takes place when the clock flips in such a way that I've pulled the obscuring object out and I've made it bright again. It goes over, 8; it goes over, 9; it goes over, it's a decade counter; swings back to 0; 1, 2, 3, 4, 5, 6, 7, 8, 9, and back to 0. You could imagine interrupting a light beam periodically by people walking by, by products going down an assembly line, or whatever, and you get a count of how many objects there've been. That's a real nice example of a lot of the electronics we've developed in this course.

That's the end of my discussion of counters, but if you want to do more with counters, do the project. The project is more sophisticated than the counters I showed in the lecture. It's to first explore your circuit simulators built in J-K flip-flop, and ask yourself: Does it work on the rising or falling edge and verify its truth table? Then simulate the 3-bit synchronous up/down counter shown below and verify that it counts both up and down. You'll have slightly different operations here, depending on which of the circuit simulators you use. By the way, if you want to get real fancy, you'll find that DoCircuits allows you to connect an actual display and decoder and actually see the numbers that your counter is producing.

But this is a sophisticated counter. It can go both up and down, and it's the kind of counter we'll need a couple of lectures hence.

Digital to Analog
Lecture 22

These days, electronics is increasingly digital. It's easier to store and process digital information; digital information gives us more bandwidth; and we can deal with more information at once in digital form. But we need to convert signals from the analog world—sound, light intensity, electrical voltages, speed, temperatures—to digital. In this lecture and the next, we'll look at how to do that conversion, starting with the easier topic, digital to analog. Key topics we'll cover include the following:

- Phones, music, and more: why we need analog/digital conversions
- Digital-to-analog converters (DACs)

- A weighted-resistor DAC
- A ladder DAC
- In your phone: the delta-sigma DAC.

Phones, Music, and More

When you talk on a cell phone, analog sound comes out of your mouth, is picked up by the microphone in the phone, and goes through an analog amplifier. Then, an analog-to-digital converter converts the sound intensities into a series of digital numbers. That information goes to a modulator, which uses it to alter a radio frequency signal produced by a transmitter inside the phone. That signal goes to the phone's antenna and out over the airwaves to a cell tower.

© Purestock/Thinkstock.

The cell tower picks up the signal, cleans it up, amplifies it, and rebroadcasts it, using filters to distinguish it from nearby frequencies of other incoming phone calls. The radio receiver in the recipient's phone picks up the signal and sends it through a demodulator, which extracts the stream of 1s and 0s that represents the digital information. That bitstream goes through a digital-to-analog converter (DAC), whose output is an analog signal—a continuously varying voltage. That signal then goes through an audio amplifier and on to a speaker or headphones, which produce analog sound. Almost all modern phones use this digital conversion and exchange data in digital form.

Digital-to-Analog Converters (DACs)

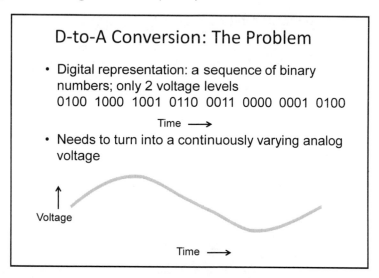

The digital representation of a quantity, such as a time-varying sound intensity or a time-varying voltage, is a sequence of binary numbers. A CD, for example, stores sound intensities as 16-bit numbers. It records these 16-bit numbers at the rate of about 44,000 times every second. Playing the sound requires a digital-to-analog converter. The DAC produces a continuously varying analog voltage that is amplified and played through loudspeakers or headphones.

The problem here is that there are always gaps in the conversion of analog information to digital. We know, for example, the sound intensity at one time and the sound intensity at a later time, but we don't know it in between. We have to assume that those gaps are filled in continuously. We thus turn the string of digital numbers into an appropriate analog waveform. In the figure above, that waveform, resulting from the string of 4-bit numbers shown, is approximately sinusoidal.

A Weighted-Resistor DAC

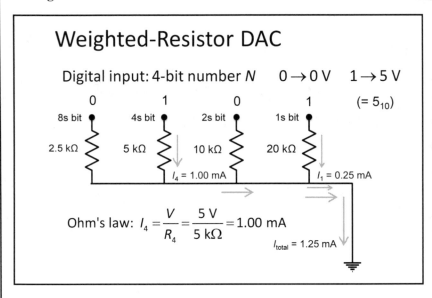

One of the easiest ways to perform digital-to-analog conversion is with *weighted currents*: We use each bit in the binary number to produce a current, and we weight those currents by the significance of the bit. The 1s bit is the least significant bit (LSB), and the maximum number of bits (e.g., the 8s bit when we're dealing with 4-bit numbers) is the most significant. We then sum those weighted currents and convert the result to an analog voltage, which corresponds to the binary number originally fed into the device. This process is implemented with precision-weighted resistors, which make different currents flow for the same voltage.

In the DAC shown here, the 8s bit goes to ground through a 2.5-kΩ resistor; the 4s bit goes to ground through a 5-kΩ resistor, which means that a 1 at the 4s bit will produce half the current of a 1 at the 8s bit. The 2s bit goes through a 10-kΩ resistor to ground, and the 1s bit goes through a 20-kΩ resistor.

It's crucially important that these resistors all go to ground. If there were any resistance in the common ground connection, then the current in each resistor would be influenced by other currents. Also, it's important to feed the DAC with precise voltages, either 0 V or 5 V in the figure shown. Using the outputs of logic circuits (which we'll use in order to simplify our circuits) isn't as accurate because there are voltage ranges, rather than precise voltages, for each of the two logic levels.

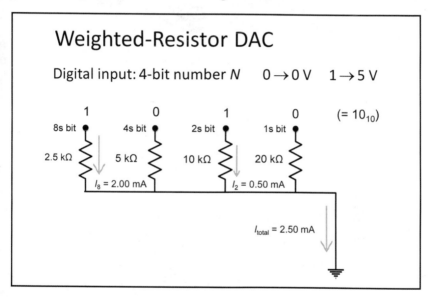

What happens when we put in the decimal number 10, represented in binary as 1010? That puts 5 V across the 2.5-kΩ resistor, causing 2 mA to flow. It puts 5 V across the 10-kΩ resistor, causing 0.5 mA to flow. Those two currents merge, giving a total current of 2.5 mA. This is a DAC that produces a current, in milliamps, whose value is ¼ of the input number ($N/4$).

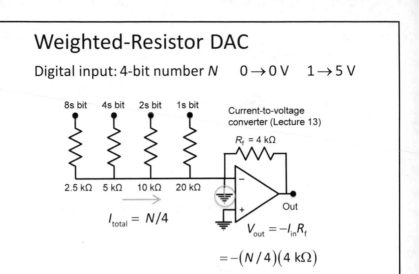

We can turn that into a combined current-to-voltage converter. As we saw in Lecture 13, $V_{out} = -I_{in}R_f$. In this case, the current coming in is $N/4$; with the 4-kΩ feedback resistor, this yields a voltage whose numerical value is equal to the negative of the input binary number: $-N$.

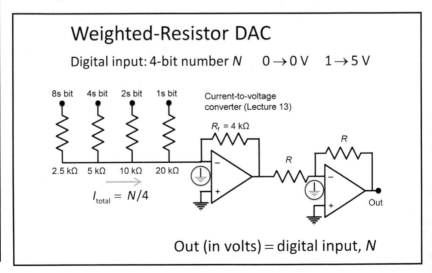

If we want to eliminate the negative result, we can add a unity-gain inverter. As you recall, this circuit produces a gain of $R_f/-R_{in}$; here, that ratio is −1. So the output in volts is equal to the digital input number, N.

A Ladder DAC

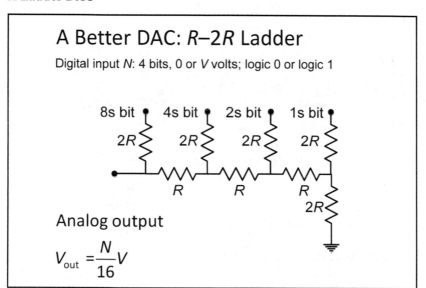

A Better DAC: R–$2R$ Ladder

Digital input N: 4 bits, 0 or V volts; logic 0 or logic 1

8s bit   4s bit   2s bit   1s bit

2R   2R   2R   2R

R   R   R

2R

Analog output

$$V_{out} = \frac{N}{16}V$$

Weighted-resistor DACs are not the best DACs because it's hard to match resistors that have very different values. Using a 20-kΩ resistor and a 2.5-kΩ resistor and getting the two resistances right is very difficult. Further, the more bits to be converted, the wider the range of resistors that must be used. A better DAC is the R–$2R$ ladder. This device is much easier to match because it uses only two different resistor values. Its analog output is a voltage rather than a current.

A Better DAC: *R*–2*R* Ladder

Digital input *N*: 4 bits, 0 or *V* volts; logic 0 or logic 1

$N_{in} = 0$

0   0   0   0

8s bit   4s bit   2s bit   1s bit

2R   2R   2R   2R

$V_{out} = 0$

R   R   R

2R

Analog output

$$V_{out} = \frac{N}{16}V = \frac{0}{16}V \quad ✔$$

Because this device is a linear circuit—all the components have currents and voltages that are linearly related—mathematically, all we need to do is determine how it works for two values, and then we'll know it works linearly for all other values.

The output of this device is *N*/16 volts; we need to show that this is true for two input values, and then, because of linearity, it will be true for all values. If we input 0000, we get 0/16, which is 0.

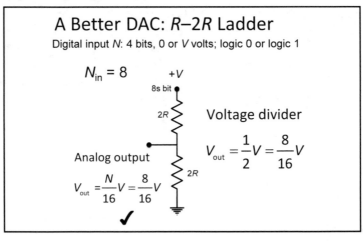

A Better DAC: *R*–2*R* Ladder

Digital input *N*: 4 bits, 0 or *V* volts; logic 0 or logic 1

$N_{in} = 8$ +*V*

8s bit

2R   **Voltage divider**

$$V_{out} = \frac{1}{2}V = \frac{8}{16}V$$

Analog output

$$V_{out} = \frac{N}{16}V = \frac{8}{16}V$$

2R

✔

For a second value, consider inputting the number 8 (1000). By going through multiple reductions of series and parallel resistor combinations, we end up with a voltage divider with a division by 2. The output, then, must be half the input, or $\frac{8}{16}V$, where V is the supply voltage.

In Your Phone: The Delta-Sigma DAC

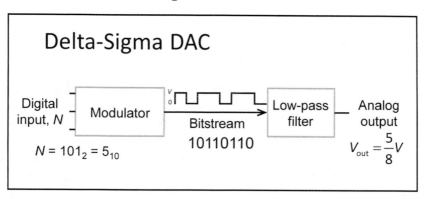

The delta-sigma DAC has a digital input corresponding to N bits and a modulator. The modulator is a circuit that looks at that digital number and produces a bitstream (a stream of 1s or 0s), making the relative proportion of 1s in that stream proportional to the value of the input number. The digital output is then passed through a low-pass filter, which smooths out the variations and produces an analog output that is basically the average of all those bits; the more 1s, the higher that average This is how the DACs in most common electronic devices work.

Suggested Reading

Introductory
Shamieh and McComb, *Electronics for Dummies*, 2nd ed., chapter 7, p. 166.

Advanced
Horowitz and Hill, *The Art of Electronics*, 3rd ed., chapter 13, sections 13.1–13.4.

Scherz and Monk, *Practical Electronics for Inventors*, 3rd ed., chapter 12, section 12.10.5.

DAC Simulation
Simulate the 4-bit weighted-resistor DAC described in this lecture, including the two op-amps. Test with one of the following options:

*   Apply all 16 different combinations of binary inputs and run a DC simulation for each combination.
*   Connect the DAC inputs to the outputs of the counter simulated in the project for Lecture 21.
*   Simulate a counter with four clocks or square-wave generators running at different frequencies.

Questions to Consider

1. Why is a weighted-resistor DAC less practical than an R–$2R$ ladder DAC?

2. Why are the inputs of a DAC labeled 1, 2, 4, 8 rather than 1, 2, 3, 4?

3. When a counter is used to feed the inputs of a DAC, the output of the DAC as viewed on an oscilloscope resembles a staircase. Why?

Digital to Analog
Lecture 22—Transcript

Here we are in the midst of a series of lectures about digital electronics, but we still live in an analog world. Our next two lectures are going to deal with the necessity of converting between the digital electronics world and the analog world we live in. How do we do those conversions? Why do we need to do them at all?

We do them largely these days because electronics is just increasingly digital. It's easier to process digital information. It's easier to store digital information. Digital information gives us more bandwidth. We can deal with more information at once. Digital information is error-free, at least in terms of what the digital part does; it may not be error-free in terms of what's coming into it. Digital is better in many ways. Increasingly, as you'll see in the final lecture, we actually use little miniature computers to do almost all of our digital processing. We need to convert signals from the analog world into digital. Sound, light intensity, electrical voltages, speed, temperature, all kinds of things we want to convert to digital. How do we do that?

In the next two lectures, we'll deal with that question. We'll start here with digital-to-analog conversion, although that's not what we need to do first to get into electronics from the analog world. We want to do that because that one's easier, and we'll be then doing analog-to-digital in the next lecture. That's going to be very sophisticated electronics, combining a whole lot of what we've already done.

Let me begin with some examples, and then pose the problem that digital-to-analog conversion imposes on us. Let's begin by thinking about a phone call. What happens when you make a phone call on your cell phone? The first thing that happens is you talk, and analog sound comes out of your mouth in the form of vibrations of the air due to the vocal cords vibrating and sounds resonating in your vocal tract and so on. That's picked up by the microphone in your phone, and that goes through an amplifier—an analog amplifier, like the sorts of amplifiers we dealt with much earlier in the course—but the next thing after the amplifier is an analog-to-digital converter. The analog-to-digital converter converts the sound intensities into a series of digital numbers that represent those sound intensities. It's got to do it pretty fast, as we'll talk about in the next lecture. That then goes to a device called a modulator, which basically takes those 1's and 0's and uses them to alter in some way, turning on and off or more likely altering slightly, the frequency of a radio frequency signal that's produced by a tiny little radio transmitter inside your phone. Then that goes into your phone's antenna, which is also built into the phone, and goes out over the airwaves and gets to a cell tower. That's the speaking end of things.

The cell tower picks up the signal. It cleans it up if need be. It does some things to it, and rebroadcasts it—by the way, using filters to distinguish it from other nearby frequencies of other phone calls coming in—and it broadcasts it out to the recipient of your phone call. The recipient of your phone call has a radio receiver in her phone, and that radio receiver picks up the radio signal. It goes through a demodulator, which extracts the stream of 1's and 0's that represents the digital information. That goes through a digital-to-analog converter, the subject of this lecture, also called a DAC, D-A-C, D-to-A converter, DAC; I'll use the word DAC quite frequently. The output of the DAC is an analog signal, a continuously varying voltage. That goes into an amplifier and goes into a speaker or headphones, and that produces output analog sound.

That's the anatomy of a phone call. In a modern phone call, almost all phones these days use this conversion to digital, and so we have to convert the analog sound into digital, do various things to get it out on a radio wave, pull it back down as a radio wave, demodulate it, and get it back into analog, and then send it out through a loudspeaker. That's the anatomy of a phone call.

What about music? Music is similar. Somebody sings into a microphone. Analog sound comes out of their mouth. Vibrations of the sound waves go through the air, go into the microphone. That's amplified into an analog-to-digital converter. In the old days, that might've been recorded to magnetic tape or burned directly to some kind of device. But nowadays, there's an analog-to-digital converter, and that analog-to-digital converter does the same thing it does in the phone: It converts that to a stream of 1's and 0's that represent the intensity—or a stream of binary digits, binary numbers—

of that sound over time. Nowadays, that might be put on a server, namely some kind of long-term storage device of the type that we discussed in the lecture on computer memory. Maybe that server is owned by an online store, and you connect to the online store, and those bits come out to you over the internet. By the way, they go through parallel to serial conversion because the internet deals with data in a serial fashion. It comes over the internet into your device, your MP3 player, your phone, whatever you store your music on. It's stored in Flash memory after undergoing a serial-to-parallel conversion. When you want to play your music, the Flash memory dumps its contents sequentially, representing the intensities of sound at different times into a digital-to-analog converter. That goes into an amplifier like we talked about in early lectures, an analog amplifier, and that comes out your loudspeaker and makes the analog sound that you can hear.

Two examples where analog-to-digital and digital-to-analog conversion are essential. We're going to abstract away from these now and talk in the rest of this lecture about how we do the process of D-to-A conversion. How do we make a DAC? How do we make a digital-to-analog converter?

Here's the problem we face: The digital representation of a quantity—a time-varying sound intensity, a time-varying temperature, a time-varying voltage, whatever it is—is a sequence of binary numbers. Here, for example, I've shown a sequence of 4-bit binary numbers, which can distinguish 16 possible levels. A CD, for example, has 16 bits in each of its numbers. It records a 16-bit number, and it does so at the rate of about 44,000 times every second. I'll talk more about that criterion in the next lecture. This is a very simplified D-to-A converter that's only going to convert with the precision of 16 possible different levels, which is what you get out of 4 bits because there are 16 different numbers from 0, 0, 0, 0 to 1, 1, 1, 1, from 0–15. They're representable by 4-bit binary number, and we need to convert this into a continuously-varying analog voltage.

We've got a little bit of a problem, because we know the sound intensity at this time, and we know it at a later time, but we don't know it right in between. The more rapidly we sampled that sound, the more rapidly we converted analog information to digital, the more accurately we know what was going on. But there are always gaps, and so we have to assume that those gaps fill in kind of continuously. We want to turn that string of digital numbers you see at the top into this roughly sinusoidal-looking waveform we see at the bottom, a time-varying, continuously-varying voltage. I picked those numbers so they roughly represent the levels on that curve. You can see the 0, 0, 0, 0, which is the lowest number we've got. The lowest number possible in 4-bit binary is right at the bottom of the curve, and so on.

How do we do that? How do we make that conversion? Here's what we do: One of the easiest ways to do it is to use what are called weighted currents. With weighted currents, we take each bit in the binary number and we use it to produce a current, and then we weight those currents by the significance of the bit. The 1's bit is the least significant; it carries the least weight. If you have $1.27, the 7 is less significant than the 2, and the 2 is less significant than the 1. The 1, the hundreds, is the most significant digit, and the 7 is the least significant digit in the decimal number 127. Similarly, in a binary number, the 1's bit is the Least Significant Bit, sometimes called the LSB. Whatever maximum number of bits you have, that's the most significant. In the example I just gave you, the 8's bit was the most significant bit.

Here's how we're going to do it. We're going to weight the currents by the significance of the bit; so a 1 at the 1's bit is going to produce only 1/8 the current of a 1 at the 8's bit. Then we're going to take those currents, one for each bit, weighted by the significance of that bit, which in binary means a factor of 2 between each one, we're going to sum those currents, and we're going to convert it to a voltage. The result is going to be an analog voltage corresponding to the binary number, the 4-bit binary number, which we fed into this device.

How are we going to implement that? If you're thinking, "How do we make different currents flow for the same voltage?" Resistors. We're going to implement this with precision-weighted resistors. We're not going to just grab random resistors off the shelf; we're going to buy real expensive, very high-precision resistors that have very accurate values and we're going to use those to weight the currents that come from our device.

Let's go over to our big screen and look at an implementation of what's called a weighted-resistor DAC, because it's going to use weighted resistors to weight the currents and that will produce the analog output we're looking for. Here's a very simple example of a weighted-resistor DAC. It's sort of a useless one right now because it dumps all its currents to ground, but bear with me, we'll change that. In this particular DAC, I've got the 8's bit going to ground through a 2.5-kilohm resistor. I've got the 4's bit going to ground through a resistor that's twice as big, and that means a 1 at the 4's bit will produce half the current of a 1 at the 8's bit; that's the weighting. The 2's bit goes through a 10,000-ohm resistor to ground, and the 1's bit through a 20,000-ohm resistor; 20,000 is, let's see, it's twice 10,000, it's 4 times 5,000, and it's 8 times 2.5 thousand. These are weighted according to the significance of these bits in the binary scheme of things.

It's crucially important that these resistors all see ground at their bottom ends. If there were more resistance here, then the current flowing in the 20-ohm resistor and the voltage at the bottom of the 20-kilohm resistor would be influenced by any other currents coming along. It's really crucial that they all dump to ground, and that's why I've drawn them all connected together, yes, but more importantly all connected to ground.

There's another issue I want to bring up right away because in the circuits I'm going to show you, I don't obey this rule. You really need to be careful. If you want a 0 at one of these inputs, you've really got to have exactly 0 volts. If you want to make it be a 1, you've got to have it be exactly a particular voltage, like, for example, 5 volts in the standard logic scheme. I'm not going to do that in the circuits I build. I'm going to use the outputs of commercial counting chips and other devices. One time I'll use switches connected to 5 volts, and then we'll be doing this accurately. But normally, if you build a DAC, you'd feed it with numbers that were precisely 0 volts or, say, 5 volts. You wouldn't connect it directly to the output of a TTL-integrated circuit whose 0 level could be anywhere between 0 and 0.8 volts and whose 1 level is typically around 3.5–5 volts, but could be even lower. Don't do what I'm going to do; you really have to put some extra circuitry in that carefully switch in exactly 0 and exactly 5 volts. The DAC I build you and show you won't be quite as accurate as it could be.

Here's our weighted-resistor DAC, and let's see what happens for several cases. Let's consider the case where the digital input is a 4-bit number n

(that stands for the 4-bit number); 0 corresponds to 0 volts, and 1, in this case, is going to correspond to 5 volts. I'm going to assume I've got a nice wave getting in accurately, either 0 or 1 volt coming in. Let me start with an example: There's 0, 1, 0, 1. No 8's, one 4, no 2's, one 1. That's 4 and 1 is 5, so that's 5 base 10. Let's see what comes out of this DAC, or let's see what currents are generated in that case. Let's think about what happens when we put 5 volts across a 4-kilohm resistor; 5 volts here, 0 volts here because it's at ground. We know how to do that, it's Ohm's law. I4, the current in the bit-4 resistor is V, 5 volts over 5 kilohms; that's exactly 1 milliamp. That's the current that comes flowing through the 4's bit resistor, and that current continues on its merry way to ground. No current in the 2's bit resistor. No current in the 8's bit resistor because there's 0 volts at the top and 0 volts at the bottom, so there are no currents there. But we do have 5 volts at the top of the 1's bit. That 1 means 5 volts; that's the logical level. Five volts across 20 kilohms is 5/20 or 1/4 of a milliamp, so we have 0.25 milliamps. Remember, these currents don't know about each other. They simply join in this wire—all those electrons flowing along, electrons going the opposite way, but we won't worry about that—and so those two currents merge, and we have a total current of 1.25 milliamps flowing in this weighted-resistor DAC.

Great. That's what happens when we put in the number 5. Let's put in a different number; let's put in the number 10, the decimal number 10. One 8, no 4's, one 2, no 1's; that's 1, 8, that's 8, plus 1, 2, that's 2. Add them up, you get 10. That's the binary representation of the decimal number 10; let's see what happens. You put 5 volts across only 2 and 1/2 kilohms, you get twice as much current as we did last time when we put it across 5 kilohms; you get 2 milliamps flowing. You put 5 volts across 10 kilohms, 5/10, you get a 1/2 a milliamp flowing. Remember how nicely it works out when you're using kilohms and volts, the currents come out automatically; and milliamps, you don't need to worry about factors of thousands and stuff. There we have 2 milliamps, and those two currents happily flow to ground. They don't know about each other, they simply combine to make a total current of 2.5 milliamps flowing to ground.

Let's do one more. Let's put the binary number 4 in, corresponding to decimal number 4 base 10. We've got no 8's, one 4, no 2's, no 1's. That's the number 4; 0, 1, 0, 0 is the number 4 expressed as a 4-bit binary. What happens here? No current in any of the resistors except the 4's bit resistor, and that current is, again, 5 volts over 5 kilohms, which is 1 milliamp. We get a total current of just 1 milliamp.

We've looked at three cases here, and let's just tabulate them. Input number is 4, number, we got 1 milliamp. Input number is 5, we got 1.25 milliamps. Input number was 10, we got twice what we got with 5, namely 2.5 milliamps. If you look at that, the current is 1/4 of the digital input number n: 4 gives us 1; 5 gives us a quarter of 5, or 1.25; 10 gives us a quarter of 10, or 2.5 milliamps.

This is a D-to-A converter that produces an output that's 1/4 of its input in current. But we can't do anything with that current yet, so now we want to take our weighted-resistor DAC; same thing, I've just drawn it a little further over to the left here so we can add some more circuitry. Remember, it was crucial that we dumped everything into ground so these currents wouldn't interfere with each other, and we have this recognition that the total current is, in milliamps, is $n/4$.

This doesn't do us much good. The currents are just all being lost to ground. We could put an ammeter here or something, but that'd be a little bit clunky. Let's take a circuit that we know about from earlier on when we were dealing with operational amplifiers: Let's take a current-to-voltage converter. Remember how this works? We've got a virtual ground here, so the resistors are all happy. They think they're seeing ground. It's not really ground; it's held at ground voltage by the feedback, the magic of feedback, in this operational amplifier. Furthermore, I put a 4-kilohm resistor here. That means if I send 1 milliamp in, we'll get a 4-volt drop across that resistor; so 1 milliamp will give us minus 4 volts at the output. There's a current-to-voltage converter back from lecture 13, and V out is minus I in, the current that's flowing in, times R feedback. In that case, the current that was coming in was $n/4$, the binary number coming in divided by 4, and we're multiplying by 4 kilohms, and so that's minus n.

The output voltage of this thing in volts is equal to the binary number. We didn't have to have that equality, it's just nice for this purpose. But you could obviously scale it by any amount you want. Put in an 8-kilohm resistor, I'd still get the scaling right. If you don't like that negative, let's just do one more thing. Let's put a unity-gain inverter on the end of it, two R's the same. Remember, this circuit, it produces a gain of feedback resistor divided by input resistor with a minus sign, and so we flip that minus sign and the output in volts is equal to the digital input number n. That's a 4-bit weighted-resistor DAC.

Now I'd like to show you what a weighted-resistor DAC actually looks like and how it works. Let's go over to our demo table and talk about what we have here. Here I have a circuit that consists of four resistors. You can see four resistors right there; those are my weighted resistors, the 2.5–20 kilohm resistors. This black chip, the integrated circuit, is a 747. It's a dual 741; two 741 op-amps in the same package. The resistors are the feedback resistor for the first one to make it a current-to-voltage converter and the two resistors that make the inverter at the end. The output of that I'm connecting over to here by this long wire over to here, and we're going to feed it later into the oscilloscope, but right now we're feeding it into an analog voltmeter. I really wanted to use my old-fashioned analog voltmeter here, because what's the point of doing a digital-to-analog conversion and then hooking up a digital voltmeter to it and converting it all back to digital again? To be pure, we've got this old-fashioned analog voltmeter.

Where are those resistors connected? Those resistors currently are connected to four switches. The rightmost switch, rightmost from your point of view, represents the 1's bit, the 2's bit, the 4's bits, that's the 8-bit, and they're all currently off. That means we have 0 volts coming into the inputs of all those resistors in the weighted-resistor DAC, and indeed, the voltage is 0. If I turn the least significant bit on, turn on the 1 bit, you'll see I get a voltage that's on the order of 1 volt. Remember, it's not going to be perfect, and partly that's because I've adjusted it also to work with integrated circuit outputs approximately. But it'll be sort of approximate. There's 1. There's 2. I get about 2 volts, a little bit more. Bring on the 1's bit also, I get a little bit over 3 volts. Drop it down to 4 and drop the other two off and go up to the 4's bit on more; 5, 6, 7, and it keeps going up. I can't go up all the way to the highest, to having all the switches on on the scale of that voltmeter. But you see an output voltage that's proportional to the binary number represented by the positions of those switches. We do indeed have a D-to-A converter.

There's a very simple demonstration of that D-to-A converter using an analog voltmeter. Before we move away from this simple circuit, let me pause and reconfigure it a little bit to show you some neat things you can do with simply a DAC. Here we are. We have the same DAC exactly; I haven't done anything to the DAC. I've connected its output. It was actually connected before to the oscilloscope, but I've now got the oscilloscope set to record that output. I've still got the input going to the light so you can see what binary number is coming in. I've still got the output also coming to the voltmeter, the analog voltmeter. The only difference is the input. The binary inputs to the DAC, the 4 bits of input, are now connected to the output of a counter, the same counter I introduced in the previous lecture. This is the 74190 decade counter, so it counts 0–9. It puts in the binary digits 0–9, and if you watch the voltmeter you can see it jumping up in steady increments. Again, it's not perfect. That's partly because my resistors aren't perfectly weighted. It's also because the outputs of the counter aren't required to be exactly 0 and 5 volts, but somewhere between 0 and 0.8, or about 3 and 5. We don't have quite nice even steps like we should. Again, if we did more sophisticated circuitry and used some kind of electronic switching, maybe MOSFET transistors, to switch between 0 and 5 volts depending on whether the output was a logic 1 or a logic 0, the output of the counter, things would work better. But that's a nicety.

You can see that happening. Everything's looking pretty much like it did before, except that the counter is now counting with a clock generated by the square wave generator in this protoboard here. That's causing the count to continually advance, causing the binary numbers to change. If you watch that, there we are at 0, 1, 2, 3, 4, 5, 6, 7, 8, 9, and so on, back to 0. That's working; the analog meter is following that.

Let me now turn up the frequency by a factor of about 100, and we get this beautiful pattern on the oscilloscope, this nice staircase. That tells you immediately it's an up counter; we're counting up. We start out at 0, and each of these steps represents an increase in the binary number at the output of the counter by 1. Correspondingly, the DAC produces an output voltage, which increases correspondingly.

Notice even though it's a digital-to-analog converter, it's not able to produce a continuous range of analog outputs. Because it's only a 4-bit DAC, it only has 16 possible states. In fact, since it's driven by this counter, it has only 10 possible states right now. We're having a little triggering issue with the oscilloscope, and so that's why, and so we're jumping around a bit. But there it is. You can see that nice beautiful staircase.

By the way, if we had a down counter, what would you see? You'd see the same thing but going down. That gives you an idea for how you can make a waveform generator of arbitrary waveform. How do you do that? You put some arbitrary set of binary numbers into a memory, and then you dump the memory sequentially to a DAC and the output is an analog voltage corresponding to those binary numbers coming in. For example, the greeting you record on your phone when you're not there—"Hi, this is me"—what's that? That's ultimately stored in a memory in your phone, and that set of numbers are dumped to a DAC in the phone and that turns that into an analog voltage, which goes runs to the loudspeaker, and that's your out-of-office greeting, or not-home greeting. You can make arbitrary waveform generators by dumping digital memory into a DAC.

Having said all that, let's talk a little bit more about digital-to-analog conversion before we wrap up, because weighted resistor DACs aren't necessarily the DACs that are always used, they're also not the best DACs. Here's why: One reason is it's very hard to match precisely resistors that have very different values. Getting a 20-kilohm resistor and a 2.5-kilohm resistor and getting those two numbers right is very difficult. Furthermore, the more bits you want to convert, the wider the range of resistors. A better DAC is called the R-2R ladder DAC. I'm not going to analyze it in all its detail, but I'm going to convince you that it works. This is a device that uses only two different resistor values, much easier to match. Here's what it looks like. Its analog output is going to be a voltage rather than a current. There it is, and we're going to see how it works. Because this device is a linear circuit—all the components have currents and voltages that are linearly related—mathematically all we need to do is determine how it works for two values, and then we'll know it works linearly for all other values.

Let's consider what happens if we connect 0, 0, 0, 0 into this DAC. I've shown that by grounding, with upside down ground symbols, the upper resistors. We got 0 everywhere, and the output better be 0. The output I'm going to claim is $n/16$ in volts. Indeed, in this case, it's $0/16$, which is 0, and so that works. We've confirmed one case, 0 at the input. Let's look at another case. Let's put it in the number 8. We'll put a 1 on the 8's bit, 0 on the 4, 0 in the 2, 0 on the 1, and let's see what happens. Here we have a lot of fun with parallel and series resistors, which you'll remember from early lectures. If you look at the rightmost two resistors, you can flip the upper one around and they're, in fact, in parallel. I've given you the formula there, but remember that two resistors of equal resistance in parallel give you half the resistance. Now we have two 2R resistors in parallel; that constitutes a resistance R. It's in series with another resistance R. That gives us another 2R. Now we're back to that same configuration again with two 2R resistors to ground. They're in parallel, so that gives us a resistance R to ground. That's in series with another resistance R, gives us 2R. Now what we've got is two 2R resistors, one connected to plus the supply voltage and one connected to ground. That's a voltage divider with a division by 2. The output in that case has got to be half the input, and half the input is $1/2$ or $8/16$. We confirm that for this particular input, this DAC also works. Because it's linear, it will then work for any other combination. You can go through that complicated analysis if you'd like for any other combination.

A R-2R ladder is a better DAC, and you can actually buy very precision laser-trimmed circuits with the resistors all built into them that are basically R-2R ladders to use in building DACs. But don't do that; don't build a DAC. You go out these days, and you buy a simple DAC.

Let me wrap up with one last example of a DAC. Before we do that, let's look at what we'd do to our DAC to get the voltage out of it. If we tried to draw current out of that DAC, we'd change the currents through those resistors, and that would change the DAC's operation. Instead we put a voltage follower—from our lectures on analog circuits, operational amplifiers, in particular—just to allow the output to draw no current.

Let's look now at one other kind of DAC. There are many other kinds; but here's what's actually probably in your smartphone, in your MP3 player, in your CD player. It's a DAC called the delta-sigma DAC. I'm not going to describe it in all its gory detail, but I'm going to look at it, and here's what it has. It has a digital input corresponding to some number n, some number of bits. It's got a thing called a modulator, and the modulator is a circuit that looks at that digital number and produces another digital thing called a bit stream, which is a stream of bits that could be 1's or 0's. What it does is make the relative proportion of 1's to 0's proportional to the value of the input number. I still have 1, 0, 1, 0, on-off, on-off coming out; that's not analog. But then I pass it through a low-pass filter—a resistor capacitor network that we talked about very early in the course—and that smoothes out those variations and produces an analog output voltage that's basically the average of all those bits. The more bits are 1, the bigger that analog voltage.

Here, for example, the input number is 1, 0, 0, base 2; that's 4 base 10. In this particular DAC, which is going to go up to 8, it's a 3-bit DAC, the bit stream is going to have half the bits 1 and half the bit 0. The analog output is going to be on average half of whatever the voltage level is at the 1's. We'll just call it D; 2 1/2 volts if it were 5 volts. If the input is 1, 0, 0, 1, then the bit stream has only one high level in it. When we low-pass filter it, we get only 1/8 of whatever the voltage is. If the bit stream has 1, 0, 1, which is 5 in decimal, we have 5/8 of the bits are high, and we get an output that's 5/8 of the voltage.

That doesn't describe all its detail, but that's how, in fact, the DACs in most common, everyday devices work. There are DACs in almost all the devices you have that produce analog outputs because they're processing electronics digitally.

Now you can receive that phone call because we understand at least the receiving end, the D-to-A conversion. In the next lecture, we got to get to the other end, the analog to digital. If you want to know more about DACs, build one. Simulate the 4-bit weighted-resistor DAC that I described in this lecture, including both op-amps. If you want to test it, I give you the three possible ways to test it: You could apply individually all 16 different combinations of binary inputs, a bit tedious; and you could run a DC simulation for each combination and see what you get at the output. You can do that. You could connect the DAC inputs to the outputs of the counter. If you did the previous lecture's project and made a counter, that simulated counter could drive your DAC and you could look at the analog output. Or you could simulate your counter with four different clocks or square wave generators running at different frequencies to act like the output of a counter. I'll let you figure out how to do that.

If you're working with CircuitLab, you can actually connect your DAC to digital circuits. If you're working with DoCircuits, you may or may not be able to. At the moment I'm recording this, DoCircuits doesn't have that capability. But option three will still work, and DoCircuits may have that capability by the time you're doing this project.

Analog to Digital
Lecture 23

In this lecture, we will build the most sophisticated circuits we'll work with in this course. They'll combine material from throughout the course, including op-amps, counters, DACs, and more. We'll put these all together as we work on analog-to-digital conversion, the more difficult part of the process of moving from the analog world into digital electronics and back to analog again. Key topics include the following:

- Characterizing analog-to-digital converters
- Sampling and the Nyquist criterion
- Flash converters
- Integrating converters
- Feedback converters.

Characterizing Analog-to-Digital Converters

Analog-to-Digital Specifications

- ## Speed
  - How much time to convert?
  - How many conversions per second?
  - Are time/rate fixed or dependent on input value?

- ## Precision
  - How many bits or digits does the A-to-D converter produce?
    - CD: 16 bits
    - Typical camera: 3 colors, 8 bits each
    - Digital scales: 3–4 digits

Analog-to-digital converters (ADCs) are characterized by speed and precision.

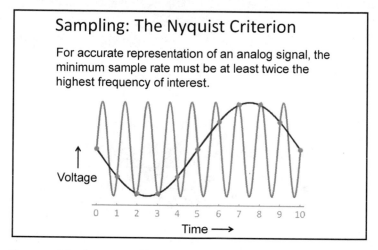

Sampling: The Nyquist Criterion

For accurate representation of an analog signal, the minimum sample rate must be at least twice the highest frequency of interest.

Voltage

0 1 2 3 4 5 6 7 8 9 10

Time ⟶

The Nyquest criterion states: For accurate representation of an analog signal, the minimum sample rate must be at least twice the highest frequency of interest. If the rate is lower, reconstructing the analog signal from the digitized values may result in an entirely different waveform, as suggested in the figure above. A CD, for example, samples at 44.1 kHz. The ADC in the recording system determines the analog sound intensity 44,100 times per second and stores each result as a 16-bit binary number representing one of 2^{16} possible sound levels. The 44.1 kHz is chosen because the upper limit of what we can hear is on the order of 20 kHz; thus, we're digitizing at somewhat more than twice that rate, enabling us to accurately capture the highest audible frequencies.

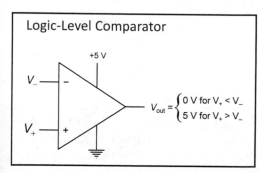

Logic-Level Comparator

+5 V

V_-

$V_{out} = \begin{cases} 0 \text{ V for } V_+ < V_- \\ 5 \text{ V for } V_+ > V_- \end{cases}$

V_+

Logic-Level Comparator
A logic-level comparator resembles a regular op-amp except it runs off a 5-V power supply if it's being used with TTL. The other end of the power supply is not −15 V but ground. As a result, the comparator's

output swings between 0 V and 5 V. If we run it open loop—that is, with no feedback—we get 0 V at the output if V_+ is less than V_- and 5 V, or logic level 1, at the output if V_+ is greater than $V-$. There's some latitude in those numbers, but they're good enough to determine definitively whether we have logic 0 at the output or logic 1. We'll make use of the logic-level comparator in several ADC designs.

Flash Converters

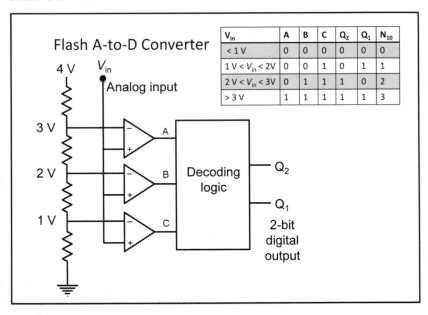

V_{in}	A	B	C	Q_2	Q_1	N_{10}
< 1 V	0	0	0	0	0	0
1 V < V_{in} < 2V	0	0	1	0	1	1
2 V < V_{in} < 3V	0	1	1	1	0	2
> 3 V	1	1	1	1	1	3

The flash ADC compares an analog input against a range of discrete voltages that are built into the converter. The process of conversion in this circuit is almost instantaneous. If there are just a few bits and we don't need high precision, this is a simple converter to use, but it gets complicated quickly if we have many bits.

In this flash ADC, we have three comparators whose outputs are three binary digits that tell us whether the input voltage is greater or less than the threshold voltage for each comparator. We then need a logic circuit that decodes that set of bits into a binary representation of the input voltage.

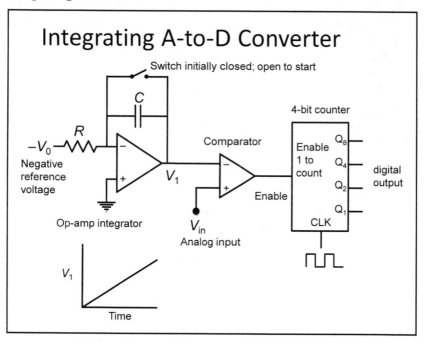

Integrating A-to-D Converter

Switch initially closed; open to start

C

R

4-bit counter

$-V_0$

Negative
reference
voltage

Comparator

Enable
1 to
count

Q_8
Q_4
Q_2
Q_1

digital
output

Enable

Op-amp integrator

V_1

V_{in}
Analog input

CLK

V_1

Time

We talked about integrators when we put capacitors together with op-amps. In particular, we saw that if we applied a steady voltage to an integrator, we got a voltage at the output of the op-amp that rose or fell at a steady rate. What integrating ADCs do is charge the capacitor at a steady rate, using a known reference voltage or an unknown analog voltage.

If we charge the capacitor at a steady rate using a known reference voltage, run a counter while the capacitor is charging, and have a comparator that tells the counter to stop when the capacitor voltage reaches the analog input, then the time that the counter has been running will be proportional to that analog input voltage, and thus, the output of the counter will be a digital number proportional to the analog input.

In this integrating ADC, we have an integrator to which we apply a negative reference voltage. When we open the switch across the capacitor, the capacitor voltage starts to rise at a steady rate. When the capacitor voltage rises above the analog input voltage, the comparator has a larger voltage at its negative input, and its output goes to zero, disabling the counter. The count is then a digital representation of the input voltage.

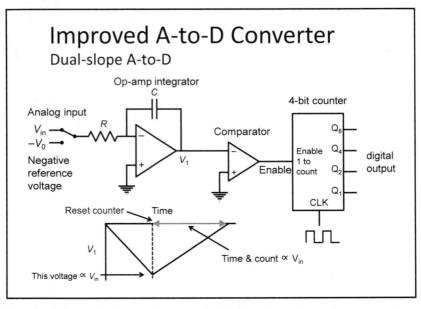

The dual-slope ADC uses the same basic technique, but it integrates, first, the unknown voltage until the counter is completely full. Then, it switches and integrates a reference voltage. The time it takes to do that depends on what that initial input voltage was, and the count that comes out at the end is, therefore, proportional to the input voltage. Dual-slope integrators are widely used in applications where precision is required, in part because small variations in resistance, capacitance, or clock frequency don't affect the conversion.

Feedback Converters

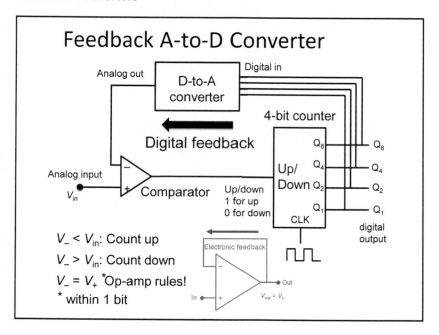

With a feedback ADC, we essentially make an ADC by putting a DAC in a feedback loop. We see here an analog input coming into the noninverting input of a comparator. We also see a 4-bit up/down counter. We then put a DAC into the feedback loop of the op-amp. Part of the feedback circuitry is the digital output of the counter going in to the digital inputs of the DAC. The analog output of the DAC goes to the inverting input of the comparator. This gives us a complete feedback ADC.

This circuit should remind you of the simple voltage follower we developed (also shown in the figure above) in which we took the output of an op-amp and connected it directly via electronic feedback to the negative input of the op-amp. The output voltage followed the input voltage. According to the op-amp rules, the two voltages at the positive and negative were essentially the same. This more complicated circuit for the ADC does essentially the same thing. It has an op-amp that is a comparator connected with a digital feedback loop. With our understanding of negative feedback, we could guess that this circuit tries to keep the inverting input—the output of the DAC—at the same voltage as the unknown voltage we're bringing in. For that to happen, the input to the DAC must be the binary equivalent of the analog input voltage. This simple circuit "bobbles" back and forth between two binary outputs whose values are either side of the actual analog input. This is like the servo voltmeter demonstrated in Lecture 11, whose needle kept "bobbling" back and forth.

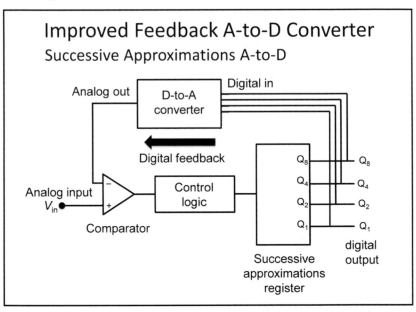

Improved Feedback A-to-D Converter
Successive Approximations A-to-D

A *successive approximations ADC* is an improved version of the feedback ADC whose conversion time is independent of the analog input voltage.

Suggested Reading

Introductory
Shamieh and McComb, *Electronics for Dummies*, 2nd ed., chapter 7, p. 166.

Advanced
Horowitz and Hill, *The Art of Electronics*, 3rd ed., chapter 13, sections 13.5–13.7.

Scherz and Monk, *Practical Electronics for Inventors*, 3rd ed., chapter 12, section 12.10.6.

Feedback ADC Simulation

Simulate a 3-bit version of the feedback ADC described in this lecture. Use your DAC/3-bit counter combination if you built it in Lecture 22. Add a comparator (CL: under Digital Gates). Plot the DAC output on one graph and $Q_{1,2,4}$ on another. You can add 6 V and 12 V to Q_2, Q_4 for clearer graphics. Verify operation for several DC input voltages. As a challenge, change the input to AC (but offset for positive only); adjust clock frequency (to what value, compared with AC input frequency?); and verify that the digital output follows the AC input.

Questions to Consider

1. Name at least six devices in your household or among your possessions that must contain ADCs.

*2. A 10-bit ADC has how many distinct digital levels?

3. What makes a flash ADC faster than an integrating or feedback type of converter?

*4. A digital kitchen scale shows three decimal digits. What's the minimum number of bits in the ADC in this scale?

Analog to Digital
Lecture 23—Transcript

Welcome to what's the penultimate lecture of this series on understanding electronics. In this lecture, we're going to build the most sophisticated circuits we'll work with in the course. They'll combine material from all the way back to the beginning, operational amplifiers, counters, D-to-A converters, you name it. We're going to put them all together because we're going to be working on analog to digital conversion, the other and harder part of that process of getting from the analog world into digital electronics, and then as we saw in the previous lecture, back to analog again.

Let me remind you of some examples of where we need to do this. Here are the examples I gave you last time of the phone: inputting information into the phone; that is, analog sound coming in, going through various processes, and then hitting an analog to digital converter. Similarly, when we took music and put it in the online store or whatever, we took the music and analog sound. We had to convert it to digital. We did it with an analog to digital converter. Many, many, many other situations use analog to digital conversion. It's not necessary, but most of our instruments these days are digital. We have digital fever thermometers. We have digital speedometers in our car. We step on digital scales. By the way, it doesn't mean they're more accurate or anything; it means they use digital electronics and digital displays. I have a scale at home that reads to a tenth of a pound, but it says it's only accurate to two pounds. I have a thermometer that reads to a tenth of a degree, but it's only accurate to two degrees. That doesn't make a lot of sense, but we like things digital; and in many cases we have to have them digital. Here's a digital thermometer, a digital speedometer, a digital scale. We need analog to digital conversions to make these things work.

Let's talk a little bit about analog to digital conversion; what we want to talk about when we're trying to specify what we want out of an analog to digital converter, an ADC. What do we want? One thing we want is speed. How much time does it take to look at an analog input voltage—and it's almost always a voltage; we've converted whatever it is, sound, intensity, temperature, whatever into a voltage. How much time does it take to convert that analog voltage into a digital number, a digital binary, representation of a number? Equivalently, how many conversions can we make every second? A slightly more subtle related concept is: Are these times or rates fixed? Or if you put five volts in does it take a different time to convert it than if you put 10 volts in? You probably don't want that latter situation. You probably want to know how long it's going to take when you're going to get the output.

Let me give you one example of the speed of a digital circuit. Here we have the oscilloscope I've been using throughout this lecture. There's a little number up here—actually two numbers—that are specifications for this scope. One of them says one ghz, one gigahertz, which means that this oscilloscope can look at signals that are oscillating about a billion times a second. The more interesting number is the number underneath. That number says five gs per s. What does that mean? It means five giga samples per second. That means this oscilloscope can convert five billion analog values, applied at its inputs, to digital numbers every second. It stores them in its internal memory, and processes them, and displays them, and does whatever we want. This is a five giga sample per second oscilloscope. That's fast; five billion conversions every second.

Now the other question we have about these conversions is how precise are they? How many digits does the analog to digital converter produce? Does it produce only four bits of binary, which would be one decimal digit? Does it produce five decimal digits? What does it do? That depends on the application. A CD, for example, is a 16-bit medium. Every sound intensity is recorded as one of 2^{16} possible values; 2,016 is about the 16 is about 64,000, 65,000, and so that's a lot of levels of gradation. That's for a CD. A typical camera has three colors, and each color is recorded to eight bits of precision. What's the intensity of red? What's the intensity of blue? What's the intensity of green? Eight bits each. So 2^8 is 256; 256 levels there. That's a typical camera; 24 bits total for the camera.

Here's a digital scale. It's reading 148.6, so it has a precision of four decimal digits. Three to four digits in this case, whether that last one goes 6, 8, 10, or just 6, 7, 8, 9, depends. We can specify the speed of an analog to digital converter; we can also specify the precision. How many gradations of digital numbers does it give us?

There's a big caveat in trying to do analog to digital conversion, and it's something called the Nyquist criterion. It says the following: If you want to represent an analog signal that's varying with time—and typically we do—we're interested in things that vary. You talk into your phone, and your voice level is varying with time. Otherwise, you wouldn't be saying anything; you wouldn't be sending any information.

Here I have a picture of a yellow sine curve that represents some varying voltage on the horizontal axis this time, and I'd like to represent that by digitizing it at periodic intervals because, again, digital information isn't continuous. We have the sound intensity or the voltage level now, we have it a little bit later, we have it a little bit later, we have it a little bit later. Here's what happens if I don't sample fast enough: I've shown you a series of white dots along there, and those white dots represent the individual times I'm going to sample. It's once every time unit; I don't care whether it's a second, or a millisecond, or whatever. But the point is if I sample the voltage at each of those times and then I say "OK, these are my voltages, let me connect them by a nice smooth curve," I get that red curve, which has nothing to do with the yellow curve; it's not the yellow curve at all. That's a mistake in sampling, it's called aliasing, and we end up with something that doesn't look even remotely like the signal we're trying to sample.

The Nyquist criterion says to get an accurate representation, you've got to sample at least twice the highest frequency of interest. You've got to pick off, at least in the case of a sine wave, two samples per cycle so you'll be sampling at twice the frequency of the sine wave. If you go higher, you're better off. For example, a CD sample's at 44.1 kilohertz, 44,100 times a second; and 44,100 times a second, the analog to digital converter, in something that is recording information to a CD, records one of 2^{16}, one of 65,000 roughly possible levels for the sound wave, digitizes it as a binary number, and stores it, or does whatever it wants to do with it. That 44,000 is chosen because the upper limit of your ear is on the order of 20 kilohertz, 20,000 cycles per second, and so we're digitizing at a bit more than twice that speed so we can capture accurately the highest audible frequencies. That's something to bear in mind with analog to digital conversion. How fast do you need your ADC, your analog to digital converter, to be? That depends on how rapidly varying is the signal you're looking at. You've got to have an ADC that converts at least twice as fast as the signal varies.

There are many, many, many ways to make ADCs in the stable of electronic instruments. I'm just going to show you three just to give you a sense of how it's done. We're going to build one of them as the project and I'll demonstrate one on my board, the same one, but I'll talk about two others in some detail.

Let's begin with the easiest. Before I begin with the easiest, I have to introduce a new device. It's not really a new device; it looks just like op AMP. It's a logic level comparator. I actually used it in that counting circuit that I built for you before. Here it is; it looks just like a regular op AMP except it runs off a five volt power supply, if you're using TTL logic with it. The other end of the power supply is not minus 15 but ground; and as a result, it's output swings between 0 and 5. If you run it open loop—that is, with no feedback—you get 0 at the output if V+ is less than V- and 5 volts, or logic level one, at the output if V+ greater than V-. There's some latitude in those numbers, but they're good enough to determine definitively whether it's a logic 0 at the output or a logic 1. We're going to make use of this logic level comparator in several of the analog to digital converters designs I'm going to show you. There's a logical level comparator.

Let's talk first about a flash A-to-D converter. What the flash A-to-D converter does is to compare the analog input against a range of discrete voltages that are built into the converter. I'll show you in a minute how we make them using techniques from the very first few lectures. A flash A-to-D converter is called that because it's very fast. The process of conversion is almost instantaneous but for the propagation time through a few gates and things. So it's very fast. It doesn't take time to do some complicated process. You put in an analog input and bingo, you get a digital output. If there are just a few bits, if you don't need a lot of precision, it's pretty simple. But if you need a lot of precision, it gets complicated very fast, exponentially fast, because it requires very complicated decoding circuitry to make it work.

Let's see how a flash A-to-D converter works. By the way, before I show you the flash A-to-D converter, which is going to have several comparators in it, the comparator I just showed you, you could even think of the comparator as the simplest A-to-D converter. It's a one bit converter. It just says "Is the voltage bigger than the voltage at the plus input of the comparator, or is it less?" That's all the information it gives you with a 1 or 0 in its output. But that is, in principle, an analog to digital conversion. Not a very precise one.

Here I'm going to consider a situation where I have a four volt supply, and I'm going to put four identical resistors between that power supply and ground. That constitutes a voltage divider. The middle point in that divider is going to be at two volts. The lower point, the junction between the lower two resistors, is going to be a one volt. The junction between the upper two resistors is going to be at three volts because they're equal resistances, they've got the same current flowing through them, they've got the same voltage across them, and no matter what those resistors are, they divide the four volts into four steps. We have one, two, and three volts. Those are the discrete voltages that are built into this converter.

Now I'm going to add three comparators, the logic level comparator I had before. I'm going to connect their plus inputs to an unknown analog input voltage, the voltage I want to convert to a digital quantity. What each comparator is doing is comparing that input voltage against the voltage it's connected to in that string of resistors. The lower comparator asks the question, "Is the input voltage greater or less than one volt?" If it's less than one volt, it outputs a 0; if it's greater than one volt, it outputs a 1. The middle comparator asks, "Is the input voltage greater or less than two volts?" If it's greater than two volts, it outputs a one. If it's less than two volts, it outputs a zero. The upper one asks the same question about three volts. We have coming out of these comparators three binary digits that tell us whether the input voltage is greater or less than the threshold voltage for that particular comparator. In some sense, we already have in these comparators—which I've labeled A, B, and C—some sort of digital representation of the analog voltage. The trouble is it's not a nice convenient binary form to work with; and there's where the hard work comes in.

What we have to do is add some kind of logic circuit that decodes that information. For this simple, three comparator circuit, which is going to give us a two bit digital output, here's the truth table we have to have. If A, B, and C are all 0, we've got to have Q1 and Q2, the two bits of our digital output, both be 0. If the input voltage is between 1 volt and 2 volts, we're going to get comparator number C, the lower one, giving us a 1 and the other two giving us 0, and we've got to develop logic that will make that into the binary number 01 and so on. We go through the possible values. If we're above three volts, all three of them are giving 1's; that has to convert into 1, 1, three volts. Again, this isn't precise because it's only got 2 bits. We know that if it's above 3, it's somewhere between 3 and 4; we only know it's above 3 actually. If it's giving us a binary 2, that's telling us that it's somewhere between 2 volts and 3 volts and so on. But this is a flash A-to-D converter, and it works very fast.

You could imagine building one with 10 bits or whatever, but that decoding logic is going to get extraordinarily complicated extraordinarily fast. Flash A-to-D converters are used either where you've got a lot of money and you really need speed, or where you don't need a lot of precision and can get away with just a few bits and the relatively easy decoding logic that's involved in that case. That's a flash A-to-D converter.

Let's take a look now at a second class of an A-to-D converter; let's look at integrating A-to-D converters. The process of integrating is the process of adding things up. We talked about integrators when we put capacitors together with operational amplifiers, and in particular if we applied a steady voltage to an integrator, we got a voltage at the output of the op-amp that rose at a steady rate. What integrating A-to-D converters do fundamentally is to charge the capacitor at a steady rate. They may do it using a known reference voltage, or they may do it using an unknown analog voltage. It doesn't matter. We'll see several examples.

But basically, if we charge the capacitor at a steady rate using a known reference voltage, and we run a counter while the capacitor is charging, and we have a comparator that says to the counter, "Stop when the capacitor voltage reaches the analog input", then the time that the counter has been running counting some steady clock signal will be proportional to that analog input voltage. Make the analog input voltage bigger and it takes more time for the capacitor to charge to that voltage and for the counter to stop. You'll remember on some of the counters I developed in the lecture on counters, I was able to build what was called an enable input that you had to set to 1 or maybe 0, it depends, in order for the counter to count at all. We're going to exploit that in a circuit like this.

Let's take a look at a diagram of a simple integrating A-to-D convertor. On the left I've got an integrator; it's a circuit with a resistor at the input and a capacitor in the feedback that we talked about before when we put capacitors with op-amps. I'm putting a negative reference voltage, some fixed voltage that's negative, at the input to that resistor. When I open this switch across the capacitor—I'll keep the switch initially closed—as soon as I open it, the capacitor voltage starts to rise. It rises because this is a negative reference voltage and we're in an inverting configuration. What happens is that voltage rises at a steady rate. I have the output of that comparator connected to the enable of the counter, and I have the analog input coming into the plus input of the comparator. Let's now figure out what happens.

Initially, the counter is counting. It's counting a fixed rate clock; the counter is enabled to count, and on we go. On the graph I show you the voltage at the output of the integrator is the curve that's rising steadily. The unknown analog input voltage that I'm trying to convert to a digital signal is shown as a straight horizontal line because it isn't changing. At the time the capacitor voltage gets above the unknown voltage, the comparator now has a bigger voltage at its minus input and its output goes to zero. That disables the counter, and I've shown the logic level of the enable on the counter up above. If we went to a higher voltage, it would take more time before the capacitor voltage reached the analog voltage. The counter would be enabled for longer, and we'd end up with a bigger count.

That's the simple integrating A-to-D converter. It doesn't really work very well, and the reason it doesn't is because it's very sensitive to the values of the capacitor, the resistor, and the clock rate. If you actually have integrating A-to-D converters—and you often do—they may well be built into the digital volt meters I've been using in this course because a very nice technique uses something like integration, but it's called dual slope. I'm not going to spend a lot of time on this, but I'm simply going to say this device uses the same basic technique that's integrating, but it integrates first the unknown voltage until the counter is completely full and then it switches and integrates a reference voltage. The time it takes to do that depends on what that initial input voltage was, and the count that comes out at the end of that is therefore proportional to the input voltage. The beautiful thing about this circuit, if you look at it closely, is it doesn't matter, to a first approximation, what the values of R, and C, and the clock rate are. If they change slightly, the way the circuit works, by integrating the input voltage until the counter is full at its maximum state and then integrating the negative reference voltage to go the other way until you get to zero, you'll find cancels out the effects of small variations in the resistor, and the capacitor, and the clock rate. This is called a dual slope because we integrate one way and then we integrate the other way. Dual slope integrators are widely used where you need precision. They do take time to make the conversion because they've got to run the counter through full counts. The more precision you want, the more counters you have to have strung together to make a high precision counter. That's a better version of the integrating A-to-D converter.

Now I'm going to get to my favorite A-to-D converter. I'll say right at the outset this isn't a good one. It's not one you'd actually use, although I'll show you at the end a variant of it that does work. But the reason I like this one so much and the reason I use it at the culmination of my electronics courses that I teach at Middlebury College is because it really ties together everything we've done. This is a feedback A-to-D converter because we're going to put in the feedback loop of an op-amp basically a D-to-A converter. Whoa, we're going to make an analog to digital converter by putting D-to-A converter in the feedback loop.

Let's look at what we've got. Here we see the analog input coming into the plus input of a comparator, and nothing else is connected yet. We see, in this case, a four bit counter with its digital outputs, a clock coming in, and this is an up down counter. I mentioned when I talked about counters in the counter lecture that you could make counters that went either up or down. If you did the project for that lecture, you actually built an up, down counter and by changing a particular bit you could count either up or down. This particular up, down counter that I'm talking about goes up if it's a 1 and down if it's a 0. The counters I'm actually using are the opposite, but that doesn't matter at all.

We haven't put this thing together yet; we need feedback. First thing we're going to do is take the output of the comparator. It's a logic level comparator; it's either a zero or a one. Logically, it's either zero volts or five volts. If it's a one, the counter counts up. If it's a zero, the counter counts down. Then we're going to take a D-to-A converter and put it in the feedback loop of the op-amp and we're going to connect. Part of the feedback circuitry is the digital output of that counter going in to the digital inputs of the D-to-A converter, the binary inputs; and the analog output of that D-to-A converter comes out and goes to the minus input of the comparator. Here we have the complete feedback A-to-D converter. Wow.

Before you get lost in circuit, let's think about sort of what it looks like. It's a circuit that has digital feedback. We've built circuits, or seen circuits, that have other kinds of feedback. For example, this circuit should remind you a lot of the very simple voltage follower we developed in which we took the output of an op-amp and connected it directly electronic feedback to the minus input of the op-amp. The output voltage followed the input voltage, and if you remember the op-amp rules, they kept the two voltage at the plus and minus essentially the same. This more complicated circuit I have for my A-to-D converter does essentially that. It has an op-amp; that is a comparator. It has a feedback loop that's a digital feedback loop. But if we believe in the magic of negative feedback, without going into the details— but we'll go into details—we'd guess that maybe this circuit tries to keep the minus input, that is the output of the D-to-A converter, at the same voltage as the unknown voltage we're bringing in. For that to happen, the input to the D-to-A converter needs to be the binary equivalent of that analog input voltage. So we ought to have on the output of those of that counter a binary representation of the analog input voltage. Wow.

Before we go on and look at detail at how that circuit works, let me show you something that this should also remind you of. Let's take a break and set up a little circuit over here. Way back when I was introducing this idea of negative feedback, I showed you this crazy makeshift servo voltmeter I built. Remember how it worked? It had an op-amp or comparator. The output of the op-amp drove the motor. It drove it one way if the output was plus 15 volts and the other way if it was minus 15 volts. The feedback consisted of a mechanical linkage between that motor and a potentiometer that dialed off some voltage and sent that voltage into one of the inputs of the op-amp. The unknown went into the other input of the op-amp and what happened ultimately, without going into every tiny bit of detail, is the circuit strove to keep the two inputs of the op-amp the same, and in the process it ran the motor the right way to make that happen.

If I, for instance, turn up the voltage, the needle goes for a while. It's driven in the clockwise direction until the potentiometer turns enough to tap off a voltage that gets a little bit above the voltage at the unknown input, the voltage that I'm supplying here. Then the motor goes the other way. But as soon as it goes the other way, it's too low and it goes back. This thing is bobbling back and forth all the time about what's basically the correct voltage, something, a needle position that corresponds to the voltage I'm inputting to the circuit. I want you to keep that volt meter in mind and the idea that it bobbles back and forth throughout the rest of this lecture. I'd love to have been able to do that while I show you the circuit I'm actually going to build, the analog to digital converter, but it turns out the switching on and off of that motor kicks up some electrical noise that messes up my digital circuits. That's a good lesson in when you build digital circuits: Things can go wrong, and things like spurious pulses associated with things like motors turning on and off can send extra pulses into the counter and the counter gets all screwed up. I couldn't do that. I'm showing you this now, and I'm now going to turn it off and disconnect it from my circuitry. We'll talk more about the feedback A-to-D converter, and then I'll demonstrate it.

Keeping that crazy servo voltmeter in mind, which is bobbling back and forth, it's comparing two voltages and running the motor one way if one is higher and the other way if the other is higher, let's look in detail at the workings of our feedback A-to-D converter. Let's focus on the left hand end there where the minus input of the op-amp is connected to the output of the DAC and the plus input is connected to the voltage, the unknown voltage, the input voltage, which we want to convert to a digital number. If V- is less than V-in, the minus input of the op-amp, the comparator, is less than the plus input, then the plus input rules and the comparator outputs five volts or a logic one. This particular counter counts up on a logic one, so if V- is less than V+, we count up. As we count up, up goes the binary number at the output of the counter. That binary number is fed back into the D-to-A converter, and so the analog output goes up also. Eventually, it goes up enough that V- is greater than V-in, and that takes the comparator and says, "Uh-oh, the minus input is now bigger." That drives the comparator output to zero; that switches the counters up down to down and the counter counts down. The counter counts up until its output digital number when converted back to analog is just above the input that you want to convert to a digital number. As soon as it gets there, it counts down. As soon as it counts down, it's a little too low, and so it counts up. It's doing digitally exactly what that crazy needle was doing when it bobbled back and forth. The motor went this way, then it went that way, then it went this way, then it went that way. I tell you that to prepare you for what you're going to see when you see this device in operation in just a moment.

But let me end with this visual description of this circuit by pointing out that what's going to happen is that V- and V+ are going to be essentially equal. They aren't going to be instantaneously equal like they were in an electronic feedback circuit, and they aren't instantaneously equal in my crazy voltmeter because it's always a little too high or a little too low. I put a little caveat on there that they're going to be equal within one bit; that's the closest we can do.

Let's now take a look at this circuit as it actually operates. What do I have here? Stuff that I've had on the board for a long time, but let me show you what it is in particular. Here I have the 74190 counter. It's an up down decade counter. Here I have that same quad comparator with digital logic output levels. Here I have the D-to-A converter that we used in the last lecture, the D-to-A converter, and the D-to-A converter inputs are coming from the four bits of output of that counter. Here's the output of the D-to-A converter. The output voltage goes to one input of the comparator. The other input of the comparator comes from this potentiometer, which I'm simply using as a variable voltage source. Had it worked, it would also have gone into here. But I can't make that happen.

I'm going to turn this circuit on, and I just chose for convenience to use the built in display that this board has. I could've easily hooked up it up to my own display that I'd already wired. Let's see what happens as we run this thing. Right now when I turn it on, the voltage, which is reading off this voltmeter, the input voltage, is toggling between 0 and 1. That's the bobbling back and forth just like this thing did. We get too high, we get too low, we get too high, we get too low, we get too high. The counter counts up, the counter counts down. What if I bring the voltage up a little bit? Bring it up a while, and for a little while the counter just counts up until it starts bobbling between two other voltages, in this case four and five. That tells me we're somewhere between four and five volts. I've got to run the clock slow enough, by the way, that you can see those numbers. I'll slow it down a little bit just to make it a little bit easier to see that bobbling. If I turn the clock up to high you won't see anything because the numbers will be just going back and forth all the time. But there's four and five bouncing back and forth. I'll turn up the voltage a little bit more, and there's eight and nine bobbling back and forth. It's somewhere between eight and nine. Turn it down a little bit and I've got four and six. I'm reading it upside down; having a little trouble reading it. But there it is, four and is that five? Yes. I can crank it up. I got eight and nine, and six and seven is in there somewhere if I'm careful.

There's a rather crude analog to digital converter using the magic of negative feedback. It's exactly analogous to this servo volt meter, and it's almost exactly analogous to the electronic feedback simple voltage follower except that one doesn't bobble back and forth because it's so fast. I'm not going to do it, but one way to avoid that digit bobble would be to add an extra bit, which I don't bother to display, an even less significant bit than the lowest significant bit that's being displayed. Then most of the time that bit would change without actually affecting the output of the four bits of the later counter. I actually wired a second counter in there, but things are a little dicey and I'm not going to try to bother to hook it in. That would be the way to get rid of that digit bobble.

Having said that, this isn't the ideal A-to-D converter. It's great for teaching, though, because it brings all these concepts that we've worked with in: feedback, D-to-A converters, counters, comparators, op-amps, the whole gamut. One of its problems is the amount of time it takes to do the conversion depends on the voltage that you put in. It has to count up to that voltage before it starts bobbling up and down. So a better A-to-D converter, an improved feedback A-to-D converter, is what's called a successive approximations converter. These are actually in use. It works very similarly except what it has is some control logic that instead of just counting up and down, it keeps adding an extra bit; first, the least significant, then the most. It keeps adding extra bits, and it keeps trying which bit is going to make it too big. It starts by setting the eights bit to one, for example; it's too big. It resets to zero. It tries the fours bit; it's too small, it keeps it. And so on. It has a fixed conversion time, and it actually works quite well. The logic behind that successive approximations register isn't terribly difficult to construct.

Let me wrap up. You've seen a number of types of analog to digital converters. The crude ones I've showed you don't work great, but there are improvements on them. There are many others. For example, the whole delta sigma process I showed you for DACs can also be made to work in reverse for ADCs. And you may have a delta sigma ADC in your phone, for example.

If you want to do the project, this is the most sophisticated project for the course. You can move on and simulate a three bit version of the feedback ADC described in this lecture. You'll need to use a three bit DAC, but you built one if you built the DAC in the previous lecture. You'll need to use a three bit counter, and you'll have one if you built one in the counter lecture. You'll need to add a comparator. If you're using CircuitLab, you'll need to add that comparator under digital gates. It's got a digital comparator that has logic level outputs. You'll need to plot the DAC output on one graph and the Q1, 2, 4 on another graph. I tend to add the 6 and 12 volts so you can see the graphs displaced and CircuitLabs displaced. You can verify the operation for several input voltages. If you want a real challenge, change the input to AC very slow and make the clock a lot faster. Watch what happens and you'll see the operation of the circuit as the digital signal tries to follow the analog signal.

Now unfortunately, at the time I'm recording this series, DoCircuits doesn't yet work for combining analog and digital, which is what's needed in this circuit. They probably will soon, and they may well by the time you're doing this. If you use DoCircuits, you'll be able to add their decoder and display, and you'll actually have a nice seven-segment display of the voltage. There's the most sophisticated circuit you'll have built in this project, in this course.

Your Future in Electronics
Lecture 24

In this lecture, we will look at two aspects of your future in electronics—as a consumer and as a creator. Even if you're not interested in building electronics, as a consumer, you will buy more electronics in the future than you have in the past because more electronic devices will be developed, and they will be, for the most part, faster and cheaper. If you've done some of the projects in this course, you may also become a creator of electronics. We've learned a number of basic building blocks, including amplifiers, flip-flops, counters, op-amps, and more. And we've seen how you can build these up from simpler circuits, ultimately based in transistors and other components. In this lecture, we will explore where your new knowledge of electronics will take you in the future.

Moore's Law Revisited

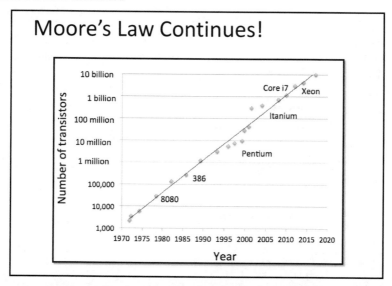

As we saw in an earlier lecture, in the 1960s, Gordon Moore, an engineer at Intel, predicted that the number of transistors on a given integrated circuit would double every 18 months. Actually, since 1965, it's been approximately true that the number of transistors on an integrated chip has doubled about every two years. But we may be headed to some sort of limit around the year 2020, when transistors are shrunk down to about a nanometer (one-billionth of a meter).

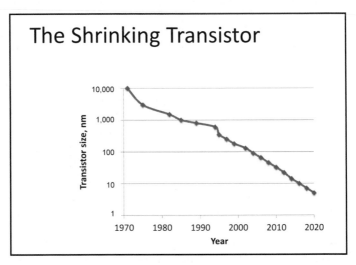

We can also think about Moore's law in terms of the size of transistors. As the graph shows, that size is heading down toward the range of 10 to 1 nanometers. Of course, as transistors continue to shrink, we can fit more on a chip, which enables us to store and process more information and perform complex operations faster. Ultimately, we will develop nanoscale-size transistors.

A Quantum Limit?

- Gordon Moore: "When [during a 2005 visit to Intel] Stephen Hawking was asked what are the fundamental limits to micro-electronics, he said the speed of light and the atomic nature of matter."
- Limiting transistor size:
  - Compton wavelength of the electron $h/m_e c = 0.0024$ nm
  - Extrapolate: will reach in year 2036

According to Stephen Hawking, fundamental limits to microelectronics are imposed by the speed of light and the atomic nature of matter.

Future Electronic Devices

Self-driving cars have already driven hundreds of thousands of miles on the public highways of California and may be for sale around the year 2020. Such cars are an example of the kind of complexity enabled by the miniaturization of electronic

© yuanyuan xie/iStock/Thinkstock.

circuits and computing power and by the miniaturization and reduced cost of sensors of all types. The future of electronics will see greater numbers of smart systems of all kinds.

Your Future as a Builder of Electronics

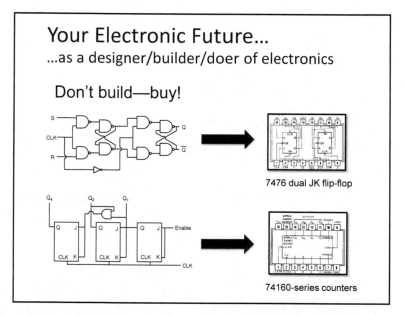

One of the most important lessons we've learned throughout this course is to buy—not build—basic circuits. For example, we saw how to build up from a simple bistable circuit to an RS master-slave flip-flop, but if you're actually building your own electronics, you should buy an integrated circuit that already has flip-flops on it, such as a 7476 dual JK flip-flop.

Microcontrollers

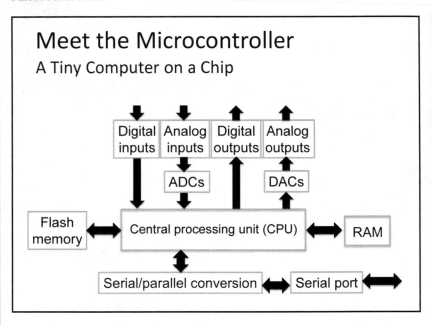

A microcontroller has a central processing unit, consisting of logic gates set up to compare, add, and perform other functions with binary numbers. It has flash memory, which we can think of as its hard drive, where it stores data and programs. It also has RAM and a serial-parallel converter and a serial port that can bring data in or out. Typically, it has both digital inputs and analog inputs that go through analog-to-digital converters. It also has digital-to-analog converters to provide analog outputs.

Arduino Code

```
int sensorPin = A0;    // input pin for the potentiometer
int ledPinred = 13;    // pin for red LED
int ledPingreen=12; // pin for green LED
int sensorValue = 0;  // value coming from the sensor
void setup() {
  // declare the ledPin as an OUTPUT:
  pinMode(ledPinred, OUTPUT);
  pinMode(ledPingreen, OUTPUT);
}
void loop() {
  // read the value from the sensor:
  sensorValue = analogRead(sensorPin);
  if (sensorValue > 100)
  // turn the red ledPin on
  {digitalWrite(ledPingreen, LOW);
  digitalWrite(ledPinred, HIGH);
  // stop the program for <sensorValue> milliseconds:
  delay(sensorValue);
  // turn the red ledPin off:
  digitalWrite(ledPinred, LOW);
  // stop the program for for <sensorValue> milliseconds:
  delay(sensorValue);
  }
  else
  {digitalWrite(ledPingreen, HIGH);
  }
}
```

Electronics hobbyists around the world use microcontroller development boards, such as the Arduino microcontroller, to build many kinds of devices. These development boards have power receptacles, a USB input, a power input, five or six 10-bit ADCs, and digital and analog inputs. They are programmable with a rather simple coding language that you can learn very quickly. One approach to learning this programming is to download the software for your microcontroller and adapt its simple code examples for your own purposes.

Summing Up Our Course

A device built by a student at Middlebury College to track the Sun by looking for the brightest point in the sky serves as a fitting end to our course. This device is an example of electronics in the service of our planet, a way to maximize our ability to extract clean, free, nonpolluting energy from the Sun.

© Igor Kirillov/iStock/Thinkstock.

Suggested Reading

Introductory

Lowe, *Electronics All-in-One for Dummies*, book VII, especially chapter 1.

Platt, *Make: Electronics*, chapter 5, experiments 34–36.

Advanced

Horowitz and Hill, *The Art of Electronics*, 3rd ed., chapter 15.

Scherz and Monk, *Practical Electronics for Inventors*, 3rd ed., chapter 13.

Questions to Consider

1. What are the ultimate limits to the shrinking size of electronic components in integrated circuits?

2. What is a microcontroller, and what relation does it bear to your laptop computer?

Your Future in Electronics
Lecture 24—Transcript

Welcome to the final lecture in this Great Course's series on electronics. This lecture is going to be a little different than the others. I'm not going to spend as much time developing new circuit concepts, although I will introduce one important new device. I'm going to instead look toward the future. In particular, I want to look toward your future in electronics; your electronic future.

There are really two aspects of that electronics future. Even if you're not interested in electronics, you as a consumer are going to be buying more electronics than you have in the past because more and more electronic devices will be developed. They'll be faster electronics and they'll by and large be cheaper electronics. Electronics will be incorporated into nearly everything. Already your car has numerous electronic devices and sensors and computers that help the car operate at top efficiency. It probably has a navigation system. It has many other electronics. They'll be incorporated into nearly everything. Even your toaster probably has some electronics in now. More significantly, perhaps, is that entirely new technologies are going to be enabled by the complexity that the miniaturization, the continuing miniaturization, of electronics permit.

Now that's just you as a consumer. But you may also choose, especially after taking this course, to be your own creator, builder, maker of electronic devices. You're ready, after this course, to make your own electronic circuits. You've probably had some practice in that, if you've done some of the projects building simulated circuits with web-based circuit simulators. I've given you a number of basic building blocks: amplifiers, flip-flops, counters, operational amplifiers, all kinds of other things. We've seen how you could build these up from even simpler circuits, ultimately based in transistors and other components.

But now I'm going to start with a mantra that you'll hear again and again in the course of this lecture. That is, "buy, don't build" these basic circuits. Don't go out and build a flip-flop out of NAND gates, for example. Buy an integrated circuit that does flip-flops; or if you need lots of flip-flops to make a counter, by the counter circuit. I'll explore that "buy, don't build" a little bit later. You can combine these basic circuits to make very complex and practical circuits. Two aspects of your electronics future: as a consumer and as a creator, builder, of electronics.

Let's look briefly at the future of electronics and in the context of the past before we move on. Here's a graph I actually showed you very early on in the course. This is Moore's Law. Gordon Moore was an engineer at Intel in the 1960s—Intel, the maker of electronic chips—and he made a proposition. He said, "I think the number of transistors"—again, that basic on/off switching device that's the heart of all modern electronics—"on a given integrated circuit (one of these silicon integrated circuits that has many, many, many transistors and other components on it) will double every 18 months." Moore wasn't quite right, but he was close. Since 1965, it's been approximately true that the number of transistors on an integrated circuit chip has doubled about every two years.

Where are we headed? We may be headed to some sort of limit around the year 2020 when we get transistors down to about a nanometer—a billionth of a meter, about 10 times the size of an atom in size—but we don't know that for sure. We've got a ways to go to get to that. This graph shows you the year on the horizontal axis, and I've projected up to about the year 2020. On the vertical axis is the number of transistors, again, on a given single integrated circuit chip, and that's a logarithmic scale. What that means is we start with 1,000; the next division up is 10,000; the next division up as 100,000; a million, 10 million, 100 million, a billion, and 10 billion. That's a logarithmic scale. Each division is a factor of 10 increase.

If exponential growth is occurring—and to double every two years is exponential growth; exponential isn't just a word that means "big," it has a particular mathematical meaning, it means you go up by a certain percentage in a certain time—and this says we increase by 100%; we double every two years. Getting a straight line graph on a logarithmic scaled plot tells you that you've got exponential growth. The straight line you see here, on that line I've labeled some individual chips, particularly some microprocessor chips that were used in personal computers. You may recognize some of those names, and those are where those particular processor chips lie in the plot of the year they were developed versus the number of transistors on a chip. You can see that we're up in the mid-second decade of the 21st century. We're up at about somewhere between one and 10 billion transistors on a chip. About as many transistors on the microprocessor chip at the heart of your computer as there are people on this planet; that's quite remarkable.

Let's look at Moore's Law in one other way of thinking about it: We can also think about it as the size of the transistor. If you go to semiconductor manufacturers, they'll tell you what size they're using for their basic feature; 32 nanometers, 14 nanometers, 64 nanometers, what is it? That size has been shrinking over the years; it's heading down toward the 10–1 nanometer range. Again, we're on a logarithmic scale. The size of the transistor is on the vertical axis in nanometers. Again, in the current year, we're down somewhere between 110 and heading toward somewhere between 10 and 1 nanometers as these transistors shrink. Again, somewhere around 2020, we may run into troubles as we approach the one nanometer range, but that isn't clear yet. The transistors have been shrinking, and shrinking, and shrinking and will continue to shrink and what that means, of course, is that we cram more transistors on a chip. We can store more information if it's a memory chip. We can process more information. We might yet again increase our word lengths. We can do more complex operations faster. That's what's happening because of the shrinking transistor.

Nothing's new about the transistor. I shouldn't quite say that; there are innovations even in the transistors. But they're basically transistors that work like we've seen throughout this course. They just get smaller and smaller and more of them get crammed into more complicated circuits. Ultimately we're looking at transistors that are nanoscale size systems. Here are a couple of images of some nanoscale size transistors. On the left, we have a carbon nanotube transistor. It's made with a carbon nanotube, a very hot thing in today's world of nanoscale electronics. I have two colleagues at Middlebury College who are condensed-matter physicists who actually work on devices like this. The left one is the scale size of the carbon nanotube, which is in the middle at the intersection of those four things coming in, which are actually electrodes that connect to the carbon nanotube. That's on the scale of a large molecule. The transistor at the right is a single-electron transistor. At the bottom, that rectangular structure pointing upward is the gate. You put charge on that gate or not and that determines whether a single electron is trapped in that channel up above the gate. This is a single-electron transistor; it works by a single electron.

What is the ultimate limit to the scale size of these things? Gordon Moore has a quote about this. Gordon Moore says, when during a 2005 visit to Intel—that's when this happened—Stephen Hawking, the famous physicist, was asked, "What are the fundamental limits to microelectronics?" He said, "The speed of light and the atomic nature of matter." The speed of light is a limit because the speed of light, although very fast—186,000 miles per second, wow—the speed of light is nevertheless finite. It's about, if you work it out in feet, one foot per nanosecond. That means in a billionths of a second, light goes this far.

The speed of light is the ultimate speed limit in the universe. The speeds of electrical signals in wires turn out to be on the order of the speed of light, typically two-thirds of it, half of it, something like that, depending on the wires and their configuration. They're close to, but not quite, the speed of light. What that means is if you want to build a fast computer—and modern computers operate with clock speeds of several gigahertz, several billion cycles per second—you can't have a computer that's this big because the time for an electrical signal to get from this part of the computer to that part of the computer is much longer than the cycle time of the clock. The speed of light is a real limit, and that's forced us to make computers smaller and smaller and smaller. One of the problems with that, as computers get more and more powerful and more complex and have more transistors on them, you're cramming more and more power dissipation and you're generating a lot of heat. One of the limitations—and one of the really clever things about today's modern laptop computers particularly—is they've managed to make very fast computers that nevertheless don't dissipate so much heat that they burn themselves up, although sometimes we get pretty close to that limit.

The speed of light is one of the limitations. The other is the atomic nature of matter. Ultimately, the limiting transistor size—although I've suggested we may get into troubles when we get down to about a nanometer in scale—the ultimate size is certainly governed by let's call it the Compton wavelength of the electron. This is a measure, I don't want to say the electron's size because an electron is probably a true point particle, but in some instances it manifests itself with a size. The Compton wavelength is the sort of size of the electron as it appears to electromagnetic radiation coming in that scatters off the electron. That size is a small fraction of a nanometer; it's down in the x-ray wavelength range. But we certainly won't be able to make transistors that have scale sizes smaller than that, and we'll probably run into trouble well before that.

If you extrapolate the exponential Moore's Law graphs I showed you, the one of transistor size, we'll probably reach this ultimate limit in the 2030s sometime. As I suggested, we'll probably reach more practical but still quantum-related limits probably around the year 2020 or maybe a little beyond it. Who knows.

We're talking about futures. What are some examples of futures? Again, I want to emphasize: The basic circuits that I've taught you in this course and that I've been teaching to students at Middlebury College for several decades are basically unchanged. It's remarkable to me that I've taught electronics through this huge revolution in electronics but the basics have stayed the same. But what's happened is we have more transistors, they're faster, we cram billions of them on a single chip, where when I started teaching about electronics we maybe had dozens or hundreds on a single chip. What a change. At some point, that amount of complexity in miniaturization enabled entirely new technologies. Here are just a couple of examples: Google has been experimenting with a self-driving car, and self-driving cars have driven hundreds of thousands of miles on the public highways of California. At some point, year 2020 maybe, these things will be available for sale. They drive themselves. They're smarter than regular drivers, human drivers. They don't fall asleep. They don't drive under the influence. They stay attentive. They don't text while they're driving. They'll make driving safer, and in some sense easier, because we won't have to do the driving. That's an example of the kind of complexity, enabled not only by the miniaturization of electronic circuits and computing power particularly, but also by the miniaturization and increasingly less expensive sensors of all types: ultrasonic sensors that bounce a beam off the car ahead to measure the distance; sensors that tell how fast you're going; and so on and so forth. There's Google's self-driving car as an example of an electronic future. We also have here a SmartPal domestic robot, which increasingly will do domestic chores; not just running around like a big Frisbee vacuuming your floor, but doing all kinds of domestic chores. Already these things are in use in places like nursing homes and other areas that are labor intensive and robots can play a role.

Your electronic future is going to see more and more smart systems of all sorts. Sometimes you won't even know that the systems are smart, but they've got more and more smarts built into them. That's your electronic future.

I've declared my mantra, which is "Don't build, buy," or "Buy, don't build," and I'd like to talk about that mantra before I move on to the final part of this lecture. Your electronic future, remember here in the second part of your future, is as a designer, a builder, a doer of electronics. There's were my mantra applies: "Don't build, buy." I spent a lecture talking about flip-flops, and I showed you how to build up from the simple bistable circuit to an RS master-slave flip-flop, which is shown here in circuit diagram. You can go build one, if you want. My students do that as part of their electronics course, and you may have done it as part of a project. But I don't recommend doing that if you're actually going out and building your own electronics. Rather, go buy an integrated circuit that already has flip-flops on it. One example, and I used it in several of the demonstrations I showed you, is a 7476 dual J-K flip-flop. You'll remember the J-K as the most versatile of all the flip-flop types; it can do anything any flip-flop can do for 40 sensors. You can buy a 7476 dual J-K flip-flop; you've got two of them to work with in a single little integrated circuit package. Buy, don't build. Don't build, buy.

Down below, I've got a counter circuit; we talked about counters. This happens to be a 3-bit counter. What do we do with this counter? It's a 3-bit synchronous counter, in this case. I wouldn't build one, I'd buy a counter circuit that already has flip-flops built into it. I wouldn't take J-K flip-flops and make them into a counter, I'd just go buy a counter chip and explore what kind of counter you want. Here's an example of the pinout connections for a 74160 series counter. In your electronics future, don't build, buy.

Some other examples: Way back early in the course, we talked about amplifiers. We built—and maybe you constructed it if you did the project—a complete audio amplifier with two stages of common emitter preamplification followed by a complementary symmetry output stage. I actually demonstrated a tiny little integrated circuit audio amplifier chip. I did it very early on when I was talking about power supplies and why the filter capacitor was important, and I did it again when we talked about amplifiers. But if you don't like a puny fraction of a watt amplifier, you can go out and buy this circuit here, which is a 68-watt audio amplifier on a single chip. By the way, it's got that big metal tab at the top so you connect something to that to dissipate the heat that that thing generates. Don't design amplifiers and try to build them, go out and buy them as chips.

In the last lecture on analog electronics, we developed a function generator. It produced square waves and triangle waves, and we could adjust its frequency and so on. That was sort of like the function generators; the fairly expensive instruments that I'd been using to make signals for my experiments. You can go out and buy a single chip function generator. Here's a picture of a little board with that single chip function generator on it, and you can buy these boards as kits for a few tens of dollars. You attach a few knobs and potentiometers, and you've got yourself a complete function generator. Not quite as good and stable and accurate and so on as the ones I've been using from the scientific instrument companies, but perfectly good for a hobbyist experimenting with electronics and wanting to have a function generator that makes different wave forms at different frequencies and different amplitudes. Don't build, buy.

Having said that, the 23 lectures you went through this before aren't wasted because to use these devices intelligently, it really helps to know what's going on inside them. I really don't like to think of them as black boxes, but sometimes it's good to do that. Now I want to move on to the final part of the lecture and introduce an entirely new idea. Not a new idea, but an idea that's a more complex circuit than anything we've seen, and that's the microcontroller. A microcontroller is just a very tiny computer on a chip, and many electronics courses that you could take in colleges and elsewhere, in fact, start with microcontrollers and they show you how to build things with them. It's great, because if you have microcontrollers you can do all kinds of stuff right away. I wanted to show you how things work inside microcontrollers and every other electronic circuit, and that's why we spent

23 lectures going through the individual pieces that work inside computers and other devices.

Here's an example of a microcontroller. It's got a central processing unit that's just a lot of logic gates set up to compare binary numbers, to add them, to do various other things. It's got flash memory; that's like its hard drive. That's where it stores stuff for the long time, including data and programs. It's got some random access memory. It's got a serial parallel converter and a serial port that can bring data in or out. Typically, it has digital inputs that can look at voltages and say, "Is this a one or is this is zero?" It's got analog inputs going through analog to digital converters, and it's got digital to analog converters giving you analog outputs. It's a complete computer on a chip.

Here's my laptop; it's actually a two-year-old unit, so it isn't the absolutely most modern. Here's a microcontroller, that long black chip. I'm just going to do a quick comparison of them. My laptop uses 64-bit words. The microcontroller, it's an Atmel ATmega328 microcontroller, has only 8-bit words. My computer has 768 gigabytes of flash memory instead of a mechanical hard drive. The microcontroller has 32 kilobytes, 32,000 bytes; mine has 768 billion. Mine has 16 gigabytes, 16 billion bytes of RAM. The microcontroller has only 2 kilobytes of RAM. The clock speed in my computer is 2.7 gigahertz, pretty fast but not the fastest these days. The clock speed in the microcontroller is 20 megahertz. They both have serial ports that are the same, they're USB ports. The processor cost for my processor alone, the microprocessor at the heart of my laptop, is $300. The processor here costs about $3 in bulk. These are comparisons between two full computers. One just has less capability than the other. Neither of them, by the way, in that $300 includes things like the display or the keyboard or other peripherals you might need to connect.

I've said I wanted you to understand the interior of controlling of computers, of all kinds of electronic circuits; that's why I've taught you this course. On the other hand, if I were starting over today as an electronics hobbyist, I would've wanted to learn everything that I learned from this course or that you learned from this course, but I would probably start in right away playing with microcontrollers because they're easy, they're basic, and they can do everything. I want to spend the rest of this lecture giving you some examples of that.

You can actually go out and buy little boards that have a microcontroller and several other components on them. This particular board is called an Arduino. There are actually several models of Arduino. Arduino is an open-source product that was developed first in Italy. It's widely available. Electronics hobbyists all over the world play with these things and use them to do all kinds of work, all kinds of devices. This particular one is an Arduino UNO board, and the long chip at the bottom, the long black thing, is the microcontroller. That is, in fact, an Arduino board over here.

There's the microcontroller. Right now, you see one LED blinking. By the way, you may have observed this thing moving around in the course of this lecture and you may have clunking noises as it moved. That's because when I stood in front of it, it's a light-seeking device—and I'll get to it soon—and it moved to a different source looking for the brightest light it could find in the sky. That's what that's doing. We'll get to that shortly.

Here's an Arduino microcontroller. You can buy this thing for about $30, and you can pay a little bit more for some peripherals. You can power it off a little power adapter, or you can plug it into your computer's USB port and power it off that, or you can plug it into the little adapter you have with your phone charger that plugs into the wall; it has a USB port because USB, remember, supplies power as well as moving signals around.

Let me talk a little bit about this Arduino microcontroller development board. I'd tell you, if I were starting out again as electronics hobbyist, I'd start right here. But I'd want to know about all the things I've showed you in this course, and the thing that'll make you a more powerful user of microcontrollers is knowing how to wire other small circuits that might be controlled by or do controlling of the microcontroller. This microcontroller, by the way, this whole board, has, in addition to a couple of power receptacles, a USB input, and a power input, it's got five or six 10-bit ADD converters. Remember the ADD converter in the previous lecture was the most sophisticated circuit we built. This has five or six of them on here. It also has digital inputs, it has digital outputs, it has analog inputs. They're all in these little connectors at the sides of the board.

It also, in addition to getting power from a USB, the beautiful thing about this is it's a computer. It's programmable with a rather simple coding language that you can learn very quickly. If you buy one of these, you can get free software and you can have examples of simple circuits that you can build and then modify the code and see how it works. These are computers. No longer are you just connecting wires to do electronics, you're also programming. You can program on your laptop, you can connect the USB cable to your laptop, you can upload the program to the microcontroller, and then it resides in the microcontroller's flash memory and the microcontroller will execute that program; that is it'll do what you're asking it to and it'll continue to do that without needing to be connected to your computer. It just needs power.

Let me give you an example of that: Here's the Arduino microcontroller. I've connected it up so two digital outputs—which produce 5 volts or 0 volts, depending on whether they're 1's or 0's—are going to two LEDs, a red one and a green one. I've connected one of the analog inputs to a potentiometer, a variable resistor again that I can tap off the middle of. That's put across the 5-volt supply; so 5 volts at one end, 0 at the other, and I can dial any voltage between 0 and 5. What I did was download the Arduino software, go to the software, and look at a simple example that made an LED blink with an interval between blinks that depended on the voltage. The bigger the voltage, the longer the interval between blinks. I'm going to turn the dial on this potentiometer up, and you can see that the LED, the red LED, is blinking more and more slowly. That was one of the simple examples of a code you can write and an utterly simple circuit you could build with this thing.

I did that and I said, "I want to do a little bit more. I want to learn a little bit about this." So I said, "Let me modify this. Let me say, suppose the input voltage is less than half a volt. Let me not try to blink that LED, it's going to blink so fast you won't be able to see it." I'm going to turn the voltage down and the LED blinks faster and faster, but then I said, "If I get below half a volt, let me turn on a green LED instead," and that's what it does. The green LED is on as long as the input voltage is less than half a volt, and if I go up over about half a volt the red LED starts blinking. The rate at which it blinks get slower as the voltage gets greater. With a little more coding, you could've made it blink faster as the voltage went greater. You can do all that.

Here's an example of what a piece of Arduino code looks like. If you've never seen computer code, you might be overwhelmed by this, but don't be. Many of the lines are simply things that initialize the Arduino microcontroller and get it ready to work. Many of the lines are simply comments about what the program is doing; they don't actually do anything themselves. But I've highlighted some lines here that have to do with things like comparing the input number that's coming in from the ADD converter and doing one thing, namely lighting the green LED if it's too low and just keeping it lit; or if it's higher, blinking the red LED and some of these commands. You can just read them out in English and you can almost see what they're saying; you can see what they're saying. They say, "Turn on this LED," and then it says, "Pause for a while," and that tells me how long the pause interval is going to be before I turn it off again, and so on. The highlighted lines are really doing the meat of this program, and they're not at all hard to understand. You, too, can get an Arduino and work with it and connect it to many of the other circuits we've developed in the course of understanding modern electronics here.

Let me end with something I'm really proud of: an example of something my student has done with microcontrollers, which I think is really a beautiful example of electronics, of everything we've learned in this course. To begin with, Middlebury College, where I teach, recently installed a field of 34 solar-tracking units that produce a total of 143 kilowatts in bright sunlight. These trackers don't actually follow the sun. What they do, instead, is they have GPS units on them and they do a calculation. This calculation is perfectly accurate because we know exactly where the earth is in relation to the sun and so on, based on the time of day and the season and so on. They calculate where the sun is in the sky, and they run motors that tilt these panels so they're pointing directly at the sun. Even if there's a cloudy day, there's no sun visible, they're still pointing at where the sun would be if the clouds weren't there. My student reasoned this way: He said, "Maybe that isn't the best thing to do. On a bright day with no clouds, it's the best thing to do; point right at the sun. But what if it's a cloudy day? What if there are clouds covering the sun and some clouds are reflecting light and there's actually more energy coming from a different part of the sky? What if it's completely overcast? Maybe you're better off putting these things flat." He set out to build a device that would literally track the sun by looking for the brightest point in the sky, the point where the most energy was coming from. Here's my student, Misha Gerschel, I'm very proud of him, a Middlebury physics major who went on in a joint program with Dartmouth and studied engineering at Dartmouth. Here he is on the roof of our science center with his device, which you can see in some of the close up pictures.

But I'd like to show you the device because it's sitting right here. What we have here, on the device, is a thing called a pyranometer. This has nothing to do with the circuitry. This is a device that actually measures the intensity of sunlight. This is a $5,000 instrument, by the way, which is very precise and gives us a very precise measure of solar energy, and it's just going along for the ride.

Remember back in the first lecture where I introduced feedback, I had that crazy demonstration I called Funny Face. I shined a flashlight, and Funny Face had two little eyes that were photoconductive cells, photo resistors, and it followed the flashlight. It had a motor and it turned in one dimension to follow the flashlight. Misha's apparatus is a much more sophisticated version of Funny Face. It has four eyes, and they're down here at the bottoms of these channels created by this big cross-shaped thing, and that's designed to cast very sharp shadows. What this does is the device looks at the voltage levels coming in from a voltage divider consisting of these photoconductive cells, these photo resistors, and another resistor and it compares them. It has two motors, one that moves the apparatus in this direction and one that moves it in this direction, and it tells those motors to move in a way that optimizes the amount of light coming into here. It points at the brightest source of light around.

Let me demonstrate it now. I have a bright light—I'll go behind here—and right now, it's found the brightest lights in the studio. The reason it was the other way at first is because when I was talking about my laptop and standing in front of it, I was blocking those lights and it went and found a brighter light that's behind me. But now I have a very bright light that I'm going to turn on and this will be an artificial sun. I'm going to hold this light up, and you immediately see the device starting to move. There it is. The total light on it is now a maximum. It's measuring not just the light from this but all the studio lights and everything else and it's optimizing that light. Notice it moved mostly around its vertical axis, but it knows about the heart of the other axis also. If I raise this light up higher, it's going to move both axes in such a way that it'll optimize, maximize, the amount of light coming in.

That's a very clever device, and Misha wrote a code for the Arduino. The Arduino is right inside there; you can see the blue Arduino board if you look closely. The other board, the white board, is containing several controllers for the motors. The motors take more power than the Arduino can supply, and so it sends signals to the motor controllers, and the motor controllers, which are powered by this heftier power supply, then tell the motors how to move. There's Misha's tracker.

Let me end this demonstration, and the entire course as well, with a movie that shows you time-lapse photography of this thing over the course of an entire day. Misha took this movie. He set up a camera that took a picture every 10 seconds through the entire course of the daylight hours of the day. It's a beautiful movie that shows what's going on with this tracker, on the top of the Middlebury College Science Center, throughout the course of an entire day. Let's take a look at that movie.

Here we are on the roof of Middlebury College's Science Center, Bicentennial Hall. We're looking east at sunrise. You can see the tracker that I just demonstrated; it's on the cart at the left center of the frame. We're going to show you now a movie that's taken at one frame every 10 seconds throughout the entire course of the day. Let's roll the movie. People coming and going; sun beginning to rise; up it comes. Watch the tracker; you'll see it beginning to move. Get a close-up soon. There it is, pointing towards the sun, some clouds coming by. It's looking for the brightest spot in the sky, even if the clouds are still there. It may not be right where the sun is. Up it goes as the day advances, into the morning. Here we are midday. You can see the tracker pivoting around as the sun goes through the sky. We sped it up a little bit there. The device to the right, by the way, is a radio telescope built by another Middlebury student for his project; it can image the sun in radio waves. Now you can see the sun in the sky and the tracker following it beautifully. As the sun begins to drop down in the afternoon, a few clouds come by. Down it goes, lower and lower, the tracker following it beautifully on a crisp, clear day where it really is seeing the brightest sky from the sun. Here we are, finally late afternoon towards sunset, looking towards the Adirondack Mountains in the west as the sun is setting lower and lower. We actually see the tracker move up a little bit as the glow of the sunset rises into the sky. Finally, evening comes, it darkens, and eventually the tracker goes to a rest position pointing straight down.

That was quite an impressive movie and quite an impressive project. All the more so, I like to think, because it's an example of electronics in the service of our planet. What could be a better thing to do than to figure out how to maximize our ability to extract clean, free, nonpolluting energy from the sun, which is supplying us energy at the rate of a thousand watts, a kilowatt, on every square meter of earth's surface in bright direct sunlight? Electronics in the service of the planet.

I want to end this lecture and this course, as I have every other one, with a project. But the project for this lecture is much more open-ended. I want you to do two things, or at least one of these two things. The first one is to go out and enjoy your evolving electronic devices. I hope you'll enjoy them all the more because you have some appreciation and knowledge of what's going on inside those devices. If you're more adventurous, build your own electronics. You're coming out of this course with the knowledge and understanding to put together the basic building blocks to make electronic devices of your own design to do what you want them to do.

I very much enjoyed teaching this course because I so much enjoy working with electronics, and I hope you have, too. Get out there and enjoy your electronics.

Answers to Selected Questions to Consider

Lecture 1
2. (a) 6 A; (b) 1440 W

Lecture 2
1. R_7, R_8, R_{12} all in parallel; R_{10}, R_{13} in parallel; R_{14}, R_{15} in series.

2. 5.7 V

Lecture 3
1. Neither measurement is successful. The ammeter's zero resistance "shorts out" R_3, so the voltage across it becomes zero, and that's what the voltmeter reads. The current through R_3 is also zero, but that's not what the ammeter reads, because the meter isn't in series with R_3. It reads the current through the series combination of R_1 and R_2, which is $V/(R_1 + R_2)$. The ammeter will be damaged (or blow its fuse) if this current exceeds the maximum current the meter can handle.

2. Time.

Lecture 5
1. (a) It acts as an open circuit or very high resistance; (b) it acts as a short circuit or very low resistance.

Lecture 7
3. 5.8 kΩ

Lecture 8
1. −15

3. (a) The load resistance; (b) the supply voltage.

Lecture 11
2. Mechanical, optical, thermal.

Lecture 14
1. $V_{out} = V_1 - V_2/2$

Lecture 16
2. $\bar{A} \cdot B$

A	B	Out
0	0	0
0	1	1
1	0	0
1	1	0

Lecture 23
2. 2^{10}, which is 1024.

4. The minimum number is 10 because a three-digit scale reads from 0–999, and $2^{10} = 1024$, the next higher power of 2. But if the conversion itself is done in binary-coded decimal, for instance, by using BCD counters, then 12 bits will be needed—4 for each decimal digit.

Bibliography

Notes: *Understanding Modern Electronics* is designed to be a standalone course introducing you to the concepts and devices behind modern electronics. Nevertheless, you can increase your knowledge of electronics with additional readings. I've chosen six books for suggested readings, although there are plenty of other good books available that you may also find helpful. You don't need to read any of the suggested books, and you certainly don't need to read them all. However, if you want to do some supplementary reading, I suggest starting with one of the introductory-level books; then, if you want to go deeper, look for one of the more advanced books.

The books I recommend here fall into two categories. First are four introductory-level books, Keith Brindley's *Starting Electronics*, Charles Platt's *Make: Electronics*, and two books in the *For Dummies* series. (Don't be put off by the *Dummies* titles; these are actually quite good introductions to electronics.) These beginning books don't go into as much detail on how electronic components work, and most don't get as far as the more sophisticated circuits discussed in the course. However, they're good for their simple analogies that help you understand electronic devices and circuits, and they get you quickly into building practical circuits without a great deal of theoretical background. They're especially helpful with advice on practical aspects of electronic circuit construction—tools, soldering, circuit boards, sources of electronic components, and so on. Because this course doesn't cover those topics, the books can be especially helpful if you're interested in pursuing actual construction of, and experimentation with, electronic circuits—something I hope you'll be inclined to try after viewing the course. (You'll also get some experience "building" circuits with software simulation if you undertake the projects associated with each lecture.)

None of the books follows the same sequence as this course, which means that the suggested readings jump around somewhat, and you'll sometimes find the same suggested readings repeated because a given chapter in the introductory books covers the material of several lectures in a way that makes it hard to break out specific sections to associate with specific lectures. Not every book includes suggested readings for every lecture.

In addition to the introductory books, I've included two more advanced books: Paul Scherz and Simon Monk's *Practical Electronics for Inventors* (3rd edition) and Paul Horowitz and Winfield Hill's *The Art of Electronics* (3rd edition). Both are oriented toward practical circuit design but without the construction tips included in the introductory books. Both go into electronics at significantly greater depth than does this course. Scherz and Monk's book is generally at a slightly lower level than that of Horowitz and Hill and may be easier to read cover to cover (and it's less expensive). For decades, Horowitz and Hill has been a standard reference tool for doing practical electronic circuit design, and I recommend it highly for serious electronics experimenters—especially if you can find a used edition. That may be a bit difficult given that the third edition was published in 2014, although the 1989 second edition remains a valuable reference book despite the fact that some of the individual components it describes have become obsolete.

Finally, I've included in the bibliography several more introductory-level books that you might find helpful, although they aren't listed specifically in the suggested readings.

Boysen, Earl, and Harry Kybett. *Complete Electronics: Self-Teaching Guide with Projects*. Indianapolis, IN: John Wiley & Sons, 2012. As the name implies, this book embodies a self-teaching approach to electronics. It covers only analog electronics, corresponding roughly to Lectures 1 through 15 of the course, although it goes deeper into some topics, such as resonant circuits. The book incorporates numerous self-tests, quizzes, and do-it-yourself projects.

Brindley, Keith. *Starting Electronics*. 4th ed. Oxford, UK: Newnes Publishing, 2011. An introductory-level book that covers, approximately, the material through Lecture 18 of this course. A good mix of practice and theory. Because the book is published in England, its lists of parts suppliers aren't particularly useful to readers outside the United Kingdom.

Horowitz, Paul, and Winfield Hill. *The Art of Electronics*. 3rd ed. Cambridge, UK: Cambridge University Press, 2014. The definitive reference guide to the theory and practice of electronic circuit design. Emphasizes practical considerations and real devices, not just underlying theory. Great graphic collages of good circuit ideas, as well as "bad circuits" that illustrate how not to design circuits. A great reference tool for the serious electronics enthusiast.

Jung, Walter. *IC Op-Amp Cookbook*. Indianapolis, IN: H. W. Sams, 1986. Ancient by electronics standards, this book is nevertheless a delightful and practical introduction to op-amps and the dazzling array of circuits they enable.

Lancaster, Don. *TTL Cookbook*. Indianapolis, IN: H. W. Sams, 1974. Even more dated than the op-amp cookbook and based on an early and now obsolete logic family, *TTL Cookbook* remains a classic introduction to digital logic. May be difficult to find.

Lowe, Doug. *Electronics All-in-One for Dummies*. Hoboken, NJ: John Wiley & Sons, 2012. Billed as "8 books in 1," this is a thorough and practical look at a wide range of electronic circuits. It covers most of the basic components used in the course and goes further into such topics as radio and infrared devices.

Mims, Forrest M. *Getting Started in Electronics*. 4th ed. Niles, IL: Master Publishing, 2012. Billed as "a complete electronics course in 128 pages," this book reads less like a narrative than a series of introductions to electronic components and circuits you can build with them. Charmingly produced, with hand-lettered text and hand drawings. Includes many practical tips and some components not covered in the course.

Platt, Charles. *Make: Electronics*. Sebastopol, CA: MakerMedia, 2009. Another good hands-on introduction to electronics; at a fairly elementary level but encompassing some reasonably sophisticated projects, including a microcontroller-driven robot relevant to Lecture 24.

Shamieh, Cathleen, and Gordon McComb. *Electronics for Dummies*. 2nd ed. Hoboken, NJ: John Wiley & Sons, 2009. Another electronics book in the *For Dummies* series. Less comprehensive than Lowe's *Electronics All-in-One* but still a good, quick introduction with a nice chapter on integrated circuits.

Scherz, Paul, and Simon Monk. *Practical Electronics for Inventors*. 3rd ed. New York: McGraw-Hill, 2013. Designed to give practical advice to inventors wanting to incorporate electronics into their inventions, this book also serves nicely as a general introduction to electronics at a level somewhat more advanced and more comprehensive than this course.